Springer Series in Materials Science

Volume 300

T0171798

The Springer Series in Materials Science covers the complete spectrum of materials research and technology, including fundamental principles, physical properties, materials theory and design. Recognizing the increasing importance of materials science in future device technologies, the book titles in this series reflect the state-of-the-art in understanding and controlling the structure and properties of all important classes of materials.

More information about this series at http://www.springer.com/series/856

Kamal K. Kar

Editor

Handbook of Nanocomposite Supercapacitor Materials I

Characteristics

 Springer

Editor
Kamal K. Kar
Advanced Nanoengineering Materials
Laboratory, Department of Mechanical
Engineering and Materials Science
Programme
Indian Institute of Technology Kanpur
Kanpur, Uttar Pradesh, India

ISSN 0933-033X   ISSN 2196-2812   (electronic)
Springer Series in Materials Science
ISBN 978-3-030-43011-5   ISBN 978-3-030-43009-2   (eBook)
https://doi.org/10.1007/978-3-030-43009-2

This Springer imprint is published by the registered company Springer Nature Switzerland AG
The registered company address is: Gewerbestrasse 11, 6330 Cham, Switzerland

Dedicated to my wife, Sutapa, and my little daughter, Srishtisudha for their loving support and patience, and my mother, late Manjubala, and my father, late Khagendranath

Preface

The search for sustainable and renewable means of energy production leads to the development of various energy conversion technologies such as wind, solar and fuel cells. No doubt, these are highly efficient and environment-friendly technologies but carries certain demerits too. As a result, the energy researchers have understood the need of various energy storage technologies such as batteries and supercapacitors to fulfil the energy demands. Various advanced batteries like Li-ion, Li-air, etc. have been developed but still suffers with very low power densities. It becomes mandatory to use another energy storage device such as supercapacitor in various integrated power modules since they can deliver extremely high power when compared to that of batteries. Various types of supercapacitors have been developed such as electrochemical double-layer capacitors (EDLCs), pseudocapacitors (or, redox capacitors) and capacitors. Supercapacitors store charges electrochemically and they exhibit high power densities, moderate to high energy densities, high rate capabilities, long life and safe operation. EDLCs are known for their high power capabilities, whereas pseudocapacitors are the best for obtaining high energy densities. In order to obtain both high power and energy densities, nanocomposite electrodes are being used. It includes combining EDLC materials with pseudocapacitive materials. The electrochemical performance of nanocomposite electrodes is superior when compared to their individual counterparts since their performance is the sum of individual components. But the performance of devices is still challenging in terms of capacitance, flexibility, cycle life, etc. These deciding factors depend on the characteristics of materials used in the devices. The book "Handbook of Nanocomposite Supercapacitors with a theme of Characteristics of Materials" focuses on the various characteristics of prospective materials. This book will be useful to the graduate students and researchers from various fields of science and technology, who wish to learn about the recent development of supercapacitor and to select the material for high-performance supercapacitor.

Capacitors consist of one or more pairs of conductors separated by an insulator and are mainly used to supply power in several electronic systems. Performance of capacitors depends on the conductors, dielectric (insulator) medium and other several basic parameters. Therefore, Chap. 1 provides the fundamental aspects of a capacitor, different types of capacitors and their applications.

As the supercapacitors are the electrochemical charge storage devices used in a variety of applications such as electronics industry, hybrid electric vehicles and power supplies, Chap. 2 gives a comprehensive overview of the basics of a supercapacitor, its components, types, characterization and parameters that decide its performance.

Transition metal oxides (TMOs) are a versatile group of compounds, finding applications in nano and flexible electronics, gas sensors, energy devices, bio-imaging and medicines. In Chap. 3, synthesis and properties of TMOs are discussed in details.

Chapter 4 gives a comprehensive overview on the synthesis of activated carbon from biomasses, the parameters controlling process of activation with a special emphasis on the characteristics of activated carbon in view of supercapacitor application. The electrochemical, structural and surface characteristics of activated carbon have been discussed to understand the underlined correlation.

Graphene, the thinnest two-dimensional material, has grabbed the attention of researcher of different fields due to the excellent combination of mechanical, electrical, optical and thermal properties. Chapter 5 reviews structure, synthesis, properties and application of graphene/reduced graphene oxide. This interesting material has been explored in the field of medicine, electronics, energy devices, sensors, etc.

Carbon nanotubes (CNTs), the interesting nanomaterials of recent ages, are the potential performer in biomedical, energy devices and storage, high strength composites and many more. Chapter 6 discusses the structure, properties, synthesis, purification, application of CNTs along with the underlined challenges and future prospect of this magical form of carbon.

Owing to their unique properties such as high corrosion resistance and excellent mechanical strength, carbon nanofibers (CNFs) are potential candidates for advanced applications including electrochemical power generation and storage, high strength composites and sensors. Chapter 7 reviews the CNFs in terms of their structure, properties and application advancements.

Chapter 8 provides a short yet inclusive overview of the common conducting polymers including the brief description of different types of conducting polymers, mechanism of conduction, their synthesis strategies, properties and applications.

This chapter emphasizes the common electrode materials that are utilized or under consideration for the development of high-performance supercapacitor devices. The characteristics of efficient electrode materials towards the development of high-performance supercapacitors have been pointed out in Chap. 9.

Chapter 10 demonstrates the types of electrolytes used in electrical double-layer, asymmetric and hybrid supercapacitors in detail. The effect of electrolyte types on the cell voltage and specific capacitance has also been discussed to understand the role of electrolytes in modern supercapacitor devices.

Chapter 11 deals with the importance of the separator materials for the fabrication of high-performance supercapacitor devices. The essential parameters along with the characteristics of separators have been included for the proper understanding of the selection of a suitable separator material.

Chapter 12 focuses on the importance of current collector materials for the fabrication of efficient modern generation energy storage devices. An emphasis has been given to understand the essential parameters for the selection of most suited current collector material based on the material properties required for supercapacitors.

Chapter 13 provides an overview of the recent applications and future prospective of supercapacitor devices.

The editor and authors hope that readers from materials science, engineering and technology will be benefited by reading of these high-quality review articles related to the characteristics of materials used in supercapacitor. This book is not intended to be a collection of all research activities on composites worldwide, as it would be rather challenging to keep up with the pace of progress in this field. The editor would like to acknowledge many material researchers, who have contributed to the contents of the book. The editor would also like to thank all the publishers and authors for giving permission to use their published images and original work.

There were lean patches when I felt that one would not be able to take time out and complete this book, but my wife Sutapa, and my little daughter Srishtisudha, played a crucial role to inspire us to complete it. I hope that this book will attract more researchers to this field and that it will form a networking nucleus for the community. Please enjoy this book and please communicate to the editor/authors any comments that you might have about its content.

Kanpur, Uttar Pradesh, India   Prof. Kamal K. Kar

Contents

11 Characteristics of Separator Materials for Supercapacitors 315

Kapil Dev Verma, Prerna Sinha, Soma Banerjee, Kamal K. Kar, and Manas K. Ghorai

Editor and Contributors

About the Editor

Dr. Kamal K. Kar is Professor, Department of Mechanical Engineering and Materials Science Programme, and Champa Devi Gangwal Institute Chair Professor since 2019 at the Indian Institute of Technology Kanpur (IIT Kanpur), India. He is the former Head of the Interdisciplinary Programme in Materials Science from 2011 to 2014 and Founding Chairman of Indian Society for Advancement of Materials and Process Engineering Kanpur Chapter from 2006 to 2011.

Prof. Kar pursued higher studies from Indian Institute of Technology Kharagpur, India, and Iowa State University, USA, before joining as Lecturer in the Department of Mechanical Engineering and Materials Science at IIT Kanpur in 2001. He was a BOYSCAST Fellow in the Department of Mechanical Engineering, Massachusetts Institute of Technology, USA, in 2003.

Prof. Kar is an active researcher in the area of nanostructured carbon materials, nanocomposites, functionally graded materials, nanopolymers and smart materials for structural, energy, water and biomedical applications. His research works have been recognized through the office of (i) Department of Science and Technology, (ii) Ministry of Human Resource and Development, (iii) National Leather Development Programme, (iv) Indian Institute of Technology Kanpur, (v) Defence Research and Development

Organisation, (vi) Indian Space Research Organization, (vii) Department of Atomic Energy, (viii) Department of Biotechnology, (ix) Council of Scientific and Industrial Research, (x) Aeronautical Development Establishment, (xi) Aeronautics Research and Development Board, (xii) Defence Materials and Stores Research and Development Establishment, (xiii) Hindustan Aeronautics Limited Kanpur (xiv) Danone research and development department of beverages division France, (xv) Indian Science Congress Association and (xvi) Indian National Academy of Engineering.

Prof. Kar has published more than 215 papers in international referred journals, 135 conference papers, 5 books on nanomaterials and their nanocomposites, 35 review articles/book chapters and more than 55 national and international patents. He has guided 18 Ph.D. students and 77 M.Tech. students. Currently, 17 Ph.D. students, 10 M.Tech. students and few visitors are working in his group.

Contributors

Soma Banerjee Advanced Nanoengineering Materials Laboratory, Materials Science Programme, Department of Mechanical Engineering, Indian Institute of Technology Kanpur, Kanpur, Uttar Pradesh, India

Pankaj Chamoli Department of Physics, DIT University, Dehradun, Uttarakhand, India;
Advanced Nanoengineering Materials Laboratory, Materials Programme, Indian Institute of Technology Kanpur, Kanpur, India

Jayesh Cherusseri Advanced Nanoengineering Materials Laboratory, Materials Science Programme, Department of Mechanical Engineering, Indian Institute of Technology Kanpur, Kanpur, India

Bibekananda De Advanced Nanoengineering Materials Laboratory, Department of Mechanical Engineering, Indian Institute of Technology Kanpur, Kanpur, Uttar Pradesh, India

Manas K. Ghorai Department of Chemistry, Indian Institute of Technology Kanpur, Kanpur, India

M. Jaleel Akhtar Department of Electrical Engineering, Indian Institute of Technology Kanpur, Kanpur, India

Kamal K. Kar Advanced Nanoengineering Materials Laboratory, Materials Science Programme, Department of Mechanical Engineering, Indian Institute of Technology Kanpur, Kanpur, India

P. K. Manna Indus Institute of Technology and Management, Kanpur, India

Tanvi Pal Advanced Nanoengineering Materials Laboratory, Material Science Programme, Indian Institute of Technology Kanpur, Kanpur, India;
A.P.J. Abdul Kalam Technical University, Lucknow, India

K. K. Raina Department of Physics, DIT University, Dehradun, Uttarakhand, India

Raghunandan Sharma Advanced Nanoengineering Materials Laboratory, Materials Science Programme, Indian Institute of Technology Kanpur, Kanpur, UP, India

Prerna Sinha Advanced Nanoengineering Materials Laboratory, Materials Science Programme, Indian Institute of Technology Kanpur, Kanpur, Uttar Pradesh, India

Jitendra Tahalyani Materials Science Programme, Indian Institute of Technology Kanpur, Kanpur, India

Alekha Tyagi Advanced Nanoengineering Materials Laboratory, Materials Science Programme, Indian Institute of Technology Kanpur, Kanpur, India

Kapil Dev Verma Advanced Nanoengineering Materials Laboratory, Materials Science Programme, Indian Institute of Technology Kanpur, Kanpur, Uttar Pradesh, India

Chapter 1
Characteristics of Capacitor: Fundamental Aspects

Jitendra Tahalyani, M. Jaleel Akhtar, Jayesh Cherusseri and Kamal K. Kar

Abstract The capacitor is a passive electrical device, used to collect electrical energy by generating a potential difference. It is generally consisting of combination of two conductors placed next to each other separated by dielectric medium. The performance of a capacitor expressed in terms of the capacitance (C) depends on the dimension/geometry of the plate/electrode and the dielectric constant of the material, where the dielectric can be defined by insulating medium having permittivity, with no AC power losses or DC leakage. The capacitor shows different response to AC and DC sources. These are mainly used to supply power in several electronic and electrical systems. Therefore, this chapter provides the fundamental aspects of the capacitors and their basic properties. It emphasizes on the parallel plate model, the basic terminologies associated with the capacitors along with the equivalent circuits of the capacitor and its response to the externally applied AC and DC sources. It also describes about different types of capacitors that are being fabricated using different materials and different construction techniques. These different types of capacitors provide some unique properties.

J. Tahalyani · M. J. Akhtar
Materials Science Programme, Indian Institute of Technology Kanpur, Kanpur 208016, India
e-mail: jtahal@iitk.ac.in

M. J. Akhtar
e-mail: mjakhtar@iitk.ac.in

M. J. Akhtar
Department of Electrical Engineering, Indian Institute of Technology Kanpur, Kanpur 208016, India

J. Cherusseri · K. K. Kar
Advanced Nanoengineering Materials Laboratory, Materials Science Programme, Indian Institute of Technology Kanpur, Kanpur 208016, India
e-mail: jayesh@iitk.ac.in

K. K. Kar (✉)
Advanced Nanoengineering Materials Laboratory, Department of Mechanical Engineering, Indian Institute of Technology Kanpur, Kanpur 208016, India
e-mail: kamalkk@iitk.ac.in

© Springer Nature Switzerland AG 2020
K. K. Kar (ed.), *Handbook of Nanocomposite Supercapacitor Materials I*,
Springer Series in Materials Science 300,
https://doi.org/10.1007/978-3-030-43009-2_1

1.1 Introduction

A capacitor is a passive two-terminal electrical device, which stores electrical energy in form of an electric field. It was invented by Ewald Georg von Kleist. A capacitor is otherwise known as a condenser. The property that determines the ability of a capacitor is termed as its capacitance. If there are two conductors in a circuit separated by a dielectric material, then a capacitance exists, and in this case, the resultant capacitor can be specifically designed to provide a capacity enhancement, which is favorably used in several practical applications. The size, shape and the positions of conductors in a circuit, and the dielectric material between the conductors should be specifically chosen in order to obtain a better performance. Figure 1.1 shows a typical curve between the specific power and the specific energy for different types of energy storages devices. It can be inferred from the graph that the capacitors have high specific power and low specific energy, while the opposite is true for batteries and fuel cells.

It has been observed that the lines of force originating from a conductor usually terminate on the conductor, which is in the proximity. When a potential difference is applied (using a voltage source) between two conductors, which are placed opposite to each other, an electric field gets induced between two faces of the conductor and the direction of the field is from higher to lower potential [2]. This electric field results to the accumulation of equal and opposite charges over the two plates. Here, no current moves through the dielectric; nonetheless, there is a charge movement over the source circuit. If this condition is kept up for adequately extensive stretch of time, the current flowing throughout the source circuit halts. However, if a time-varying voltage is applied to the capacitor, the source encounters a continuous current as a result of the charging and discharging cycles of the capacitor.

Fig. 1.1 Ragone plot of various energy storage devices (redrawn and reprinted with permission [1])

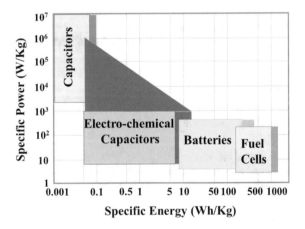

1.2 Parallel Plate Model

A capacitor is generally consisting of combination of two conductors placed opposite to each other separated by vacuum, air or insulating (dielectric) materials. The
elementary model of a capacitor as shown in Fig. 1.2 consists of two parallel plate
conductors having area A separated by distance d using a dielectric medium having
permittivity ε.

This simple model can also be used for qualitative assessment of some other
geometry of capacitors. By considering charges ($\pm Q$) on the surface of two plates,
the amplitude of the surface charge density may be expressed as $\pm\rho = \pm Q/A$ on their
surface. Assuming the area, A of the plates to be much larger in comparison to the
separation distance d, we can neglect the edge effect and approximate the uniform
distribution of electric field E between the two conducting plates. The magnitude
of electric field in this case is given by $E = \rho/\varepsilon$, where ε is the permittivity of the
medium [2, 3]. The potential (V) developed in this case may be determined by the
line integral of the $\mathbf{E}$ between the two parallel plates as [4]

$$V = \int_0^d \mathbf{E} \cdot \mathbf{dz} = \frac{\rho d}{\varepsilon} = \frac{Qd}{\varepsilon A} \qquad (1.1)$$

where V is the voltage, $\mathbf{E}$ is the electric field, $\mathbf{dz}$ is the vector along the field $\mathbf{E}$, d is
the separation distance, ρ is the surface charge density, ε represents the permittivity
of the dielectric medium, Q is the charge and A is the area.

Fig. 1.2 Schematic of
parallel plate capacitor

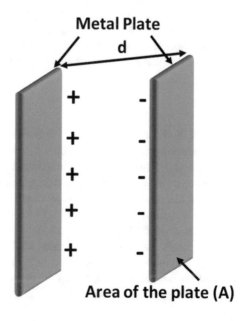

Further by solving (1.1), and using the relationship $C = Q/V$, following relationship may be obtained

$$C = \frac{\varepsilon A}{d} = \frac{k\varepsilon_0 A}{d} \tag{1.2}$$

where ε is the permittivity of medium, ε_0 is free space permittivity ($= 8.854 \times 10^{-12}$ F/m) and k is the relative permittivity of the dielectric material inserted between the two plates ($k = 1$ for free space, $k \approx 1$ for air and $k > 1$ for other natural materials).

From (1.2), it can be inferred that the capacitor basically depends on the dimensions of the structure and the medium of separation. Firstly, the capacitance can be enhanced either by increasing the cross-sectional area of conductors or by reducing the separation between two conductor plates. Secondly, the capacitance shows direct dependency on k also referred as the dielectric constant of the material, where higher the value of k leads to the larger value of the capacitance. This dependency may be attributed to the microscopic dipole moments induced in the dielectric material when it is introduced between conducting plates in order to compensate for the effect of originated charges on the plates and adjust the relation. These microdipoles, effectively, appear as if negative charges are lined up against the positive plate, and positive charges against the negative plate. A dielectric material having more number of dipoles can effectively hold higher electric field inside it and leads to higher value of k. Hence, a higher capacitance value can be obtained by fabricating the capacitor with high dielectric constant, by reducing the distance between the conducting plates or by increasing the area of conducting plates.

A parallel plate capacitor stores an absolute extent of energy until it reaches to the dielectric breakdown voltage. Every dielectric material used in the capacitor has a specific value of dielectric strength given by U_d, which decides the breakdown voltage of the capacitor as $V = V_{bd} = U_d d$.

The maximum electrostatic energy that can be stored in a dielectric media placed between the plates during charging in a linear capacitor is [2]

$$E_s = \frac{1}{2}CV^2 = \frac{1}{2}\frac{\varepsilon A}{d}(U_d d)^2 = \frac{1}{2}\varepsilon A d U_d^2 \tag{1.3}$$

where E_s is the energy stored, C is the capacitance, V is the voltage, U_d is the dielectric strength, d is the separation distance, A is the area and ε is the permittivity.

Equation 1.3 reveals that the maximum energy, which can be acquired in the capacitor, shows proportional linear dependency on dielectric volume and permittivity, and it also shows parabolic dependency on dielectric strength. Hence, by maintaining the overall volume of the capacitor, i.e., by increasing area and decreasing separation distance between two plates with the same amount, maximum amount of electrostatic energy stored remains unaltered inside the capacitor. In (1.1) and (1.2), it is assumed that the electric field is entirely concentrated in the interior of the dielectric layer placed between the two conducting plates. But practically, there are some fringing fields coming outside the dielectric material resulting from the

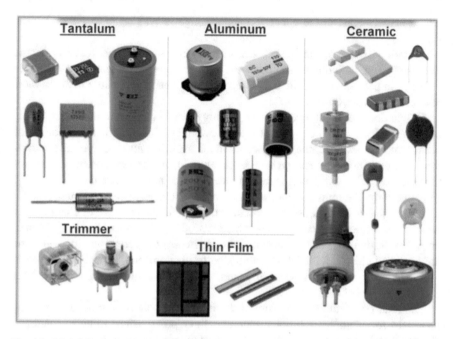

Fig. 1.3 Digital images of different types of capacitors [reprinted with permission (*image source* https://eeeproject.com/types-of-capacitors/)]

edges of the two plates, which show the enhancement in the effective capacitance value of the capacitor. This edge effect can be made negligibly small, by increasing the ratio of the area of the conducting plates to the separation distance between the plates. Figure 1.3 shows the images of various types of capacitors used in practical application of electronic device.

1.3   Dielectric Polarization Mechanism

A perfect dielectric may be defined as insulator having real value of permittivity, with no AC power losses or DC leakage. Since the dielectric does not pose any free electrons to conduct electricity, under application of electric fields the electrons (associated to molecules or atoms) get displaced from their position in distorted format, which gives rise to the polarization [5]. The polarization phenomenon can be understood using different mechanisms: the electronic polarization, the vibrational polarization, the dipolar or orientational polarization and the space charge or interfacial polarization.

Electronic polarization usually occurs in neutral atom. In this case, when an electric field is applied, the electron clouds get displaced with respect to the nucleus giving rise to a dipole, which vanishes after removing the field. This polarization would usually exist in the entire frequency range in which the capacitor is used [5, 6].

Vibrational or ionic polarization occurs mostly in ionic materials. When an electric field is applied, a dipole is induced in the ionic materials by mean of stretching or contraction between the positive and negative ions until the bonding between them restricts further movement [5–7].

Orientational or dipolar polarization mostly occurs in the polar dielectrics having permanent dipoles like polymers or dipolar ceramics. Under normal situation, these permanent dipoles are randomly oriented in the materials making zero net polarization. However, after application of external electric, the dipoles reorient themselves as per the electric field direction thus augmenting the net polarization.

Commonly, the translational polarization is correlated with the existence of the migrating charge carriers over microscopic distance in the presence of electric field, which results into the accumulation at the electrodes or interfaces like grain boundaries , voids, defects, etc. This accumulation increases the permittivity by distorting the localized electric field. Mostly, this kind of phenomenon has been observed in lower frequency regions in the multiphases or heterogeneous materials like polymer nanocomposite [8].

The total polarization is the sum of the individual polarization occurring in the materials, for example, in dipolar materials, three polarizations (dipolar, ionic and electronic polarization) are summed up to give the total polarization [7].

On the basis of polarity, the dielectric materials can be divided into two regions:

(1) Polar dielectrics possess permanent dipoles with no net polarization in the absence of electric field [2]. However, they possess net permanent polarization in the presence of the applied electric field. Materials in this category are water, HCl, NH_3, etc.
(2) Non-polar dielectrics are those, which do not possess any polarization in absence of electric field but achieve the temporary polarization in the presence of the electric field. Examples are N_2, O_2, CH_4, etc.

Using the model of parallel plate capacitor described above, it was observed that the capacitance value would be directly proportional to the dielectric constant of the insulator material. The dielectric constant usually provides an idea about the storage of electrical energy in the material when it is fully charged. This energy is stored in materials by means of polarization, which can be permanent or induced. The permanent dipoles are aligned parallel to each other in the insulator and in line with the internal electric field. The induced polarization occurs in presence of applied electric field. For the first-order dependency, the induced polarization $\mathbf{P}_{ind}$ and the permanent polarization $\mathbf{P}$ are directly proportional to the applied electric field $\mathbf{E}$ using the following expression [9]

$$\mathbf{P}_{ind} = X_{ind} \cdot \mathbf{E} \tag{1.4}$$

$$\mathbf{P} = X_e \cdot \mathbf{E} \tag{1.5}$$

P_{ind} and E vectors are parallel to each other, X_{ind} and X_e are induced and permanent dipole moment susceptibility, which can be further represented by K_p (relative dielectric constant) as

$$K_p = (1 + X_{ind} + X_e) \tag{1.6}$$

Hence, for the capacitor application in order to store more energy, the materials should have high value of K_p, which is directly related to the polarization of the materials.

1.4 Terminologies

1.4.1 Capacitance

Capacitance is described as the change in electrical charge on the conductors to the corresponding change in potential between them. The SI unit of the capacitance is the "Farad", denoted by letter F. Capacitance values of the conventional dielectric capacitors used in general electronics range are from ~1 pF (10^{-12} F) to ~1 mF (10^{-3} F). The expression for capacitance basically varies with respect to the geometry of the capacitor.

(a) For parallel plate capacitor (as shown in Fig. 1.2), capacitance C is given by [4]

$$C = \frac{\epsilon_r \epsilon_0 A}{d}, \tag{1.7}$$

where ε_r and ε_0 are the relative and vacuum permittivity respectively, A is the area of the parallel plate and d is the separation distance between the two conducting plates.

(b) In spherical capacitor, the parallel plates are replaced with the concentric charge sphere of different radius separated by a dielectric material (permittivity $\varepsilon_r\varepsilon_0$) as shown in Fig. 1.4.

The equal and opposite charge ($+Q$ and $-Q$) will be generated on the inner and outer surfaces of the sphere. The potential difference V will be generated against the two surfaces, which is given by [3]

$$V = \frac{Q}{4\pi \varepsilon_r \varepsilon_0} \left(\frac{1}{a} - \frac{1}{b} \right) \tag{1.8}$$

where Q is the charge on the surface of the sphere, a and b are the inner and outer radius of the sphere, respectively, and ε_r and ε_0 are the relative and vacuum permittivity, respectively. By comparing (1.8) with $C = Q/V$, capacitance C is written as:

Fig. 1.4 Schematic of spherical capacitor

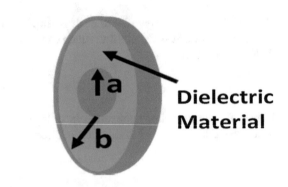

Fig. 1.5 Schematic of cylindrical capacitor

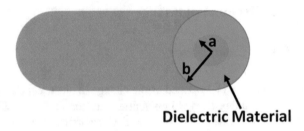

$$C = 4\pi \varepsilon_r \varepsilon_0 \frac{ab}{b - a} \tag{1.9}$$

(c) Similarly, the cylindrical capacitor consists of two coaxial conducting cylinders of same length L and having different radius separated with dielectric material (permittivity $\varepsilon_r \varepsilon_0$). The equal and opposite charge ($+Q$ and $-Q$) will be generated on the inner and outer cylinder surfaces. This is similar to the structure of the coaxial cable as shown in Fig. 1.5.

The capacitance C is given as [3]

$$C = \frac{2\pi \varepsilon_r \varepsilon_0 L}{2.3 \log_{10}(b/a)} \tag{1.10}$$

where L is the length of the cylinder, a and b are the radius of inner and outer cylinders, respectively.

(d) The two-wire capacitor consists of two wires having length L and radius a placed parallel to each other at distance d separated by a dielectric material having relative permittivity $\varepsilon_r \varepsilon_0$ as shown in Fig. 1.6.

The equal and opposite charge ($+Q$ and $-Q$) will be generated on the wires. The potential V is given by [10]:

$$V = \frac{Q}{\pi \varepsilon_r \varepsilon_0 L} \ln\left(\frac{d}{2a} + \sqrt{\frac{d^2}{4a^2} - 1}\right) \tag{1.11}$$

Fig. 1.6 Schematic of parallel wire

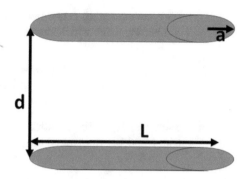

Again, by comparing (1.9) with $C = Q/V$, capacitance C is written as

$$C = \frac{\pi \varepsilon_r \varepsilon_0 L}{\ln\left(\frac{d}{2a} + \sqrt{\frac{d^2}{4a^2} - 1}\right)} \tag{1.12}$$

(e) The capacitance for a single disk capacitor can be calculated analytically as [11]:

$$C = 8\varepsilon_r \varepsilon_0 a \tag{1.13}$$

where ε_r and ε_0 are relative and vacuum permittivity, respectively, and a is the radius of the disk.

(f) Two similar conducting spheres having radius a are separated by distance d ($>2a$) with a material having dielectric constant $\varepsilon_r \varepsilon_0$. The capacitance for this configuration is given by [12]

$$C = 2\pi \varepsilon_r \varepsilon_0 a \left\{ \ln 2 + \gamma - \frac{1}{2} \ln(k) + O(k) \right\} \tag{1.14}$$

where $k = 2D - 2$ and $D = d/2a$ (>1), γ is the Euler's constant.

(g) The two parallel coplanar strip with strip width a and spacing d having dielectric material with dielectric constant $\varepsilon_r \varepsilon_0$, and length L. The total line capacitance is given by Gevorgian and Berg [13]

$$C = \frac{\varepsilon_r \varepsilon_0 L K(\gamma)}{K(\gamma_0)}, \tag{1.15}$$

where $\gamma = \sqrt{1 - \gamma_0^2}$ and $\gamma = d/(2a + d)$.

(h) Electrode geometries are also analyzed with respect to the infinite ground/conducting wall to calculate the capacitance value.

(i) Wire having length L and radius a is placed at a height h parallel to the conducting plane which shows the capacitance C approximately as [14]

$$C = \frac{2\pi \varepsilon_r \varepsilon_0 L}{\ln\left[\frac{h}{a} + \sqrt{\frac{h^2}{a^2} - 1}\right]} \tag{1.16}$$

(ii) Sphere having radius a with a charge Q and the potential V at the surface with respect to the ground are given by Natarajan [3]

$$V = \frac{Q}{4\pi \varepsilon_r \varepsilon_0 a} \tag{1.17}$$

By comparing the (1.17) with $C = Q/V$, the capacitance C is given as

$$C = 4\pi \varepsilon_r \varepsilon_0 a \tag{1.18}$$

(iii) A charged sphere having radius a is placed at a distance h from the conducting wall. Capacitance C is given by [15]

$$C = 4\pi \varepsilon_r \varepsilon_0 \left(1 + \frac{1}{2}\log(1 + a/h)\right) \tag{1.19}$$

(iv) A toroid of circular section having minor and major radius of a and b, respectively, is placed parallel to the conducting plane at the height h between the geometrical center of the toroid and the conducting plane. By using image method, capacitance C of the toroid can be written as [14]

$$C = \frac{4\pi^2 \varepsilon_r \varepsilon_0 b}{\ln\frac{8b}{a} - K\left(\gamma^2\right)\gamma} \tag{1.20}$$

where $\gamma^2 = \frac{b^2}{b^2 + h^2}$ and $K(\gamma^2)$ is the complete elliptic integral of fist kind having modulus γ^2

1.4.2 Dielectric Parameters

Dielectric parameters are quite important for various applications involving these types of materials. The basic dielectric parameters are represented in terms of the real and the imaginary parts of the dielectric constant and associated loss tangent. Dielectric constant is the property of material, which provides information about the material's ability to store or hold the electric energy. Dielectric constants of few insulating materials are given in Table 1.1. If the dielectric is not perfect, then it also

Table 1.1 Dielectric constant of few insulating materials

Materials	Dielectric constant	References
Vacuum	1	[18]
Air	1.006	[18]
Polypropylene (PP)	2	[18]
Polyphenylene sulfide (PPS)	2	[18]
Poly(tetrafluoroethylene)	1.9	[19]
Polynaphthalene	2.2	[19]
Transformer oil	2.2	[9]
Paraffin	2.25	[9]
Polystyrene	2.6	[19]
Polycarbonate (PC)	2.8	[20]
Polyquinoline	2.8	[19]
Silicon oil	2.8	[9]
Polyester (PET)	3	[18]
Polyester (PEN)	3.3	[18]
Polyester (PET)	3.3	[20]
Poly(ether ketone ketone)	3.5	[19]
Impregnated paper	2.0–6.0	[18]
Fused quartz	3.85	[9]
SiO_2	3.9–4.5	[21]
Paraelectric ceramics (Class 1)	5.0–9.0	[18]
Diamond	5.58	[21]
Mica	6.8	[18]
Aluminum oxide	8.5	[18]
Polyvinylidenefluoride (PVDF)	12	[20]
ZrO_2	25	[19]
Tantalum oxide	27.7	[18]
La_2O_3	30	[19]
TiO_2	80	[19]
$SrTiO_3$	2000	[19]
Barium titanate (Class 2)	3000–8000	[18]

has some dielectric loss associated with it. In this case, the permittivity becomes complex, which can be expressed as

$$\varepsilon^* = \varepsilon' + j\varepsilon'' = \left(\frac{1}{j\omega C_0 Z^*}\right) = \left(\frac{-Z''}{\omega C_0 \left(Z'^2 + Z''^2\right)}\right) + j\left(\frac{Z'}{\omega C_0 (Z'^2 + Z''^2)}\right)$$

$$(1.21)$$

where ε' and ε'' are the real and imaginary parts of the complex permittivity (ε^*), respectively, ω is the angular frequency, z' and z'' are the real and imaginary parts of the complex impedance (Z^*), respectively, and C_0 is the capacitance in vacuum [16]. Low frequency behavior of real part is originated from AC conductivity, whereas at higher frequency, the dipoles will not have sufficient time to orient themselves as per the applied field; hence, the real part decreases with frequency. Typically, the temperature-assisted dielectric parameters are mainly associated with the polar dielectrics as compared with the non-polar dielectrics. Dielectric parameters usually show increment with increasing temperature for polar dielectric materials. Real part ε' and imaginary part ε'' resemble to storage of energy and dissipation of energy in presence of applied electric field. The imaginary part has to be very low for an energy storage device like capacitor. This energy loss component is directly associated with the ac conductivity as [16]

$$\varepsilon'' = \frac{\sigma_{ac}}{\omega\varepsilon_0} \qquad (1.22)$$

The dissipation factor or loss tangent (tan δ) is related to dielectric losses, which represents the measure of the rate of energy loss during polarization and depolarization of dielectric materials [17]. Loss tangent is calculated as the ratio of the imaginary to the real part of the permittivity [16]

$$\tan \delta = \frac{\varepsilon''}{\varepsilon'} \qquad (1.23)$$

In general, the losses are due to dipolar, interfacial and conduction losses and associated with different polarization mechanism. Loss tangent is more favorable at matching of the applied electric field frequency and relaxation time. This energy loss in form of heat not only decays the energy storage capacity of the capacitor but also increases the equivalent series resistance (ESR), which is coupled with the capacitor. Hence, for capacitor application, it is desirable to control the dielectric parameters such that ε_r is maximum and loss tangent is minimum. Polarization and relaxation are usually found to be dependent on frequency of the applied field as well as on temperature of the system. Hence, the ε_r and dissipation factor show great dependency on temperature, voltage and frequency of the applied voltage, which affect the capacitance value [17].

1.4.3 Breakdown Voltage

The dielectric of the capacitor becomes conductive after applying a specific electric field, which is termed as the dielectric strength of the material E_{ds}. The applied voltage at which this phenomenon happens is known as the capacitor breakdown voltage, V_{bd}. The expression for breakdown voltage in a parallel plate capacitor is

$$V_{bd} = E_{ds}d \qquad (1.24)$$

Hence, the maximum energy storage is restricted by breakdown voltage of the capacitor [22]. The dielectric strength of few insulating materials is given in Table 1.2.

Very thin layers of dielectric are used in capacitors, and hence, absolute breakdown voltage of capacitors is thus limited. In general electronic applications, the

Table 1.2 Dielectric strength for some of the insulating materials

Materials	Dielectric strength (kV/mm)	References
Polyetherimide film (26 μm)	486	[23]
Fused silica (SiO_2)	470–670	[24]
Polytetrafluoroethylene (PTFE) film	87–173	[25]
Micas (muscovite, ruby, natural)	118	[25]
Transformer oil	110.7	[26]
Boron nitride (BN)	37.4	[25]
Silicone rubber	26–36	[25]
Neoprene	15.7–27.6	[25]
Polypropylene (PP)	23.6	[25]
Butyl rubber	23.6	[25]
Polychlorotrifluoroethylene	19.7	[25]
Polymethyl methacrylate (PMMA)	19.7	[25]
High-density polyethylene (HDPE)	19.7	[25]
Epoxy resin	19.7	[25]
Polystyrene (PS)	19.7	[25]
Polyetherketone (PEK)	18.9	[27]
Polyethersulfone (PES)	15.7	[25]
Ethylene-tetrafluoroethylene copolymer	15.7	[25]
Polycarbonate (PC)	15	[25]
Alumina (99.9% Al_2O_3)	13.4	[25]
Polyvinyl chloride (PVC)	11.8–20	[27]
Silicone oil (Basilone M50)	10–15	[28]
Aramid paper	12.2	[25]
Zirconia (ZrO_2)	11.4	[29]
Perfluorinated polyethers		
Galden XAD (Mol. wt. 800)	10.5	[28]
Transformer oil Agip ITE 360	9–12.6	[28]
Room-temperature vulcanized silicone rubber	9.2–10.9	[30]
Polyvinylidene fluoride	10.2	[25]
Aluminum silicate (Al_2SiO_5)	5.9	[25]
Air	1.41	[31]

typical ratings for capacitors used are ranging from a few V to 1 kV. The thickness of the dielectric must be high to make high-voltage capacitors, which have larger charge storage per capacitance than those rated for lower voltages. The most predominant factor, which critically affects the breakdown voltage for a capacitor, is the geometry of the plates. Pointed ends or edges might lead to dielectric breakdown, which may result into short circuiting. The short-circuit in the capacitor leads to drawing of current from the nearby circuit, which may result in explosion since the capacitor dissipates the energy [32]. When the electric field is sufficiently high, it can pull electrons from the atoms of the dielectric material, which leads to the dielectric into conducting state resulting into breakdown of the capacitor. There are many other possible scenarios, for example, impurities present inside the dielectric materials, the nature of the dielectric (amorphous, crystalline, etc.), imperfections in the crystal structure. The above-mentioned facts may cause an avalanche breakdown, which are mostly observed in the case of semi-conducting devices like diodes and transistors. The various parameters that affect the breakdown voltage of the capacitor are humidity, pressure and temperature [33]. The dielectric materials commonly used are paper, glass, ceramic, mica, plastic film, and oxide layers. For high-voltage applications, the capacitor is fabricated having vacuum between their plates and they exhibit low losses, but the disadvantage is that their capacitance is very low.

1.4.4 Equivalent Circuit

A capacitor can be described practically by a lumped circuit consisting of an ideal capacitance value C in series with an equivalent series resistance (ESR) and an equivalent series inductance (ESL). By considering DC leakage resistance of the dielectric media, a shunt resistance is introduced in parallel to C. The equivalent circuit is shown in Fig. 1.7 [9], where the series resistance Rs represents the resistance offered by contact and electrodes of the capacitor and shows dependency on the peak current as well as the applied frequency.

Leakage resistance R_p shown parallel to C arises in the capacitor due to the resistivity, and it represents the dielectric loss, which are much larger than the frequency-dependent loss. Inductance L mainly depends on the geometrical parameters, and it has very less dependency on frequency and voltage.

Fig. 1.7 Circuit model of a practical capacitor (redrawn and reprinted with permission [34])

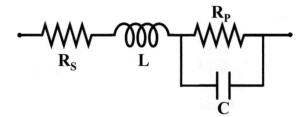

1.4.4.1   Equivalent Series Resistance

Ideally, a capacitor charges and discharges, but it does not dissipate any energy, which is not true when it is used practically. All the fabricated capacitors display some inadequacy due to the fabrication technique, defects in materials used for fabrication, humidity, temperature, etc. These flaws create some resistance, which is termed as the equivalent series resistance (ESR) for a capacitive component. The ESR is described as the AC resistance offered by capacitor, R_s and R_p combined together, to reveal a single resistance R_{ESR} in the equivalent circuit, which shows all the resistive losses in the capacitor. The reactance of the capacitor X_c shows dependency on frequency represented as $X_c = 1/j\omega C$ (imaginary), and the ESR adds up as a real component to the overall complex impedance Z_c and is mathematically expressed as [9]

$$Z_c = X_c + R_{ESR} = \frac{1}{j\omega C} + R_{ESR} \tag{1.25}$$

where X_c is the reactance offered by capacitance, ω is the angular frequency, C is the capacitance value of capacitor and R_{ESR} is the equivalent series resistance of the capacitor.

The capacitive reactance approaches to zero at infinite frequency and hence the ESR becomes significant, which leads to the overall impedance to become completely resistive. When the reactance becomes negligible, the power dissipation is given by

$$P_{RMS} = \frac{V_{RMS}^2}{R_{ESR}} \tag{1.26}$$

where P_{RMS} is the root mean square (RMS) power, V_{RMS} is the RMS voltage and R_{ESR} is the equivalent series resistance of the capacitor.

1.4.4.2   Equivalent Series Inductance

Similar to ESR, an equivalent series inductance (ESL) exists in the capacitor at higher frequency, i.e., the reactance offered by induced inductive component dominates in the capacitor. The reactance of this inductive component is given by $X_L = j\omega L$ and usually adds up in the series with the capacitor [9]. Hence, the overall impedance is now modified for angular frequency ω as

$$Z_c = R + j(\omega L - 1/\omega C) \tag{1.27}$$

where R is the resistance, ω is the angular frequency, L is the inductance and C is the capacitance.

The inductive reactance is positive showing increment with respect to frequency; at higher frequency, this inductive reactance dominates over the capacitive reactance.

By continuously lowering the frequency, a point will arrive, where both inductive and capacitive reactances are equal and cancel out each other's effect. At this point, the frequency is called as the self-resonating frequency (SRF) and the impedance is purely resistive ($Z = R$) at this point. Hence, to avoid the large RMS currents inside the capacitor, one needs to use the device far below SRF. If the conductors are isolated by a material of a little conductivity contrasted with an ideal dielectric, at that point a little spillage of current flows among them, and subsequently, the capacitor has a limited parallel resistance [34]. The capacitor gradually discharges after some time, and the discharging time is changed depending on the anode material.

1.4.5 Power Factor

The power factor is the ratio of ESR to its impedance, which gives electrical losses of capacitor. In practical capacitor, due to termination or impurities in dielectric media, the voltage is not lagging by exact 90° [9]. The power factor PF is represented in percentage and is usually below 10%, and PF have same values as dissipation factor, DF.

$$\%PF = \frac{R_{ESR}}{Z_C} \times 100\% \tag{1.28}$$

where PF is the power factor, Z_C is the overall impedance and R_{ESR} is the equivalent series resistance.

1.4.6 Dissipation Factor

Dissipation factor (DF) is the ratio of ESR to the capacitive reactance, i.e.,

$$\%DF = \frac{R_{ESR}}{X_C} \times 100\% \tag{1.29}$$

where R_{ESR} is the equivalent series resistance and X_C is the reactance offered by the capacitor. In advanced capacitor, the DF is described with respect to the frequency, temperature, biased voltage as it varies for different dielectric materials [9]. Generally, the DF shows the increment with respect to frequency and is a measure of the power loss in the capacitor.

1.4.7 Q Factor

The quality factor otherwise known as Q factor of a capacitor is a figure of merit for a capacitor and is reciprocal of DF. It is defined as the ratio of the reactance to its resistance at a certain frequency, which gives a measure of its efficiency [9]. The large value of the capacitor's Q factor implies that the capacitor approaches closer to the ideal capacitor (i.e., zero loss) behavior. The Q factor of a capacitor is expressed mathematically as

$$Q = \frac{X_c}{R_c} = \frac{1}{\omega_c R_c} \tag{1.30}$$

where Q is the quality factory, X_c is the capacitive reactance, R_c is the resistance and ω_c is the angular frequency.

The Q factor can be applied in capacitor discharging application to evaluate the reduction in life of capacitor due to the net applied reversal voltage. In this case, results are totally dependent on the nature of insulating media.

1.4.8 Leakage Current

The leakage current is generally representing the average flow of DC current through the capacitor in AC biasing after certain stabilization time (typically varies from few minutes to few hours) [9].

1.4.9 Insulation Resistance

This basically depends on the DC voltage and the leakage current, defined as the ratio of applied DC biasing voltage over capacitor terminals to the leakage current [9]. This resistance is strongly dependent on the time and having values near to megaohms.

1.4.10 Ripple Current/Voltage

The ripple voltage and current are the maximum allowed RMS value of AC current and voltage, when superimposed to the same DC voltage level. Ripple current is the AC segment of a connected power source in which the power source is having either steady or fluctuating frequency [9]. Ripple current leads to the production of heat inside the capacitor, since the dielectric material generates some heat due to the dielectric losses resulting from the applied frequency-assisted field through the resistive supply lines. These phenomena can be modeled with the ideal capacitor values by inserting ESR in the circuits. Precaution of the maximum ripple current/voltage ratings specified by the manufacturer is to be taken care while using the capacitor in any circuit for design as exceeding the ripple current leads to the failure or explosion of electrical component/circuits. If different types of capacitors are used, for example, the solid tantalum electrolyte-based capacitor uses manganese dioxide having lower value of ripple current, and then, it requires high ESR. Ceramic capacitors have no limitation of ripple current and have lower value of the ESR. However, exceeding certain value of ripple current may lead to the degradation in this case even though the ESR rating is low.

1.4.11 Capacitance Instability

Generally, the value of capacitance deteriorates due to aging of capacitor. For example, in case of ceramic capacitors, the degradation of dielectric leads to decrement of the capacitance value. The most significant factors affecting the aging process of the capacitors are the nature of dielectric, ambient operating temperature, storage temperature and operating voltage. By heating the component above the Curie point, the aging process can be reversed. Aging happens fastest at the beginning of life of the component, but after some time, the capacitor achieves stability. In the case of electrolytic capacitors, aging happens as a result of the evaporation of the electrolyte component [4]. The capacitance generally shows linear dependency on the temperature and sometimes shows nonlinear dependency at some extreme temperature region. The temperature coefficient can be positive or negative and depends on the nature of materials used for capacitor fabrication. This temperature dependency is usually indicated as parts per million (ppm) per °C.

1.4.12 Power Dissipation

To transfer the charges from one plate to other conducting plate, some work must be done in the direction opposite to the electric field of charged plates. In the absence of this external supply, the separation of charge continues in the electric field and the energy stored inside is likely to release, and the charges are permitted to get back to their equilibrium state. The work done (W) in this electric field is established, and the amount of stored energy is [35]

$$W = \int_0^Q V(q)\mathrm{d}q = \int_0^Q \frac{q}{C}\mathrm{d}q = \frac{1}{2}\frac{Q^2}{C} = \frac{1}{2}CV^2 = \frac{1}{2}VQ \qquad (1.31)$$

where W is the work done, Q is the total accumulated charge, q is the individual charge, V is the potential and C is the capacitance.

For time-varying voltage $V(t)$, the stored energy will also vary, and hence, the power must either flow in or out of the capacitor. This power mathematically represents the derivative of the stored energy with respect to time and is given by

$$P = \frac{dW}{dt} = \frac{d}{dt}\left(\frac{1}{2}CV^2\right) = CV(t)\frac{dV(t)}{dt} \qquad (1.32)$$

where P is the power, W is the work done, C is the capacitance, V is the voltage and t is the time.

The losses of a real capacitor can be modeled as an ideal capacitor with the ESR, which dissipates the power during charging or discharging cycle of the capacitor. For an AC input voltage, the dissipated power in the ESR is mathematically represented as

$$Pd_{rms} = \frac{V_{rms}^2 R_{esr}}{R_{esr}^2 + \left(\frac{1}{2\pi fC}\right)^2} \qquad (1.33)$$

where Pd_{rms} is the power dissipated in the ESR, V_{rms} is the rms voltage, R_{esr} is the equivalent series resistance, f is the frequency and C is the capacitance.

1.4.13 Current–Voltage Relation

Current $I(t)$ is known by flow of charge per unit time through any electric component present in the circuit. But in the case of a capacitor, the dielectric layer is present, which restricts the charge carriers to flow through it. Due to this dielectric layer, electrons get accumulated on the cathode (negative electrode), equal and opposite charges get accumulated over the anode (positive electrode) forming a charge depleted region between the two electrodes. Hence, the charges present on the electrodes are proportional to voltage V (1.1) and are equal to the time integral of the current. To solve the antiderivative form, the constant of integration is added, which is represented by the initial voltage $V(t_0)$. The integral form of the capacitor equation is given by [4]

$$V(t) = \frac{Q(t)}{C} = \frac{1}{C}\int_{t_0}^{t} I(\tau)d\tau + V(t_0) \qquad (1.34)$$

where $V(t)$ is the voltage, Q is the charge, C is the capacitance, I is the current and t is the time.

Taking the derivative of (1.34) and multiplying by C produce the derivative form

$$I(t) = \frac{dQ(t)}{dt} = C\frac{dV(t)}{dt} \tag{1.35}$$

1.4.14   DC and AC Response

1.4.14.1   DC Circuits

Figure 1.8 shows a series circuit, which contains a resistor, a switch, a capacitor, a constant DC source of voltage V_0.

If the capacitor is initially not charged, however, the switch will be opened and closed at t_0, and it will follow the Kirchhoff's voltage law [4]

$$V_0 = V_r(t) + V_c(t) = i(t)R + \frac{1}{C}\int_{t_0}^{t} i(\tau)d\tau \tag{1.36}$$

where V_0 is the constant DC source voltage, V_r is the voltage across the resistance R, V_C is the voltage across the capacitance C and the parameter i represents the current through the series circuit.

Taking the derivative of (1.36) and multiplying by C, the first-order differential equation is obtained as shown below

$$RC\frac{di(t)}{dt} + i(t) = 0 \tag{1.37}$$

where R is the resistance, C is the capacitance, i is the current and t is the time.

At, $t = 0$, the voltage through the capacitor will be zero and the voltage across the resistor will be V_0. The initial current will be then $I(0) = V_0/R$. From this hypothesis, solving the differential equation, (1.37) becomes [4]

Fig. 1.8 Charging circuit of a capacitor

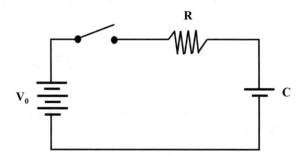

$$I(t) = \frac{V_0}{R} e^{-\frac{t}{\tau_0}} \tag{1.38}$$

$$V(t) = V_0 \left(1 - e^{-\frac{t}{\tau_0}}\right) \tag{1.39}$$

$$Q(t) = CV_0 \left(1 - e^{-\frac{t}{\tau_0}}\right) \tag{1.40}$$

where V is the voltage, I is the current, Q is the charge, C is the capacitance, R is the resistance, V_0 is the voltage at resistance R at $t = 0$, t is the time and $\tau_0 = RC$ is the time constant of the system.

When the capacitor reaches to the equilibrium with the source voltage, the voltage through the resistor and the current through the entire circuit decay exponentially. When the capacitor is discharging, the initial voltage (V_i) of the capacitor replaces V_0 [3]. Then (1.38–1.40) become

$$I(t) = \frac{V_i}{R} e^{-\frac{t}{\tau_0}} \tag{1.41}$$

$$V(t) = V_i e^{-\frac{t}{\tau_0}} \tag{1.42}$$

$$Q(t) = CV_i \, e^{-\frac{t}{\tau_0}} \tag{1.43}$$

where V is the voltage, I is the current, Q is the charge, C is the capacitance, R is the resistance, V_i is the initial voltage, t is the time and $\tau_0 = RC$ is the time constant of the system.

1.4.14.2   AC Circuits

When a capacitor is connected to a sinusoidal voltage source, it causes a displacement of current flow over it. In this case, the voltage source is $V_0 \cos(\omega t)$, and therefore, the displaced current (I) can be expressed as [4]

$$I = C \frac{dV}{dt} = -\omega C V_0 \sin(\omega t) \tag{1.44}$$

where C is the capacitance, V is the voltage and t is the time. When $\sin(\omega t) = -1$, the capacitor will experience a maximum current of $I_0 = \omega C V_0$.

Therefore, the ratio of the peak voltage to the peak current is designated as the capacitive reactance, which is symbolized by X_c

$$X_c = \frac{V_0}{I_0} = \frac{V_0}{\omega C V_0} = \frac{1}{\omega C} \tag{1.45}$$

When X_C approaches to zero, ω approaches to infinity. In this case, the capacitor looks like a short wire and passes current dramatically at high frequencies, whereas if X_C approaches to infinity, ω approaches to zero. In this case, the capacitor seems like an open circuit passes current poorly at low frequencies. The current of the capacitor can be expressed in the form of cosine to better compare with the voltage of the source,

$$I = -I_0 \sin(wt) = I_0 \cos(wt + 90°) \tag{1.46}$$

In this condition, the current is out of phase with the voltage by $+\pi/2$ radians or $+ 90°$. In other words, the current leads the voltage by 90°.

1.4.15 Self-discharge

The term self-discharge varies for different energy storage systems. For a battery, it is a phenomenon for energy or voltage reduction due to internal reactions of chemical, whereas for capacitor, without any chemical reactions, the open-circuit voltage or voltage drops or self-discharge results from the leakage of current between electrodes and decreased the energy accumulated by capacitor [36]. Self-discharge of a charged capacitor can occur when some Faradaic electron transfer process or processes take places at and below the maximum potential attained on charge. A capacitor in the charged condition is in a state of high positive free energy relative to that for the discharged or partially discharged state. The term self-discharge is sometimes related to the Faradaic reactions discharging on the surface and eliminating any physical processes that cause the voltage drop (e.g., charge redistribution) [37]. Self-discharge can be used as an umbrella term surrounding both the chemical reactions and physical processes that leads to the voltage drop. The term activation-controlled Faradaic self-discharge will signify both the rate of limitation (activation-control) as well as the fact that it is a faradaic chemical reaction [37].

1.4.16 Dielectric Absorption

Losses are observed in all the fabricated capacitors in which a loss associated with dielectric material is formally called as dielectric soaking or dielectric absorption. RC circuiting model of dielectric absorption is shown in Fig. 1.9 in which C is the ideal capacitance of capacitor and RC branches determine the dielectric absorption [38].

Even after discharging the capacitor, the charge will be cured on the surface of electrode, which is due to dielectric absorption. This phenomenon can be explained on the basis of time-dependent polarization and depolarization. During polarization

Fig. 1.9 RC circuiting model of dielectric absorption

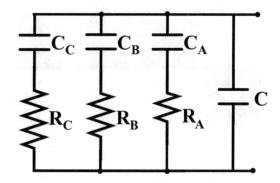

and depolarization, charges will get assembled near to the grain boundaries especially in polycrystalline materials. Typically, the dielectric time constant for different dipoles varies from nanoseconds to few minutes or even larger in case of permanent polarization. Dielectric absorption can be determined in frequency domain as well as in time domain.

In case of frequency domain, voltage and current phases are measured, which show less than 90° of difference due to dielectric absorption. These phases can be used in solving the *RC* branches of the equivalent circuit [38].

In time domain analysis, it consists of two stages, firstly, the capacitor is charged by voltage for enough amount of time, so that all the *RC* branches get charged (time for voltage applied is greater than the time constant of all the individual *RC* branch). Secondly, after charging, the capacitor is short-circuited for a small time until the ideal capacitance *C* is discharged while maintaining the almost same charge on all the *RC* branches. At the end, by opening the circuit voltage across all the branches immediately after discharging of the capacitor, the value of this voltage increases from zero to the asymptotic values [38].

1.5 Structure

The structure of capacitors for various applications is quite different, and hence, various types of capacitors depending on the physical structure are generally utilized. Yet, fundamentally, the capacitor is arranged by utilizing two electrical conductors as metallic plates isolated by a dielectric material. The electrical conductor can be a foil, sintered dab of metal, slight film or an electrolyte. The electrically protecting dielectric acts to expand the charging limit of the capacitor.

1.5.1 Capacitors in Parallel

Figure 1.10 shows that two capacitors are connected in parallel.

Capacitors connected in parallel configuration each have the same applied voltage. When they are connected in parallel, their capacitance adds up. The net capacitance, C_{Total} is the sum of instantaneous charge build-up in the individual capacitors (C_1 and C_2) [2, 4]

$$C_{Total} = C_1 + C_2 \qquad (1.47)$$

When n number capacitors are connected in parallel (Fig. 1.11), the net equivalent capacitance will be the sum of their individual capacitance. Potential difference across each capacitor is same, but the charge on each capacitor is different.

1.5.2 Capacitors in Series

Figure 1.12 shows that two capacitors are connected in series combination.

When they are connected in series, their capacitance adds up. Figure 1.13 illustrates the series combination of many capacitors.

The net capacitance, C_{Total} is given by [2, 4]

$$C_{Total} = \frac{C_1 C_2}{C_1 + C_2} \qquad (1.48)$$

When n numbers of capacitor are connected in series, the net reciprocal capacitance is given by sum of their reciprocal individual capacitor. In series combination of capacitor, charges on the capacitors are same, but voltage across each capacitor is different. Hence, while making series connection we have to ensure that the voltage across any of the capacitor should not exceed their maximum rated potential. These

Fig. 1.10 Illustration of two capacitors connected in parallel configuration

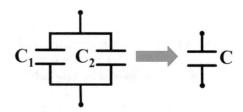

Fig. 1.11 Illustration of many capacitors connected in parallel configuration

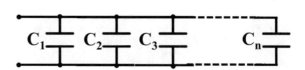

Fig. 1.12 Illustration of the series connection of two capacitors

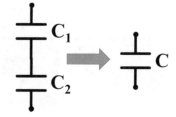

Fig. 1.13 Illustration of the series connection of many capacitors

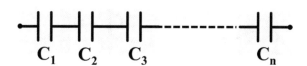

types of connections are indeed used to achieve a higher working voltage, for example, in the case of smoothing a high-voltage power supply. When the capacitance and leakage currents of each capacitor are identical, the voltage ratings get added up. In certain applications, the series strings are connected in parallel, which forms a matrix and the goal of doing the same is to maximize the energy storage of the network without overloading any capacitor. Series combination of capacitors is also utilized in bipolar AC to adjust the polarized electrolytic-based capacitors.

1.6   Capacitor Types

Different types of capacitors are fabricated based on the type of electrodes or dielectric materials and their geometries (shape, size, thickness, etc.). Figure 1.14 shows the different types of the capacitors included in this chapter.

1.6.1   Paper Capacitors

For attaining relatively high-voltage performance, paper was used as a choice of dielectric material in the past. Paper capacitor is a capacitor that uses paper as the dielectric to store electric charge. It consists of aluminum sheets and paper sheets. But the disadvantage of paper is that it absorbs moisture, which decreases the insulation resistance of the dielectric. Hence, it is covered or soaked with oil or wax to protect it from outside harmful environment. Paper capacitors are the fixed type of capacitors that means these capacitors provides fixed capacitance (capacitance means ability to hold or store electric charge). In other words, the paper capacitor is a type of fixed capacitor that stores fixed amount of electric charge. Based on the construction, the paper capacitors are classified into two types. They are (i) paper sheet capacitor and

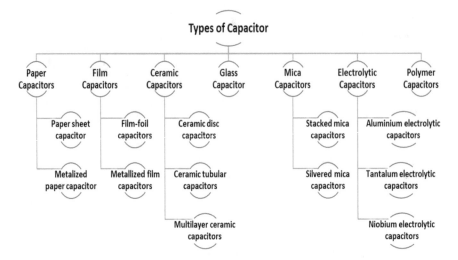

Fig. 1.14 Different types of the capacitors

(ii) metalized paper capacitor. The paper capacitors are used in high-voltage and high-current applications [39].

1.6.1.1 Paper Sheet Capacitor

The paper sheet capacitor is constructed by using two or more aluminum sheets and paper sheets. The paper sheet placed between the aluminum sheets acts as dielectric, and the aluminum sheets act as capacitor electrodes. The assembly of a paper capacitor is shown in Fig. 1.15A.

The paper sheet is poor conductor of electricity, so it does not allow flow of electric current between the aluminum sheets. The paper sheets and aluminum sheets are rolled in the form of cylinder, and wire leads are attached to both ends of the aluminum sheets (Fig. 1.15B). The entire cylinder is then coated with wax or plastic

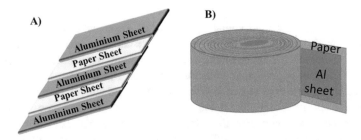

Fig. 1.15 A Assembly of a paper sheet capacitor and **B** rolling-up of paper sheet capacitor in the form of a cylinder

resin to protect it from moisture present in the air. The paper sheet capacitors are used in the high-voltage and high-current applications.

1.6.1.2 Metalized Paper Capacitor

In the case of a metalized paper capacitor, the paper is coated with thin layer of zinc or aluminum. The paper coated with zinc or aluminum is rolled in the form of cylinder. The entire cylinder is then coated with wax or plastic resin to protect it from moisture. The zinc or aluminum coated on the paper acts as electrodes, and the paper acts as dielectric. Paper capacitors that are made of zinc are easily destroyed due to chemical action. Therefore, aluminum is widely used for the construction of paper capacitors. The size of metalized paper capacitor is very small when compared with the paper sheet capacitor. In metalized paper capacitor, the aluminum is directly coated on the paper. Therefore, aluminum layer of metalized paper capacitor is very thin compared to the aluminum layer of the paper sheet capacitor. The schematic of a metalized paper capacitor is shown in Fig. 1.16.

1.6.2 Plastic Film Capacitor

The paper in the paper sheet capacitors has been largely substituted by plastic film in plastic film capacitors. Film capacitors are also known as plastic film capacitors or film dielectric capacitors. Plastic film capacitor is a capacitor, where aluminum or zinc is used as electrode that is separated by a plastic film as the dielectric to store electric charge. The plastic film capacitors offer better stability and aging performance when compared to that of the paper capacitors. The most noteworthy advantage of plastic film dielectrics over natural dielectrics (e.g., paper and mica) is that the plastic film could be manufactured, which can be made to meet particular prerequisites, such as thickness of dielectric and higher temperature resistance. Since they are non-absorbent, their moisture absorbent characteristics are predominant to those of mica. In plastic film capacitors, polyester, polypropylene, polyethylene terephthalate and polyphenylene sulfide are commonly used as dielectrics. Polyethylene terephthalate

Fig. 1.16 Schematic of a metalized paper capacitor

(PET) and polycarbonate (PC) are the more common plastic films incorporated as dielectric. Capacitors utilizing polyethylene terephthalate as the dielectric are the most widely recognized of the plastic film types available today in the market. The main preferred reason for polyethylene terephthalate dielectric capacitors is the high order of insulation resistance to temperature. They are generally used in circuits, where loss and high insulation resistance is required. Large numbers of plastic film capacitors are widely used in power factor correction circuits, suppression circuits and motor start circuits. Plastic film capacitors are classified into two categories. They are (i) film-foil capacitors and (ii) metalized film capacitors [40]. The advantages of film capacitors include high stability, low cost and low losses even at high frequencies.

1.6.2.1   Film-Foil Capacitors

The film-foil capacitor is made of two plastic films or sheets, each layered with thin aluminum metal foil or sheets, which are used as electrodes. The plastic sheets and aluminum sheets are then rolled into a jelly roll in the form of a cylinder, and wire leads are attached to the both ends of aluminum sheets using soldering technique or spraying metal technique. Polyester, polypropylene, polyethylene terephthalate and polyphenylene sulfide are commonly used as dielectric in film capacitors [39]. In plastic film capacitors, the aluminum sheets are mostly used due to their low cost and high conductivity acting as electrodes and plastic sheets acting as dielectric. Thickness of plastic film is the separation distance of the capacitor, and its operating area is determined by area of electrodes. The schematic of a film-foil capacitor is depicted in Fig. 1.17.

Plastic film capacitors are basically constructed as per their use in applications of electrical circuit, which formally requires the higher value of insulation resistance, lower values of loss factor over the wide ranges of temperature, lower dielectric absorption and where the AC component of the aroused voltage smaller than that of the DC voltage rating. Film capacitors are mostly suitable for application of A/D (analog-to-digital) converters, filters, integrators, multivibrator timing circuits and many additional applications, where the stability of the capacitance is essential.

Fig. 1.17   Schematic of a film-foil capacitor

1.6.2.2 Metalized Film Capacitors

In metalized film capacitors, the aluminum sheet or foil is replaced by a layer of metal vacuum deposited on the film layer having thickness of approximately 1/100 times that of the metal foil. This reduced thickness saves the space and provides higher resistance from metalized electrode. Connecting edges of metals to leads can be done only by spraying the metal instead of soldering. The most commonly used metal layer is aluminum or zinc that is extremely thin. The plastic film layers made of synthetic material act as dielectric, and the aluminum layers act as electrodes. The major advantage of film dielectric capacitors over natural dielectric capacitors is that the plastic film is synthetic or artificial. Therefore, it is possible to increase the thickness and heat resistance of the dielectric material. In other words, it can be said that the thickness and heat resistance of the plastic film capacitor can be tuned. The schematic of a metalized film capacitor is depicted in Fig. 1.18.

The metalized capacitors have a self-healing character called as clearing. Two different types of breakdown are found in this case: (1) momentary breakdown (complete breakdown lasting only for a small time interval) and (2) sudden decline in resistance of insulation for large period of time and then returning back to normal. Metalized film capacitors are ranked for both the continuous DC and AC operation under the limitation that the value of voltage for the AC component is smaller in comparison to the DC voltage. The circuits, which can permit the irregular momentary breakdown, can use these capacitors. Metalized PC film capacitors are basically proposed for their use in SCR commutating, bypass applications, power supply filter circuits, etc., whereas the metalized PET film capacitors can be used to replace the commercial general-purpose capacitor and can be used in industries electronic equipment requiring coupling as well as decoupling.

Fig. 1.18 Schematic of the construction of a metalized polymer film capacitor winding without the end connections (redrawn and reprinted with permission [41])

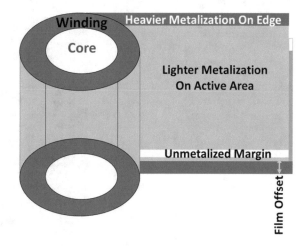

1.6.3 Ceramic Capacitors

In a ceramic capacitor, the ceramic material is used to construct the dielectric and conductive metals are used to construct the electrodes. Ceramic material is chosen as dielectric because ceramic materials are poor conductors of electricity, and hence, they do not allow flow of electric charges through them. The schematic of a ceramic capacitor (i.e., ceramic materials are sandwiched between metal electrode) is shown in Fig. 1.19.

The ceramic capacitors are usually little, reasonable and valuable for high recurrence applications; however, their capacitance shifts compellingly with the voltage and temperature, and they deteriorate partially. These capacitors additionally experience the effects of the piezoelectric. Ceramic capacitors are commonly grouped into two classifications: One is characterized as class 1 dielectrics that have certain variation of capacitance with respect to the temperature. The other is characterized as Class 2 dielectrics that can work at higher potential. As of now, multilayer ceramic capacitors are additionally used, which are practically very small; however, a few of them have shown naturally wide esteem resilience and microphonic issues and are commonly brittle. Ceramic capacitors are essentially planned for use, where a smaller size with having large electrical capacitance along with large resistance of insulation is required. The installation of lead in ceramic capacitors makes it more convenient for application of printed circuit. Common functioning ceramic capacitors are unengaged for applications, which require high accuracy but they are convenient for application like filter/bypass or as unfussy coupling elements used in high-frequency circuits, where considerable modification in capacitance induced by variation in temperature is permitted. These types of capacitor can be used in

Fig. 1.19 Schematic of a ceramic capacitor

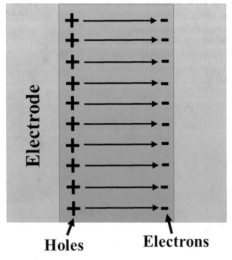

Ceramic Dielectric

Electrode

Holes   Electrons

frequency determining circuits also but some modification is recommended. Applications of ceramic capacitors include tone compensation, automatic volume control filtering, antenna coupling, resonant circuit, volume control RF bypass, lighting ballasts, etc. In these applications, frequency or voltage change, moderate changes due to the temperature and dissipation factor are not critical, and they will not affect the suitable functioning of the circuits. Variable ceramic capacitors are small-sized trimmer capacitors, which is used in application where repeatedly fine-tuning adjustments are necessary. Usually, they are used as trimming and coupling in oscillator circuits, filter and phase shifter. This type of capacitor is comparatively more stable in providing high shock and vibration resistance, which influence the changes in the capacitance value, whereas for more stability, air trimmers may be used. Based on the shapes, ceramic capacitors are divided into three sub-classes. They are (i) ceramic disk capacitor, (ii) ceramic tubular capacitor and (iii) multilayer ceramic capacitor [39]. The advantages of ceramic capacitor include high stability, low losses, high capacitance and small size.

1.6.3.1 Ceramic Disk Capacitors

The ceramic disk capacitors are manufactured by coating silver on both sides of the ceramic disk. The ceramic disk functions as the dielectric material and silver coated on both sides of the disk functions as capacitor electrodes. For low capacitance, a single ceramic disk coated with silver is used, whereas for high capacitance, multiple layers have been used. The leads are made of copper, which are attached to the ceramic disk by soldering. In order to protect from heat, a protective coating is applied to the ceramic disk capacitor. As usual, the area of a ceramic disk or dielectric and spacing between the silver electrodes also determines the capacitance of a ceramic disk capacitor. The main disadvantage of using ceramic disk capacitor is its high capacitance change with a slight variation in temperature.

1.6.3.2 Ceramic Tubular Capacitors

As the name implies, the ceramic tubular capacitor consists of a hollow cylindrical ceramic material. A coating of silver ink is applied on the inner and outer surfaces of the hollow cylindrical ceramic material. The hollow cylindrical ceramic material functions as the dielectric material, and the silver ink coated on inner and outer surfaces functions as capacitor electrodes.

1.6.3.3 Multilayer Ceramic Capacitors

The multilayer ceramic capacitor is manufactured of multiple layers of ceramic material and conductive electrodes placed one above the other. The conductive electrodes are placed between each layer of ceramic material, and these multiple layers of

Fig. 1.20 Schematic of a
multilayer ceramic capacitor
(redrawn and reprinted with
permission [43])

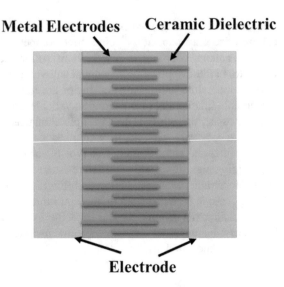

ceramic material act as the dielectric of the capacitor. The multiple layers of ceramic
dielectric and electrodes are connected by the terminal surfaces [42]. By this way,
multiple layers of ceramic dielectric and capacitor electrodes can be fabricated. Some
of these capacitors contain hundreds of ceramic layers and electrodes in which the
thickness of each layer is in micrometers. The schematic of a multilayer ceramic
capacitor is depicted in Fig. 1.20.

1.6.4 Glass Capacitor

The fabrication of glass capacitors is easy and simple like every other capacitor.
Basically, it consists of three elements like glass dielectric, aluminum electrodes and
leads for terminals [39]. These capacitors are constructed in multilayer patterns in
which leads are being welded to the electrodes, which avoid pressure connections to
come loose and melt due to soldering. The continuous glass ribbon is used as dielectric
to precisely control dimension, physical and electrical properties of the capacitor.
These capacitors are assembling as alternate layer of glass ribbons and electrode.
Further, the assembly units are sealed by applying high pressure and temperature,
which result in a single monolithic block. Although these capacitors are of monolithic
structure, they are not necessary hermetically sealed as the there is some mismatch
in temperature coefficient of block and the case.

These fabrication techniques have provided certain advantage to glass capacitors
like [39]

- Low dielectric absorption
- Higher insulation resistance

- Fixed temperature coefficient
- Used where miniaturization is demanded
- Can be operated in high temperature and highly corrosive environment
- Life can be extended to 30,000 h and even more.

Glass has certain advantage in terms of corrosion or degradation, delamination, high resistance to microfractures and many other problems, which are associated with most of the crystalline materials. Furthermore, the axial glass capacitors can be hermetically encased in glass providing a proper glass-to-metal seal at the leads. This type of construction in glass capacitor is practically immune to environmental factors like solder heat, radiation hardness, humidity containing moisture, vibration or shock, salt spray, etc. [39].

Glass capacitors, which are fabricated using the inorganic materials provide high resistance to the breakdown voltage, operating temperature as well as resist the nuclear radiations, that means these capacitors will not become a toxic hazard when exposed to radiation. Gamma irradiation may affect this capacitor by permanently changing its capacitance value but have low dissipation factor with respect to frequency. Glass capacitors can be operated for wide range of frequency and temperature. They also exhibit zero aging rate, zero piezoelectric noise regardless of component age. Hence, traditionally, this type of capacitor has shown great interest in aerospace and military applications due to its high performance. Glass capacitors have applications across the entire spectrum of electronic circuits. There are failures in many electronic systems, which are not acceptable practically examples in satellite systems, and under sea repeater. To design such systems, glass capacitors are the best choice. These types of capacitors are favorable for commercial as well as industrial application, where performance of electrical circuits is critical. Also, it can bear higher radio frequency (RF) currents over a wider range of frequency [39].

Applications, where glass capacitors have been used and proven to provide better properties, are radar/microwave application, RF circuit, where cooling through fan cannot be fully trusted, fire alarm or safety control circuits, circuits of instruments, which are used for mechanical metal cutting, space/defence application where temperature or humidity affects the most.

1.6.5 Mica Capacitors

Mica capacitor is a capacitor in which it uses mica as the dielectric material. Mica is a silicate mineral found in rocks. Different types of dielectric materials used in mica capacitors are muscovite (otherwise called white mica), ruby (otherwise called rose mica) and amber mica. Mica capacitors are very reliable and high-precision capacitors. The advantages of mica capacitors include stable capacitance, high temperature operability, high working voltages, low losses and accurate. But the disadvantage relies on their huge cost. Mica capacitors are generally constructed as per the requirement of circuits for precise frequency filtering and coupling. They

Fig. 1.21 Schematic of a
stacked mica capacitor

are also used, where close impedance limits are crucial with respect to temperature, aging and frequency, for example, tuned circuits, which control frequencies, reactance or phases. Due to the fundamental characteristics of dielectric, mica capacitors are inexpensive, quite small in size, ease in availability along with the good stability and they are high reliable. These capacitors can be used in stable low power networks and delay lines. Button-style capacitors are designed to be used in frequencies up to 500 MHz and mostly used in very-high- and ultra-high-frequency circuit applications like coupling and bypassing [39]. Mica capacitors are used in high-frequency and RF applications, coupling circuits, resonance circuits, RADAR, LASER, space, filters, etc. Depending on the configuration, mica capacitors are two types. They are (i) stacked mica capacitors and (ii) silvered mica capacitors.

1.6.5.1 Stacked Mica Capacitors

The stacked mica capacitors are manufactured with thin mica sheets, which are arranged in a manner that one upon the other, and each mica sheet is separated by thin metal sheets. Either copper or aluminum is used as metal sheets. The entire unit is further enclosed in a plastic case in order to protect it from the possible mechanical damage and moisture in the ambient atmosphere. The terminals of a mica capacitor are connected at each end. The mica sheets placed in between two metal sheets function as the dielectric material that opposes the flow of electric current and copper or aluminum sheets function as capacitor electrodes. A schematic of stacked mica capacitor is shown in Fig. 1.21.

1.6.5.2 Silvered Mica Capacitors

The silvered mica capacitors are manufactured by coating silver on both sides of the mica sheets. The coating is performed using the screening technique [39]. To achieve a desired capacitance, several silver-coated mica sheets are arranged one over other.

The silver coated on the mica functions as electrodes for the capacitor and mica sheets functions as the dielectric material.

1.6.6 Electrolytic Capacitors

Electrolytic capacitor is a capacitor that uses an electrolyte-coated conducting plate as one of the electrodes in order to achieve higher capacitance. The symbol of an electrolytic capacitor is represented either by two parallel straight lines (European standard) or a straight line with a curved line (American standard), as shown in Fig. 1.22.

The electrolyte used will be an ionic conducting liquid, which enhances the charge storage within the capacitor. In electrolytes, the ions are of two types: anions (negatively charged ions) and cations (positively charge ions). The electrical conductivity of the electrolytic capacitor increases at an increase in the temperature and decreases at a decrease in the temperature. Hence, the capacitance of an aluminum electrolytic capacitor increases when the temperature increases and decreases when the temperatures decreases, which shows a typical temperature dependent nature of capacitance of an aluminum electrolytic capacitor. Most of the electrolytic capacitors are polarized, that is the voltage applied to the terminals must be in correct polarity (positive connection to positive terminal and negative connection to negative terminal). If it is connected in reverse or wrong direction, the capacitor may be short-circuited, that is a large electric current flow through the capacitor and that can permanently damage the capacitor. Depending on the type of dielectric used to manufacture the capacitor, the electrolytic capacitors are classified into three types. They are (i) aluminum electrolytic capacitors, (ii) tantalum electrolytic capacitors and (iii) niobium electrolytic capacitors. The schematic of an electrolytic capacitor is shown in Fig. 1.23.

The advantages of electrolytic capacitors include high capacitance and low cost. The disadvantages of electrolytic capacitors are high leakage current and short lifetime. These capacitors are widely used in filters and time constant circuits.

1.6.6.1 Aluminum Electrolytic Capacitors

Aluminum electrolytic capacitor is manufactured using two aluminum foils, aluminum oxide layer, an electrolytic paper and an electrolyte. In place of electrolytic paper, a paper spacer soaked in an electrolytic liquid or solutions may be used. Liquid

Fig. 1.22 American standard symbol of an electrolytic capacitor

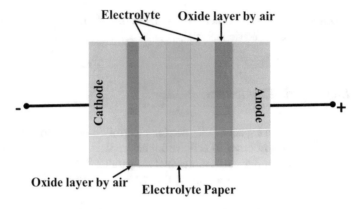

Fig. 1.23 Schematic of an electrolytic capacitor

or solid electrolytes are commonly used. In aluminum electrolytic capacitor, both the anode (positive electrode) and cathode (negative electrode) are made of pure aluminum foil. A thin layer of aluminum oxide (electrically insulating in nature) is coated with the anode aluminum foil and which functions as the dielectric of the electrolytic capacitor. The cathode- and oxide-coated anode is separated by an electrolytic paper, which is soaked in an electrolyte solution. The cathode aluminum foil is also covered with a very thin insulating oxide layer naturally formed by air. But the thickness of this oxide layer is very thin when compared to that of the one formed at the anode aluminum foil. The configuration of aluminum electrolytic capacitor looks like two capacitors connected in series with anode capacitance C_{anode} and cathode capacitance $C_{cathode}$. Hence, the expression for the total capacitance, C_{Total}, of an aluminum electrolytic capacitor can be written as [39]

$$C_{Total} = \frac{C_{anode} C_{cathode}}{C_{anode} + C_{cathode}} \tag{1.49}$$

These capacitors are mostly used in filter, coupling and bypass applications, which require large capacitance in small cases and where excessive value of capacitance can be tolerated over the nominal value. Polarized capacitors must be used only in DC circuits by observing the polarity properly. In case of pulsating DC, the sum of the peak of AC signal plus the applied DC voltage should never exceed the DC rating. The peak AC value always should be less in comparison with the applied DC voltage so as to maintain the polarity, for both negative and positive peaks, which avoids overheating and damage of the device [39].

1.6.6.2   Tantalum Electrolytic Capacitors

Tantalum electrolytic capacitors are just like other electrolytic capacitors, comprised of an anode, an electrolyte and a cathode. The design of this capacitor is such that the anode is completely isolated from cathode, which results to a very small leakage current. Three types of tantalum capacitors are there.

Foil-Type Tantalum Capacitor
In this type of capacitor, tantalum foil is used as anode, which forms a layer of oxides of tantalum after electromechanical treatment. The foil is etched in acid to enhance the effective area and provide high capacitance value followed by anodization in the solution under application of direct voltage, which produces pentoxide dielectric film over the surface of the foil. The tantalum foil capacitor is constructed in cylindrical pattern by rolling the two foils having paper saturated with electrolyte as spacer. They are more versatile and available in etched/plain foil and polarized/unpolarized format and hence make it more suitable to be used in electrical circuit application. But due to their wide tolerance, foil-type capacitor is not used in timing and precision circuits. Essential precautions should be taken when used in application of low voltages as some of these capacitors might show change in capacitance as well as in dissipation factor as low-voltage DC biasing [39].

Tantalum Solid Electrolyte Capacitor
Fabrication process of solid tantalum electrolyte capacitor is basically divided into four steps as shown in Fig. 1.24.

Firstly, the tantalum metal is first crushed into fine powder followed by high-temperature sintering in form of pellet, which produces a highly porous metallic anode to provide better capacitance value. Anode is further covered with insulating oxide, i.e., tantalum pentoxide as dielectric layer. This overall processing technique is termed as anodization. This step plays the crucial role, as the growth of oxide layer decides the dielectric thickness, which specifically validates the accurate values of capacitance as well as use to minimize the tolerance value of capacitor. Electrolyte is integrated over the anode via pyrolysis technique in the solid tantalum capacitor. Further, they are immersed in the solution followed by heating in the oven to get

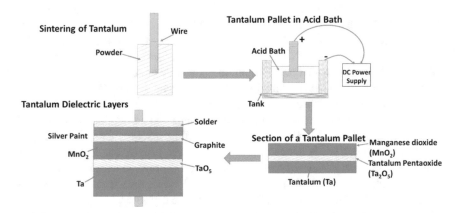

Fig. 1.24 Fabrication process of solid electrolyte capacitor

deposition of manganese dioxide. Above process is repeated to get enough thickness. At final stage, graphite or metal contacts are made to ensure proper cathode connection.

Tantalum Wet Electrolyte Capacitor

In wet tantalum capacitors, instead of solid electrolyte, a liquid electrolyte is utilized contradicting to the solid tantalum capacitors. Initial process for making anode followed by dielectric deposition is same as that of solid tantalum capacitor. After this process, the anode-coated dielectric is immersed into a cage containing the electrolyte solution. The enclosure composed of case and the electrolyte solution together perform the functioning of cathode in wet tantalum capacitors [44]. Tantalum capacitors have high gravimetric capacitance (capacitance per mass) and high volumetric capacitance (capacitance per volume) in comparison to other type of electrolytic capacitor. The main advantage of the tantalum capacitor is that they have lower leakage current, having low ESR value and can be operated at higher temperature. When it comes to costumer to make a choice out of the three tantalum capacitor, foil tantalum capacitor is chosen, where circuits are operated a high voltage or reverse voltage. Wet sintered anode or wet-slug tantalum capacitor possesses low DC leakage and hence used for the application, where DC leakage is critical. The tantalum case wet-slug units can be used up to +200 °C temperature under high ripple current and also can withstand reversal voltage of 3 V. These characteristics are more reliable over silver cases and hence can be used in military and aerospace applications [39]. Solid electrolyte designs are cheap and can be used in many applications, where their very tiny size for a given unit of capacitance is required and can tolerate maximum of 15% rated DC voltage [39]. They have advantage in performing at low temperature and provide better resistance to corrosion.

1.6.6.3 Niobium Electrolytic Capacitors

A new stream of niobium capacitor was established in the USA and Soviet Union in the 1960s, which is analogous to other available electrolytic capacitors. Niobium electrolyte capacitor consists of a solid electrolyte (cathode) on the surface of the layered niobium pentoxide (Nb_2O_5). This oxide layer acts as a dielectric material and is coated over the anode made of niobium monoxide or niobium metal. Niobium electrolytic capacitors are also available as surface-mount device (SMD) in the form of a chip capacitor, which can be directly mounted on the printed circuit board (PCB) surface. These capacitors have similar voltage and capacitance ratings as of tantalum chip capacitors [45]. Niobium electrolytic capacitors are polarized-type capacitor and hence while connecting the DC power supply, connection polarity should be taken care. For these types of capacitors, if the reverse voltage or the ripple current is supplied to be higher than the specified value, then it may damage the dielectric material and might result into failure of the capacitor. Niobium capacitor manufacturers must specify the special rules to design circuit for the user's safety as the destruction of the dielectric may have some vital catastrophic consequences.

1.6.7 Polymer Capacitors

Polymer capacitor is a stream of capacitors, which utilizes conductive polymers as the electrolyte. The main difference of polymer capacitors with the ordinary electrolytic capacitor is that the polymer capacitors use solid polymer electrolytes, whereas the latter use liquid or gel electrolytes. The problem of drying up the electrolyte is completely resolved with the help of solid polymer electrolytes. Electrolyte drying is an important factor, which determines the lifetime of electrolytic capacitors. There are numerous kinds of polymer capacitors, including polymer aluminum electrolytic capacitors, polymer tantalum electrolytic capacitor and hybrid polymer capacitors. Mostly, polymer capacitors have lower maximum rated voltage in comparison with ordinary electrolyte capacitors. Hence, we can directly replace ordinary electrolytic capacitors with polymer capacitor as long as the maximum rated voltage is not exceeding for polymer capacitor [4, 39]. Polymer capacitors have several advantages over the ordinary electrolyte capacitors such as, long lifetime, high maximum working temperature, good stability, low ESR and a much safer failure mode. But the disadvantage relies on the higher cost when compared to the wet electrolyte capacitors [46].

1.7 Color Code

In order to specify the capacitance information of a capacitor, color codes are used. Color codes are the information by which the capacitance is represented. In color coding technique, the capacitance value is marked on the body of the capacitors by using different colors. The colors painted on the capacitors body are called color bands. All the color bands painted on the capacitors body are used to indicate the capacitance value and capacitance tolerance. Each color painted on the capacitors body represents a different number. Generally, the capacitors are marked with four or more color bands. Depending on the number of color bands present on the capacitors, they are classified into two classes: (i) four-color band capacitor and (ii) five-color band capacitor [4].

1.7.1 Four-Color Band Capacitor

When the capacitors are marked with four color bands, they are called as four-color band capacitor. In a four-color band capacitor, the first and second color bands painted on the capacitor represent the first and second digits of the capacitance of the capacitor. The third color band represents the decimal multiplier, and the fourth color band represents tolerance. The four-color coding of the capacitor is schematically marked in Fig. 1.25.

Each color in the color coding of a capacitor signifies a particular value, and Table 1.3 illustrates the same.

1.7.2 Five-Color Band Capacitor

Similar to the four-color band capacitor, in five-color band capacitor, the first four color bands represent the same details. And the fifth color band represents voltage rating of the capacitor. Color codesin a five-color bandcapacitor are shown in Fig. 1.26.

Table 1.4 gives the details of the voltage ratings of the capacitor based on the color band on the capacitor.

1.8 High-Performance Capacitors

High-performance capacitors are also known as supercapacitors, also called an ultra-capacitor. It is a high-capacity capacitor with a capacitance value much higher than normal capacitors. This supercapacitor is electrochemical capacitor with relatively high energy density, typically on the order of thousands of times greater than an electrolytic capacitor. In supercapacitors, charges are stored by electrostatically and faradaic methods on the electrode materials. Electric double-layer capacitors (EDLCs)/pseudocapacitor (adsorption pseudocapacitor, redox pseudocapacitor and intercalation pseudocapacitor)/hybrid capacitor are distinct from the conventional capacitors in the terms of their charge storage. Electric double-layer capacitor (EDLC) stores charge electrostatically, whereas in pseudocapacitor, faradaic charge storage mechanism takes place. Hybrid capacitor is the combination of electrostatic

Fig. 1.25 Color codes in a four-color band capacitor table

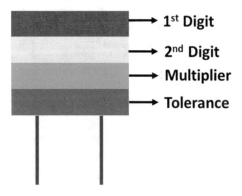

Table 1.3 Capacitor color code table

Band Color	1st digit	2nd digit	Multiplier	Tolerance (>10 pF)	Tolerance (<10 pF)	Temperature Co-efficient
Black	0	0	X1	± 20%	±2.0 pF	
Brown	1	1	X10	± 1%	±0.1 pF	-33X10^{-6}
Red	2	2	X100	± 2%	±0.25 pF	-75X10^{-6}
Orange	3	3	X1,000	± 3%		-150X10^{-6}
Yellow	4	4	X10,000	± 4%		-220X10^{-6}
Green	5	5	X100,000	± 5%	±0.5 pF	-330X10^{-6}
Blue	6	6	X1,000,000			-470X10^{-6}
Violet	7	7				-750X10^{-6}
Grey	8	8	X0.01	+80%, -20%		
White	9	9	X0.1	± 10%	±1 pF	

Reproduced and reprinted with permission. *Source* https://www.codrey.com/capacitor/capacitor-colour-code-values-with-examples/

Fig. 1.26 Color codes in a five-color band capacitor

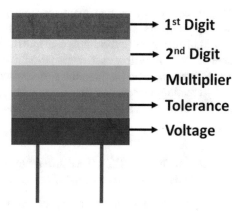

→ **1st Digit**

→ **2nd Digit**

→ **Multiplier**

→ **Tolerance**

→ **Voltage**

and electrochemical capacitor. In hybrid capacitor, asymmetric electrode materials are used for charge storage. These high-performance capacitors or supercapacitors consist of electrodes (most important is the electrical properties of electrode), electrolyte (most important is the breakdown voltage of electrolyte) and separators (prevent short circuits).

Storage of charge in these supercapacitors is a surface phenomenon, and hence, storage of charges increases with respect to the surface area of the electrodes used.

Table 1.4 Capacitor color code table

Band Color	Voltage Rating (V)				
	Type J	Type K	Type L	Type M	Type N
Black	4	100		10	10
Brown	6	200	100	1.6	
Red	10	300	250	4	35
Orange	15	400		40	
Yellow	20	500	400	6.3	6
Green	25	600		16	15
Blue	35	700	630		20
Violet	50	800			
Grey		900		25	25
White	3	1000		2.5	3

To increase the surface area, many researchers have used different materials like conducting polymers, carbon (particle, tubes, fibers, sheets, etc. [47]), metal oxide and hybrid of materials (hybrid polymer mostly used). Materials chosen depend on the ease of availability, low cost, degradation with time/environmental conditions. There are different ways to increase the surface area of electrodes by using nanomaterials or by creating nanoporous structures, where size of pores is minimized to achieve the high surface area [48]. Based on electrodes, supercapacitors have two types:

(1) symmetric or SEC's (uses same materials for electrodes as both anode and cathode).
(2) asymmetric or AEC's (uses different materials for electrodes as both anode and cathode), also known as hybrid capacitors.

Electrolyte breakdown voltage gives us the maximum attainable voltage or energy density. Power density depends on the ESR value of capacitor. It shows strong dependency on conductivity of electrolyte. There are two types of electrolytes used in these capacitors:

(1) Organic electrolytes are used mostly in commercial devices and can achieve voltages of the order of ~2.5 or 3 V. The high resistivity of electrolyte limits the power of cell.
(2) Aqueous electrolytes show comparatively lower voltages (~1 V).

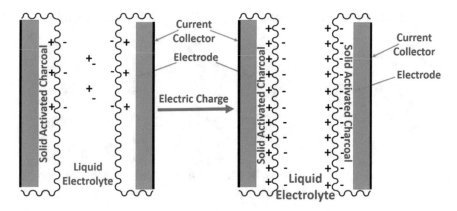

Fig. 1.27 EDLC charge storage mechanism (redrawn and reprinted with permission [48])

The selection of electrolyte and electrode influences the capacitance values, i.e., the ion size in electrolyte and nanopores on surface should be optimized. Separator functions to electrically separate the two electrodes so as to remove the chances of short circuits, but it allows ions to move allowing charges to transfer between two electrodes. Polymer/paper separators are mostly used for organic electrolyte, whereas ceramic/glasses are used for aqueous electrolyte. Separators having high electrical resistance and ionic conductance with lower thickness are used to get high performance of this class of capacitor.

The mechanism of charge storage is by means of electrochemical double-layer formation [49–52], and hence, this class of capacitors is also known as electrochemical capacitors or EDLC. The mechanism of charge storage in EDLC is shown in Fig. 1.27.

EDLCs store charges at the electrode/electrolyte interface by forming double layer of charges. These capacitors exhibit high energy densities when compared to that of conventional capacitors [53, 54]. The EDLC contains no conventional dielectric unlike ceramic or aluminum electrolytic capacitors. An electrolyte (either solid or liquid) is filled between two electrodes, instead. The schematic of supercapacitor ((a) electrical double-layer capacitor (EDLC), (b) pseudocapacitor (PC) using metal oxide and (c) hybrid supercapacitor (HSC)) is shown in Fig. 1.28. In case of EDLC, charge storage occurs at the interfaces between the electrolyte and electrodes (Fig. 1.28a), whereas in pseudocapacitors (PCs), it involves reversible and fast Faradaic redox reactions for charge storage (Fig. 1.28b). Most common example of electrode is metal oxide or any redox active molecule. When a supercapacitor stores charges by matching the capacitive carbon electrode with either a pseudocapacitive or lithium-insertion electrode (Fig. 1.28c), it is then called a hybrid supercapacitor (HSC).

The capacitance obtained is linearly proportional to the area of the surface formed by electrical double layer. Therefore, large surface area containing materials such as activated carbon is used for electrode of EDLC to achieve high capacitance.

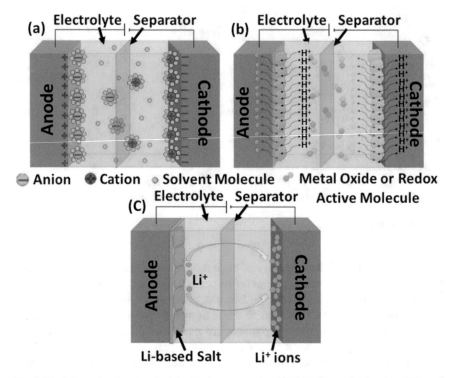

Anion   **Cation**   **Solvent Molecule**   **Metal Oxide or Redox Active Molecule**

Fig. 1.28 Schematic of **a** electrical double-layer capacitor (EDLC), **b** pseudocapacitor (PC) and **c** hybrid supercapacitor (HSC) (redrawn and reprinted with permission [55])

The charge storage process in these capacitors is completely different from that of conventional capacitor. Different models are suggested for the charging of these capacitors

(1) Classical equivalent circuit
     This is very simple and basic model suggested for supercapacitor as shown in Fig. 1.29.

Fig. 1.29 Classical equivalent circuit of a supercapacitor (redrawn and reprinted with permission [48])

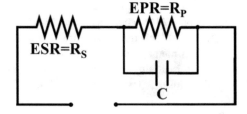

This shows that supercapacitor is practically represented by ESR and EPR and a capacitance, where EPR and ESR can be calculated as [48]

$$EPR = \frac{-t}{(\ln(V_2/V_1)C)} \tag{1.50}$$

$$ESR = \frac{\Delta V}{\Delta I} \tag{1.51}$$

where V_1 is the initial voltage, V_2 is the final voltage, t is the time, C is assumed to be equal to the rated capacitance, ΔV is the change in voltage and ΔI is the change in current.

By using the equivalent circuits, energy change (ΔE) (integral of instantaneous power) can be determined by

$$\Delta E = \left(\frac{C}{2}\right)(V_1^2 - V_2^2) \tag{1.52}$$

And hence capacitance by rewriting the above equation, we can get capacitance C as

$$C = \frac{2\int_{t1}^{t2} vi\,dt}{V_1^2 - V_2^2} \tag{1.53}$$

(2) Three-branch model

This model is given by three parallel branches of RC (as shown in Fig. 1.30) to achieve best-fitted data to the practical collected data [56].

The three branches have different time constant and are named accordingly. The first branch called as R_i branch or immediate branch mostly influences over few seconds, second one called as R_d or delayed branch shows dependence over few minutes range, and third branch, R_l or long branch influence for more than

Fig. 1.30 Three-branch model (redrawn and reprinted with permission [48])

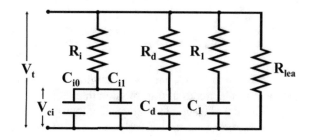

10 min. One more resistance R_{iea} shows the leakage current in the supercapacitor. Equations to determine the R_i, C_{i0} and C_{i1} are:

$$R_i = \frac{V_1}{I_{CH}} \tag{1.54}$$

$$C_{i0} = I_{CH}\left(\frac{t_2 - t_1}{\Delta V}\right) \tag{1.55}$$

$$C_{i1} = \frac{2}{V_4}\left(I_{CH}\left(\frac{t_4 - t_1}{V_4}\right)\right) - C_{i0} \tag{1.56}$$

where I_{CH} is the constant current for charging, V_1 is the voltage at time t_1, ΔV is the voltage change at time from t_1 to t_2, t_3 is charging time for V_3, t_4 is $t_3 + t_1$ and V_4 is the remaining voltage on supercapacitor after discharging for time t_1. In order to determine R_d in component in the delayed branch, discharging time for both battery and supercapacitor is calculated as t_5 for an obtained voltage of $V_5 = V_4 - \Delta V$. R_d can be calculated as:

$$R_d = \left(\frac{\left(V_4 - \frac{\Delta V}{2}\right)\Delta t}{C_{i0} + (C_{i1}V_3)}\right) \tag{1.57}$$

Another component of delayed branch C_d can be obtained by:

$$C_d = \frac{I_{CH}(t_4 - t_1)}{V_6} - \left(C_{i0} + \left(\frac{C_{i1}}{2}V_6\right)\right) \tag{1.58}$$

where t_6 is calculated as $t_5 + 3(R_dC_{i1})$ and V_6 is the voltage at time t_6. Similarly, for long-term branch, after discharging the battery and supercapacitor to time t_7, we got $V_7 = V_6 - \Delta V$. Hence, R_1 and C_1 can be obtained by equations:

$$R_l = \frac{\left(V_6 - \frac{\Delta V}{2}\right)\Delta t}{(C_{i0} + (C_{i1}V_3))\Delta V} \tag{1.59}$$

$$C_l = \frac{I_{CH}(t_4 - t_1)}{V_8} - \left(C_{i0} + \frac{C_{i1}}{2}V_8\right) \tag{1.60}$$

where Δt is $t_7 - t_6$ and V_8 is charging voltage after charging time of 30 min.

(3) Porous electrodes as transmission lines

Model of electrode as equivalent transmission lines is shown in Fig. 1.31.

Assuming, electrolyte resistance is very much higher than the bulk resistance. The capacitor acts as impedance element $Z = 1/j\omega C$ at higher frequency and

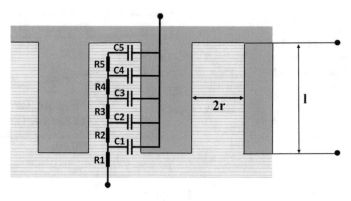

Fig. 1.31 Porous electrode representation as a five-element transmission line (redrawn and reprinted with permission [1])

there is negligible current flow in deep site of pore. The most of the current flows through $R1$ and $C1$ into the bulk of the material. The pores are assumed as cylindrical in shape (radius r, length l), for double layer capacitance, impedance can be calculated as [1]

$$Z(\omega) = \sqrt{\frac{R_w}{j\omega C}} \coth \sqrt{jwR_wC} \tag{1.61}$$

$$Z(0) = \sqrt{\frac{1}{j\omega C}} \tag{1.62}$$

$$Z(\omega \to \infty) = \sqrt{\frac{R_W}{j\omega C}} \tag{1.63}$$

Where C is the low frequency resistance, and R_w is the low frequency resistance spanning the 45° Warburg region. The total impedance is determined by adding contact resistance as well as the ionic resistance of bulk electrolyte and separator i.e. $R_{contacts} + R_{solid} + Z(\omega)$. The low frequency resistance, R_w, for an electrolyte filled porous conducting layer depends on the conductivity of the electrolyte κ and the porosity and given by [1]

$$R_W = \frac{l}{3\pi r^2 n\kappa} \tag{1.64}$$

Where l represents the thickness of active layer i.e., pore length in case of straight, cylindrical pores and r represents radius of the pore, and n represent number of pores. While charging the energy stored € by supercapacitor is given by [36]

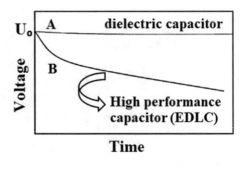

Fig. 1.32 Spontaneous voltage changes between electrodes of **a** "dielectric" or electrolytic capacitor and **b** supercapacitor, kept under open-circuit conditions (redrawn and reprinted with permission [36])

$$\epsilon = \frac{1}{2}CU^2 \tag{1.65}$$

where C is the capacitor and U is the voltage. This equation is also valid for electrolyte as well as dielectric-type capacitor. This phenomenon of voltage drops in Fig. 1.32 shows how spontaneously the changes occur in voltage for dielectric/electrolyte and supercapacitor during open-circuit condition. This voltage drop is also known as self-discharge or leakage current. It depends on capacitance, voltage, temperature, humidity and the chemical stability of the electrode/electrolyte combination. Supercapacitor self-discharge time is specified in hours, days or weeks.

Different test can be performed to determine various properties/parameter like power density, capacitor charging windows and charging/discharging cycles, AC impedance characterization, shock vibration test, lifecycle test, etc. Supercapacitors have broad applications and can be used where both high power and high energy density are required. Application is which it can be used are memory backup, electric vehicles, portable power supplies, adjustable speed drives, battery improvement, electrochemical actuators, renewable energy applications, power quality, etc. A detailed discussion on the special type of capacitor has been included in the next Chap. 2.

1.9   Concluding Remarks

A capacitor is a passive two-terminal electrical device, which stores electrical energy in form of an electric field and possesses high specific power and low specific energy. There are many basic terminologies associated with the capacitor such as the breakdown voltage, ripple current, power factor, dissipation factor, quality factor, self-discharge, dielectric absorption. Different types of capacitors are being fabricated based on different electrode and dielectric spacer and their usage for various applications. Paper capacitor uses paper as the dielectric to store electric charge for attaining relatively high voltage performance. Plastic film capacitor uses aluminum or zinc as electrodes that separated by a plastic film. In ceramic capacitors, the ceramic material is used to construct the dielectric. Electrolytic capacitor uses electrolyte-coated conducting plate as one of the electrodes in order to achieve higher capacitance. Polymer capacitor uses conductive polymers as the electrolyte. High-performance capacitors, i.e., supercapacitors (electric double-layer capacitors (EDLCs)/pseudocapacitor (adsorption pseudocapacitor, redox pseudocapacitor and intercalation pseudocapacitor)/hybrid capacitor) are distinct from the conventional capacitors in the terms of their charge storage. EDLC storage is a surface phenomenon, and hence, storage of charges increases with respect to the surface area of the electrodes used.

Acknowledgements The authors acknowledge the financial support provided by Department of Science and Technology, India (DST/TMD/MES/2K16/37(G)), for carrying out this research work. Authors are thankful to Ms Tanvi Pal for drafting few figures.

References

1. R. Kotz, M. Carlen, Electrochim. Acta **45**, 2483 (2000)
2. R. Katz Publications, H. Semat, R. Katz, *DigitalCommons@University of Nebraska -Lincoln 25-1 Capacitance of an Isolated Sphere in Vacuum* (1958)
3. R. Natarajan, *Power System Capacitors* (CRC Press, Boca Raton, 2006)
4. R.P. Deshpande, *Capacitor* (McGraw-Hill Education, New York, 2014)
5. W. Emphasis, P. Concepts, E. Processes, *Dielectiric Phenomena in Solids: With Emphasis on Physical Concepts of Electronic Processes* (Elsevier Academic Press, California, 2004)
6. D.W. Hess, K.F. Jensen, *Microelectronics Processing* (American Chemical Society, Washington DC, 1989)
7. T.W. Dakin, I.E.E.E. Electr, Insul. Mag. **22**, 11 (2006)
8. K.K. Kar (ed.), *Composite Materials: Processing Applications Characterization* (Springer, Berlin, 2017)
9. W.J. Sarjeant, in *Proceedings of Electrical Electronics Insulation Conference* (IEEE, Chicago, IL, USA, 1989), pp. 1–51
10. Harry E. Green, IEEE Trans. Microw. Theory Tech. **47**, 365 (1999)
11. H. Nishiyama, M. Nakamura, IEEE Trans. Compon., Hybrids, Manuf. Technol. **16**, 360 (1993)
12. A.D. Rawlins, IMA J. Appl. Math. (Institute of Mathematics and Its Applications) **34**, 119 (1985)
13. S. Gevorgian, H. Berg, in *2001 31st European Microwave Conference* (IEEE, London, England, 2001), pp. 1–4
14. J.R. Riba, F. Capelli, Energies **11**, 1 (2018)
15. J. Crowley, Proc. *ESA Annual Meeting on Electrostatics* 1 (2008)
16. J. Tahalyani, K.K. Rahangdale, R. Aepuru, B. Kandasubramanian, S. Datar, RSC Adv. **6**, 36588 (2016)
17. Q. Li, F.-Z. Yao, Y. Liu, G. Zhang, H. Wang, Q. Wang, Annu. Rev. Mater. Res. **48**, 219 (2018)
18. N.A.B. Zulkifli, M.A. Johar, O.M.F. Marwah, M.H.I. Ibrahim, IOP Conf. Ser.: Mater. Sci. Eng. **226** (2017)

19. P. Barber, S. Balasubramanian, Y. Anguchamy, S. Gong, A. Wibowo, H. Gao, H.J. Ploehn, H.C. Zur Loye, *Polymer Composite and Nanocomposite Dielectric Materials for Pulse Power Energy Storage* (2009)
20. L. Qi, L. Petersson, T. Liu, J. Internation, Council. Electr. Eng. **4**, 1 (2014)
21. M. Kohno, J. Photopolym. Sci. Tec. **12**, 189 (2008)
22. S.T. Pai, Q. Zhang, *Introduction to High Power Pulse Technology* (World Scientific, Singapore, 2009)
23. J.P. Zheng, P.J. Cygan, T.R. Jow, I.E.E.E. Trans, Dielectr. Electr. Insu. **3**, 144 (1996)
24. G.I. Skanavi, F. Dielektrikov, editors, *Oblast Silnykh Polei (Physics of Dielectrics; Strong Fields)* (Gos. Izd. Fiz. Mat. Nauk (State Publ. House Phys. Math. Sci.), Moscow, 1958)
25. W.T. Shugg (ed.), *Handbook of Electrical and Electronic Insulating Materials* (Van Nostrand Reinhold, New York, 1986)
26. W.R. Bell, IEEE Trans. Electr. Insu. **EI-12**, 281 (1977)
27. C.T. Lynch (ed.), *Practical Handbook of Materials Science* (CRC Press, Boca Raton, FL, 1989)
28. E. Forster, H. Yamashita, C. Mazzettii, M. Pompili, L. Caroli, S. Patrissi, I.E.E.E. Trans, Dielectr. Electr. Insu. **1**, 440 (1994)
29. R.A. Flinn, P.K. Trojan, *Engineering Materials and Their Applications*, 2nd ed. (Houghton Mifflin, 1981)
30. M.G. Danikas, I.E.E.E. Trans, Dielectr. Electr. Insu. **1**, 1196 (1994)
31. A.A. Al-Arainy, N.H. Malik, M.I. Qureshi, I.E.E.E. Trans, Dielectr. Electr. Insu. **1**, 305 (1994)
32. S. Paul, *Paul Scherz - Practical Electronics for Inventors 2 E (2006, McGraw-Hill_TAB Electronics).Pdf*, second (Mc Graw Hill, 2006)
33. J. Biird, *Electrical Circuit Theory and Technology*, 3rd edn. (Elsevier Ltd., Burlington, 2007)
34. M. Hallikainen, F. Ulaby, M. Abdelrazik, I.E.E.E. Trans, Antennas Propag. **34**, 1329 (2004)
35. P. Hammond, *Electromagnetism for Engineers*, 3rd edn. (Pergamon Press, Great Britain, 1986)
36. A. Lewandowski, P. Jakobczyk, M. Galinski, M. Biegun, Phys. Chem. Chem. Phys. **15**, 8692 (2013)
37. H.A. Andreas, J. Electrochem. Soc. **162**, A5047 (2015)
38. J.C. Kuenen, G.C.M. Meijer, I.E.E.E. Trans, Instrum. Meas. **45**, 89 (1996)
39. C.J. Kaiser, *The Capacitor Handbook, First edit* (Springer, Netherlands, 1993)
40. N. Valentine, M.H. Azarian, M. Pecht, Microelectron. Reliab. **92**, 123 (2019)
41. J.S. Ho, S.G. Greenbaum, A.C.S. Appl, Mater. Inter. **10**, 29189 (2018)
42. H. Trinh, J.B. Talbot, J. Am. Ceram. Soc. **86**, 905 (2009)
43. S. Ducharme, ACS Nano **3**, 2447 (2009)
44. K.K. Kar, A. Hodzic (eds.), *Developments in Nanocomposites* (Research Publishing Services, Singapore, 2014)
45. T. Zednicek, B. Vrana, W. Millman, C. Reynolds, *Carts-Conference* **142** (2002)
46. K.K. Kar, A. Hodzic (eds.), *Carbon Nanotube Based Nanocomposites: Recent Development* (Research Publishing Services, Singapore, 2011)
47. R. Kumar, S. Sahoo, E. Joanni, R.K. Singh, W.K. Tan, K.K. Kar, A. Matsuda, Prog. Energy Combust. Sci. **75**, 100786 (2019)
48. P. Sharma, T.S. Bhatti, Energ. Convers. Manage. **51**, 2901 (2010)
49. J. Cherusseri, R. Sharma, K.K. Kar, Carbon **105**, 113 (2016)
50. J. Cherusseri, K.K. Kar, J. Mater. Chem. A **4**, 9910 (2016)
51. J. Cherusseri, K.K. Kar, Phys. Chem. Chem. Phys. **18**, 8587 (2016)
52. J. Cherusseri, K.K. Kar, RSC Adva. **6**, 60454 (2016)
53. R. Sharma, K.K. Kar, J. Mater. Chem. A **3**, 11948 (2015)

54. J. Cherusseri, K.K. Kar, RSC Advances **5**, 34335 (2015)
55. X. Chen, R. Paul, L. Dai, Natl. Sci. Rev. **4**, 453 (2017)
56. A. Sani, S. Siahaan, N. Mubarakah, Suherman, IOP Conf. Ser.: Mater. Sci. Eng. **309** (2018)

Chapter 2
Capacitor to Supercapacitor

**Soma Banerjee, Prerna Sinha, Kapil Dev Verma, Tanvi Pal,
Bibekananda De, Jayesh Cherusseri, P. K. Manna, and Kamal K. Kar**

Abstract Supercapacitors bridge the gap between conventional electrolytic capacitors and batteries. These are capacitors with electrochemical charge storage. The basic equations used to describe the capacitors are same in the case of supercapacitors but their mechanism of energy storage is different. Various electrode-active materials such as activated carbon, mesoporous carbon, carbon nanotubes, graphene, etc., are invariably used in the supercapacitors with high performance. Both aqueous and organic electrolytes are used in supercapacitors but high voltage can only be delivered by the supercapacitors manufactured with organic electrolytes. However,

S. Banerjee · P. Sinha · K. D. Verma · T. Pal · J. Cherusseri · K. K. Kar (✉)
Advanced Nanoengineering Materials Laboratory, Materials Science Programme, Indian Institute of Technology Kanpur, Kanpur, Uttar Pradesh 208016, India
e-mail: kamalkk@iitk.ac.in

S. Banerjee
e-mail: somabanerjee27@gmail.com

P. Sinha
e-mail: findingprerna09@gmail.com

K. D. Verma
e-mail: kdev@iitk.ac.in

T. Pal
e-mail: tanvipal93@gmail.com

J. Cherusseri
e-mail: jayesh@iitk.ac.in

B. De · K. K. Kar
Advanced Nanoengineering Materials Laboratory, Department of Mechanical Engineering, Indian Institute of Technology Kanpur, Kanpur, Uttar Pradesh 208016, India
e-mail: debibek@iitk.ac.in

T. Pal
A.P.J. Abdul Kalam Technical University, Lucknow 226031, India

P. K. Manna
Indus Institute of Technology and Management, Kanpur 209202, India
e-mail: pkmanna8161@yahoo.co.uk

© Springer Nature Switzerland AG 2020
K. K. Kar (ed.), *Handbook of Nanocomposite Supercapacitor Materials I*,
Springer Series in Materials Science 300,
https://doi.org/10.1007/978-3-030-43009-2_2

the cycle life of aqueous electrolyte-based supercapacitors is high when compared with the organic ones. The present and future flexible and wearable technologies necessitate the development of flexible solid-state capacitors to supply them power. Supercapacitors are found applications in a variety of fields such as electronics industry, hybrid electric vehicles, and power supplies. The two major demerits of the present supercapacitors are low energy density and high cost. Hence, novel low-cost supercapacitors should be developed with high energy density to fulfill the needs of society. The present chapter discusses the Faradaic and non-Faradaic processes, types of supercapacitors, structure—i.e., electrode, electrolyte, electrolyte membrane, and current collector—key parameters for estimation of performance, electrochemical characterizations, etc.

2.1 Introduction

Increased energy consumption and the depletion of fossil fuel lead to the thinking of novel alternative strategies to resolve the problem. Increasing energy consumption due to the increased levels of electronic devices in the society is a major challenge in reality. As the energy resources are limited, it is indeed mandatory to think the use of sustainable, renewable, and clean energy resources. The eco-friendly nature of renewable resources is highly welcomed for their implementation in energy production. Various renewable technologies such as solar, wind, and tidal are environmentally friendly, whereas the uses of fossil fuels are extremely harmful due to the production of CO_2 and other effluents to the atmosphere. However, such energy conversion devices are facing problems of their intermittent nature of energy production. The energy can't be produced at night when used solar technologies and cannot be produced at no wind when wind technologies are used. Hence, it is mandatory to couple these types of renewable energy conversion technologies with novel electrical energy storage systems. Electrical energy storage systems such as batteries and supercapacitors are promising energy storage devices in which they can fulfill the basic requirements of storing energy from the renewable energy conversion technologies. Supercapacitors are otherwise known as electrochemical capacitors. Hence, the electrical energy will be available round the clock as they can store energy for prolonged periods of time. Batteries and supercapacitors can function thus as uninterrupted power systems and hence have achieved much demand in the present scenario. Batteries are powerful in terms of their high energy densities, whereas the supercapacitors are beneficial in their high power densities. By coupling the renewable energy conversion technologies either with batteries or with supercapacitors, one can develop novel hybrid devices which can both produce and save energy simultaneously. Such hybrid energy technologies are very promising for the future electronic technologies to supply them power. In the case of heavy-duty applications, it is mandatory to have both energy density and power density simultaneously. Hence, novel hybrid devices have been developed by using batteries and supercapacitors. The performance of battery–supercapacitor modules is excellent, and they are suitable candidates in the

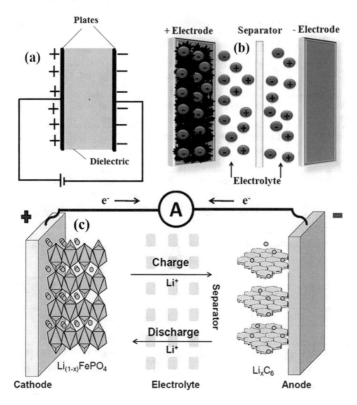

Fig. 2.1 Schematic of **a** capacitor, **b** supercapacitor, and **c** conventional Li-battery. Redrawn and reprinted with permission from [1, 2]

field of hybrid energy vehicles. But the fundamental aspects of capacitor and different types of capacitor are discussed in Chap. 1. Figure 2.1 schematically represents the conceptual difference among the three energy storage systems, i.e., capacitor, super-capacitor, and conventional Li-battery. For better understanding, Table 2.1 represents a comparative of capacitor, supercapacitor, and conventional Li-battery.

2.2 History

The working principle of electrochemical capacitors is first explained by Becker and Ferry. They have systematically studied the working mechanism of an electrochem-ical capacitor, which they have assembled, and the same is patented later although the performance of the device is not awesome [6]. In 1960, research groups have been involved in the manufacturing of electrochemical capacitors and they tried to bring the same to market however, failed as the capacitors got technical failure. The popularity of conventional capacitors was also one of the potential reasons for their

Table 2.1 Comparison among capacitor, supercapacitor, and conventional battery devices [1–5]

Parameters	Capacitor	Supercapacitor	Battery
Components	Conducting plates, dielectric material	Electrodes, electrolyte, separator, current collector	Electrodes, electrolyte, separator
Charge storage mechanism	Electrostatic	Electrostatic and electrochemical	Electrochemical
Charge stored	Between charged plates	Interface of electrode and electrolyte	Entire electrode
Points to be focused	Geometric area of electrode, dielectric	Electrode surface area, microstructure, electrolyte	Active mass, electrolyte, thermodynamics
Energy density (Wh/kg)	<0.1	1–10	20–150
Power density (W/kg)	$\gg 10,000$	500–10,000	<1000
Operating voltage (V)	6–800	2.3–2.75	1.2–4.2
Charge time	10^{-6} to 10^{-3} s	Seconds to minutes	0.3–3 h
Discharge time	10^{-6} to 10^{-3} s	Seconds to minutes	1–5 h
Cycle life (Cycles)	$\gg 10^{6}$($\gg$10 years)	>10^{6} (>10 years)	~1500 (3 years)
Operating temperature (°C)	−20 to 100	−40 to 85	−20 to 65

rejection as the dielectric capacitors were in the market more than hundred years at that time. After tremendous struggle and technical corrections, the first electrochemical capacitor came to exist in the market was the 'MaxCap™,' [7]. The different types of MaxCap™ electric double-layer capacitors are shown in Fig. 2.2.

2.3 Faradaic and Non-Faradaic Processes

In electrochemical analysis, Faradaic and non-Faradaic processes represent two fundamental modes of electrode characteristic (Figs. 2.3 and 2.4). In Faradaic process, the charge transfer takes place during redox reaction at the electrode. However, a mere presence of redox reaction at the electrode side does not imply Faradaic process. For the conduction of the Faradaic process, electronic charge injected into an electrode during redox reaction should be transferred away from it and charge should not be stored at the electrode either. Generally, Faradaic process is observed in lead–acid batteries and fuel cells. In non-Faradaic process, no charge transfer takes place. Here, ionic and electronic charges remain at or in the electrode similar to adsorption and desorption processes. In non-Faradaic process, either there will be no redox

Fig. 2.2 MaxCap™ electric
double-layer capacitors [8]

reaction at all or there will be redox reaction with the reacting species staying at the electrode material. Non-Faradaic process is observed in electric double-layer capacitor (EDLC), intercalation, and electrode with redox active surface functionalities. In ideally polarized electrode, where no charge transfer takes place across metal solution and the interface, only Non-Faradaic process takes place. In charge transfer electrode, Faradaic and non-Faradaic both processes occur simultaneously [9].

2.4 Types of Supercapacitors

Based on the mechanism of charge storage, supercapacitors are divided mainly into three categories (Fig. 2.5): (i) electric double-layer capacitors (EDLCs), (ii) redox capacitors, and (iii) hybrid capacitors. Supercapacitors are different from the conventional capacitors in terms of their energy storage. In the case of dielectric capacitors, the charge is stored electrostatically and in the case of supercapacitors; it is by chemical reactions between the electrodes of the capacitor and the electrolyte [10]. In the case of dielectric capacitors, no electrolyte is used, whereas in supercapacitors electrolytes are used between the two electrodes but both of them utilize the dielectric membrane.

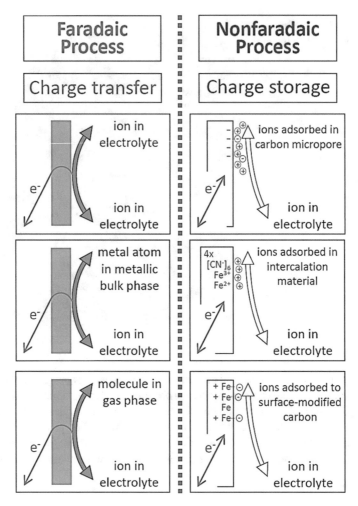

Fig. 2.3 Difference between charge transfer versus charge storage using six examples. Reprinted with permission from [9]

2.4.1 Electric Double-Layer Capacitor

In EDLC, the electrostatic force of attraction creates the electrolyte ions to adhere with the surfaces of the electrodes [11, 12]. During charging, electric charges are accumulated at the electrode/electrolyte interface and this leads to the formation of electric double-layer [13, 14]. Here, the charge storage is by the formation of electric double layer, and no charge transfer reactions such as Faradaic reactions take place. On voltage applications, one electronic layer of charges is formed on the surface lattice structure of the electrode material, and to compensate these charges further, ions of opposite polarity from the electrolyte material further accumulate on the surface

Fig. 2.4 Key differences between Faradaic and non-Faradaic processes. Reprinted with permission from [9]

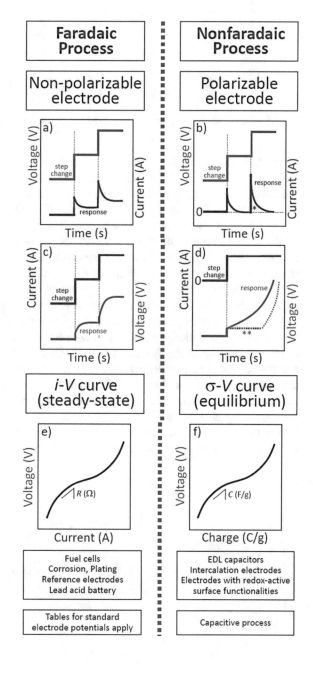

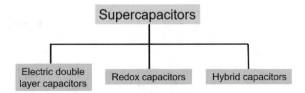

Fig. 2.5 Classification of supercapacitors based on the mechanism of charge storage

of the electrode material [15]. These two layers of charges are separated by solvent molecules called inner Helmholtz plane. Because of the size difference of electrons and ions, charge density of the electronic layer on the electrode material is higher compared to that of ionic layer. To compensate this extra electronic charge density, diffused layer of ions comes into the picture. The electrode potential varies linearly till the outer Helmholtz layer and gradually becomes exponential as approaches to the diffuse layer. EDLC provides fast charge/discharge cycle because of electrostatic charge storage mechanism. Cycle life of EDLC is higher as compared to pseudoca-pacitor due to non-Faradaic charge storage mechanism. Since there are no chemical reactions, the composition of the electrode structure remains almost constant. Hence, high reversible capacity and long cycle life have been observed here [13, 16]. The schematic representation of an electric double-layer during charging and discharging conditions is shown in Fig. 2.6. During the charged condition, the electrolyte ions are adsorbed on the capacitor electrode surfaces, and during discharging, they go back to the electrolyte solution and the electrons are passed through the load connected at the external circuit. The number of ions adsorbed/desorbed on the capacitor electrode surface during charging depends on the surface area of the active material. Hence, the electrodes with large surface area much preferred since the maximum utilization of surface area leads to higher capacitance [13, 15]. In the past, activated carbons

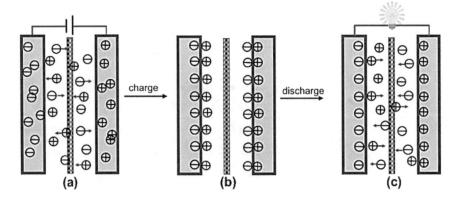

Fig. 2.6 Principle of charge storage of electric double-layer capacitors, **a** process of charging, **b** after charging and **c** process of discharging. Reprinted with permission from [22]

were used as electrode-active materials in electric double-layer capacitors, whereas the advanced capacitors utilize carbon nanomaterials with large surface area such as carbon nanotubes [17, 18], carbon nanopetals [19], and graphene [20].

Operating Principle of EDLC
Charge/discharge process of EDLC can be expressed by the following equations [21] On the positive electrode (2.1)

$$E_p + I^- \underset{\text{Discharging}}{\overset{\text{Charging}}{\rightleftharpoons}} E_p^+ // I^- + e^- \tag{2.1}$$

and on the negative electrode (2.2)

$$E_n + e^- + I^+ \underset{\text{Discharging}}{\overset{\text{Charging}}{\rightleftharpoons}} E_n^- // I^+ \tag{2.2}$$

Net charging and discharging process (2.3)

$$E_p + E_n + I^- + I^+ \underset{\text{Discharging}}{\overset{\text{Charging}}{\rightleftharpoons}} E_p^+ // I^- + E_n^- // I^+ \tag{2.3}$$

where E_p and E_n are the positive and negative electrodes, respectively. I^- and I^+ represent the anion and cation of the electrolyte. e^- stands for electron and // represents double-layer formation at the interface.

2.4.2 Pseudocapacitor

Pseudocapacitor stores energy by means of charge transfer reactions, i.e., by Faradaic reactions [23, 24]. A schematic representation of a pseudocapacitor is shown in Fig. 2.7. The reduction/oxidation happens at the capacitor electrodes, and these reactions should be reversible so that the cycle life of the supercapacitor will be high. In the case of electrodes in pseudocapacitors, they undergo continuous chemical changes during rapid charge/discharge cycling. Fast and reversible Faradaic redox reaction takes place in the pseudocapacitor. Adsorbed ions in pseudocapacitor do not undergo chemical transformation of the electrode material since charge transfer is the only process taking place [25]. Here, charge is transferred through the redox reaction, adsorption, and intercalation on the surface of the electrode material. Pseudocapacitance has been observed along with EDLC in the supercapacitor devices. Various electrode materials such as electronically conducting polymers, and transition metal oxides and also their nanocomposites are used for manufacturing supercapacitors [26–28]. These materials undergo reversible oxidation/reduction reactions for a long period of time, and hence, the energy density of such capacitors is generally very

Fig. 2.7 Schematic representation of a redox capacitor. Reprinted with permission from [29]

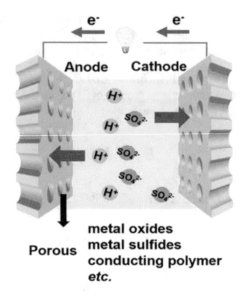

high and they exhibit high capacitance. Nanostructured materials with good porosity (preferably mesoporous structure) are highly preferred since these porous electrodes possess good diffusion of electrolyte ions through them. The performance of pseudocapacitors is excellent when compared with the electric double-layer capacitors but the power density of such capacitors remains generally low when compared with the latter one.

There are three types of electrochemical processes, which contribute to charge transfer in pseudocapacitance as mentioned below

Adsorption pseudocapacitance: Adsorption pseudocapacitance is observed due to the adsorption and desorption of ions of the electrolyte on the electrode material. Adsorption process can be expressed by the following (2.4), where lead is adsorbed on the surface of gold electrode in 10 mM $HClO_4 + 1$ mM PbF_2—[21, 30] (e.g., lead and gold system is shown here).

$$Au + xPb^{2+} + 2xe^- \leftrightarrow Au.xPb_{ads} \tag{2.4}$$

Redox pseudocapacitance: Redox pseudocapacitance predominates, when ions are electrochemically adsorbed on the surface of the electrode material with Faradaic charge transfer. For example, redox reaction of hydrated RuO_2 in acid electrolyte (e.g., $HClO_4$) can be expressed by the following (2.5) [21, 31]

$$RuO_x(OH)_y + \delta H^+ + \delta e^- \leftrightarrow RuO_{x-\delta}(OH)_{y+\delta} \tag{2.5}$$

Intercalation pseudocapacitance: In the intercalation pseudocapacitance, electrolyte ions intercalate into the Van der Waals gaps and lattice of the electrode material. For example, intercalation pseudocapacitance process can be expressed by the following (2.6) in which lithium ions of the electrolyte intercalate into the lattice of niobium oxide [21]

$$Nb_2O_5 + xLi^+ + xe^- \leftrightarrow Li_xNb_2O_5 \tag{2.6}$$

2.4.3 Hybrid Capacitor

The name hybrid capacitor implies that the supercapacitor consists of electrodes of two or more different electrode materials. The mechanism of energy storage by these electrodes will be a combination of both electric double-layer formations as well as by means of pseudo-Faradaic reactions [32, 33]. In hybrid capacitor, asymmetric electrode materials are used for electrical energy storage. Purpose behind the hybrid capacitor is to achieve high energy density. Hybrid capacitors have high potential window as well as high specific capacitance as compared to symmetric capacitor [21]. Three types of electrodes are used in general for hybrid capacitors, e.g., composite electrode, battery-type electrode, and asymmetric electrode. In composite electrode, carbon-based materials are incorporated with pseudocapacitive electrode material. In battery-type electrode, one of the electrodes will be of carbon and the other will be of battery electrode types. In asymmetric capacitor, both electrodes will be capacitive in nature; however, one will follow EDLC behavior and the other will be made of pseudocapacitive electrode material. Low self-discharge rate, high potential window, high energy density, and high specific capacitance are the key advantages of the hybrid electrochemical capacitors. A schematic representation of a hybrid capacitor is shown in Fig. 2.8. These types of supercapacitors have achieved great interest

Fig. 2.8 Schematic representation of a hybrid supercapacitor. Reprinted with permission from [29]

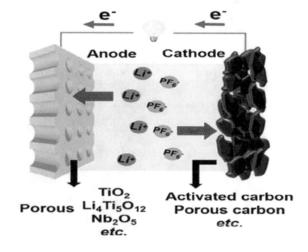

of the industries in the recent past due to their enormous charge storage capacity. Hybrid capacitors exhibit excellent characteristics when compared with the electric double-layer capacitors as well as redox capacitors. The new generation capacitors of these types are of huge potential. By combining Faradaic process with non-Faradaic processes, the hybrid capacitors are capable to store more charge, and hence, the energy density and the power density are very high. They exhibit high capacitances when compared with the electric double-layer as well as redox capacitors. Some hybrid capacitors are even capable to work in high voltages depending on the type of electrolyte used in their construction. These capacitors target the next-generation electronic devices to supply them power in a very efficient way. Hybrid capacitors have filled the gap between the electric double-layer capacitors and redox capacitors. For example, charge storage mechanism in hybrid capacitor is shown in the following (2.7) and (2.8), where highly porous graphite carbon (HPGC) is used as negative electrode and $Ni_2P_2O_7$ used as positive electrode in aqueous NaOH electrolyte [34]. At the positive electrode

$$Ni_2P_2O_7 + 2OH^- \leftrightarrow Ni_2P_2O_7(OH^-)_2 + 2e^- \tag{2.7}$$

At the negative electrode

$$HPGC + xNa^+ + xe^- \leftrightarrow HPGC(xe^-)//xNa^+ \tag{2.8}$$

2.5 Structure

The structure of a supercapacitor (Fig. 2.9) includes mainly four components: electrodes, electrolyte, current collectors, and electrolyte separator membrane [13].

Each component of a supercapacitor possesses a definite role in the operation. Supercapacitors are manufactured in various shapes and sizes. The various supercapacitor modules manufactured by the inventlab® company are shown in Fig. 2.10 (as

Fig. 2.9 Structure of a supercapacitor

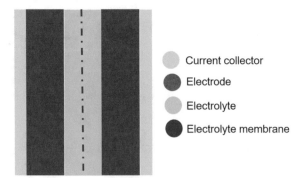

Current collector

Electrode

Electrolyte

Electrolyte membrane

Fig. 2.10 Supercapacitor modules manufactured by inventlab®. Reprinted with permission from [35]

representative for understanding). The details about each component of any standard supercapacitor are discussed in the next section.

2.5.1 Electrode

The active part of supercapacitor is the electrode, as the charge storage inside the supercapacitor depends on the type of the electrode-active materials used in them. Electrodes should have characteristics such as good electrical conductivity, large surface area, good porous structure, and good redox activity. The electrode-active material in electric double-layer capacitors does not undergo charge transfer reactions, whereas the redox capacitors possess such reactions. The electrodes in electric double-layer capacitors consist of a compact structure, whereas the redox capacitors are made up of electronically conducting polymers that are much flexible in nature. The selection of electrode-active materials is a pre-requisite for achieving the best performance. Recent progresses in the field of nanoscience and nanotechnology are helped to synthesize novel nanomaterials for supercapacitor application. The reason, why the nanostructured materials have been preferred, is because of their unique features when compared to macro- or microelectrode materials. Nanomaterials exhibit unique characteristics such as good electrical conductivity, large surface area to volume ratio, good thermal stability, and good chemical resistance. The porous structure of the electrode-active material can easily be modified by using nanostructured materials. As they exhibit large surface area, the electrolyte ions can easily be diffused through their pores and that will give rise to an enhancement in their performance. Hence, materials with comparatively larger surface area are preferred. Carbon nanomaterials include carbon nanotubes, graphene oxide, mesoporous carbons, carbon nanofibers, graphene, etc., which are widely used for the same purpose. Nanocomposite electrodes are also employed for manufacturing supercapacitor electrodes. The high performance of the nanocomposite electrodes has made them very attractive candidates in manufacturing supercapacitors with high capacitances and energy

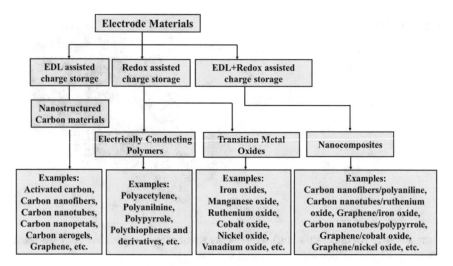

Fig. 2.11 Chart of various electrode-active materials used in supercapacitors [13]

densities. One major issue rooted in the electrode materials is the cycle life as the electrode damage happens and deteriorates the performance after say 1000 cycles. Depending on the type of mechanism of charge storage by the electrode-active materials, they can be classified as (i) materials with double-layer capacitance, (ii) materials with pseudocapacitance, and (iii) nanocomposite materials with both double-layer capacitance and pseudocapacitance [13]. A brief overview can be obtained from Fig. 2.11.

The electrode materials of a supercapacitor device can be a number of types; they can be made of carbon nanomaterials, transition metal oxides, and electrically conducting polymers. In the next section, a brief overview on different materials used as electrode materials for fabrication of a supercapacitor has been discussed.

2.5.1.1 Carbon Nanomaterials

Nanostructured carbon materials have been attracted by the supercapacitor research due to their unique properties such as decent electronic conductivity, large surface area to volume ratio, good chemical and electrochemical stabilities, and environment-friendliness [36–38]. Carbon nanomaterials are widely used invariably in electric double-layer capacitors, redox capacitors, and in hybrid capacitors. The mechanism of electrical energy storage by carbon nanomaterials is by means of electrochemical double-layer formation. The carbon nanomaterials do not take part in the Faradaic reactions and that is the reason behind very high life of the carbon nanomaterial-based electric double-layer capacitors as compared to other types of supercapacitors. The pre-requisite for a carbon nanomaterial to be used in a supercapacitor is that it should exhibit a porous structure so that the electrolyte ions can easily be diffused through these pores, and the accessibility of the electrolyte ions toward the extreme inner pores will participate in the charge storage, and thereby, the performance of the supercapacitor will be enhanced. For storing huge amount of energy in a supercapacitor for a long period of time, the porous carbon nanomaterial-based electrodes should perform reversible oxidation/reduction reactions. That means, the adsorption of electrolyte ions at the electrode/electrolyte interface at the time of charging should be reversibly desorbed during the discharging step. If the ions are not desorbed completely or if they undergo some chemical reactions with the electrode, may lead to reduction in performance and subsequently the electrode architecture fails in such a way that the specific surface area available for the ion adsorption will be minimized. Not only porous architecture, the pore size and pore size distribution are also important in the case of carbon nanomaterial-based electrodes in supercapacitor application. As the pore sizes are within the mesopore range (i.e., in between 2 and 50 nm), the performance will be higher as the diffusion of electrolyte ions will be favorable for achieving higher capacitance [34, 39]. However, in the case of micropores (the size of the pores <2 nm) and macropores (the size of the pores >50 nm), the supercapacitor performance is found to be low. Good electronic conductivity of the carbon nanomaterials helps in the transport of electrons through them even after thousands of cycles as there will be no or negligible change in morphology. Examples of nanostructured carbon materials used as electrode-active materials in supercapacitors are carbon nanotubes (CNTs), carbon nanofibers, carbon nanopetals, mesoporous carbons, graphene, etc. Both single-walled and multi-walled CNTs are used for the application. Detailed characteristics of activated carbon, graphene and reduced graphene oxide, carbon nanotubes, carbon nanofibers, polymers, electrodes, electrolytes, separators, and current collectors are discussed in Chaps. 4, 5, 6, 7, 8, 9, 10, 11 and 12, respectively.

2.5.1.2 Transition Metal Oxides

Transition metal oxides are much attracted by the attention of high capacitance supercapacitors due to their redox-type of charge storage. They are widely used in various energy conversion as well as energy storage applications due to the good mechanical, electronic, and redox properties [40–42]. Nanostructured and porous transition metal oxide-based electrodes are much preferred, as the electrolyte diffusion toward the interior pores will be high; thereby, an enhanced performance is obtained. The redox capacitance is coming as a result of their multiple valence state changes. Transition metal oxides can be classified into two categories: (i) noble transition metal oxides (e.g., RuO_2, IrO_2, etc.) [43, 44] and (ii) base transition metal oxides (e.g., MnO_2, Co_3O_4, etc.) [45, 46]. The noble transition metal oxides are relatively costlier compared to the base transition metal oxides. The low-cost and environment-friendly nature of the base transition metal oxides have attracted much for the supercapacitor application; however, performance is relatively low when compared with the noble transition metal oxides. A major demerit of low energy densities of supercapacitors can be resolved by using novel transition metal oxides/hydroxides with novel design in the electrode architecture. Nanostructuring of the electrode architecture is the best strategy adopted for achieving the best performance of the supercapacitor. Detailed characteristics of transition metal oxides are discussed in Chap. 3.

2.5.1.3 Electronically Conducting Polymers

Electronically conducting polymers attracted the attention of flexible energy conversion and storage devices in the recent past. The good electronic conductivities and easy processability have made them successful candidates in those applications. Electronically conducting polymer-based electrodes for supercapacitor application have many advantages such as the redox-type charge storage acting as performance booster, feasibility of fabrication of flexible electrodes, and thereby opening up the provision of flexible supercapacitors. Flexible supercapacitors are mandatory for application in the flexible and wearable electronic devices, which have shown an increasing trend in the recent past. The charge storage by electronically conducting polymers is by a doping process. Examples of electronically conducting polymers used in supercapacitor electrode application are polypyrrole [47, 48], polyaniline [49, 50], poly(3,4-ethylenedioxythiophene) [51, 52], etc. The conductivity in case of polypyrrole is by means of p-doping. The pi-electrons will be removed from their conjugated structure, which leaves a net positive charge. Both the postulates related with 'polarons' as well as 'bipolarons' exist in the literature. For example, the schematic representation of the conduction mechanism of polypyrrole is shown in Fig. 2.12. Detailed characteristics of conducting polymers are given in Chap. 8.

2.5.2 Electrolyte

The nature of electrolyte determines the power density of supercapacitors since the electrolyte resistance plays a major role in determining the same. The term 'electrochemical series resistance' is used to represent the collective resistances within the supercapacitor system. If the electrolyte resistance is high, the power density will be low. Different types of electrolytes are used in supercapacitors, namely aqueous, organic, ionic, etc. They are categorized into two types either liquid or solid. Different types of liquid and solid electrolytes are shown in Fig. 2.13. The selection of an electrolyte is based on the stable potential window in which it can work. Aqueous electrolytes (like acids, alkaline) offer low specific resistances and hence are suitable for manufacturing supercapacitors. Also, the price of aqueous electrolytes is very low when compared with the organic electrolytes, which is one of the major advantages. But the organic electrolytes offer high specific resistance, which in turn reduces the power density of the supercapacitor [13]. Aqueous electrolytes also have disadvantages such as instability at higher voltages, leading to electrode corrosion and are environmentally hazardous. But, on the other hand, the organic electrolytes are stable even at higher operating voltages and also very toxic and flammable. The reason,

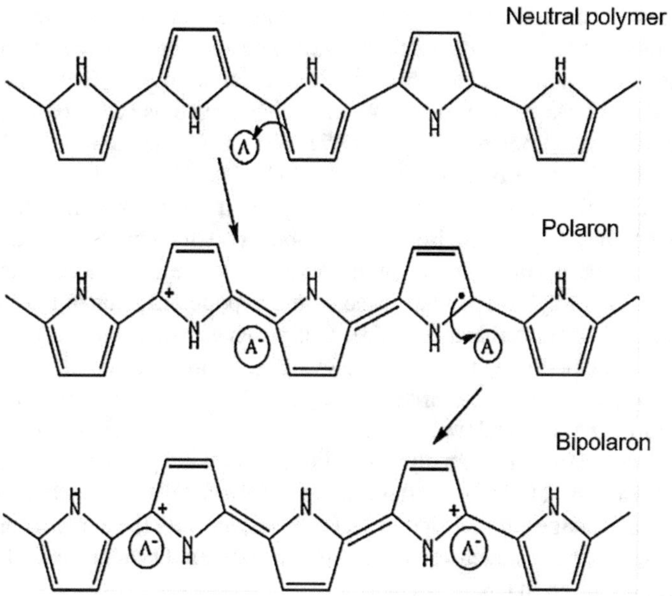

Neutral polymer

Polaron

Bipolaron

Fig. 2.12 Schematic representation of the conduction mechanism in polypyrrole. Reprinted with permission from [53]

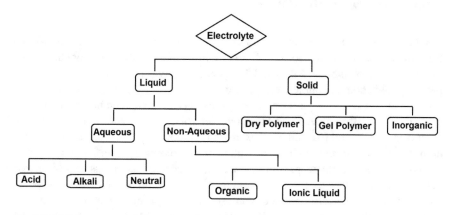

Fig. 2.13 Different types of electrolytes for supercapacitor

why the aqueous electrolytes experience small resistances, is because the protons are of high mobility and small size that lower the resistance. On the other hand, in the case of organic electrolytes, due to large sizes, the resistance has been increased. Hence, the selection of electrolyte as well as the porosity of the electrode architecture is important for best performance of the supercapacitor. As the supercapacitor operates for several thousands of cycles, the electrolyte depletion causes an increase in the internal resistance, which leads to a decrease in the capacitance and hence the energy density. This is the reason why both the aqueous and organic electrolytes are not much preferred in the commercially available supercapacitors. The selection of electrolytes in the commercial supercapacitors is ionic liquids. These special categories of electrolytes exhibit unique characteristics such as high conductivity and wide electrochemical potential window [54, 55]. Ionic liquids are non-flammable in nature, and this property makes them safe to handle. By adjusting the concentration of ionic liquids very high, electrolyte depletion can be minimized. They also possess good chemical and environmental stabilities, which have made them potential candidates for use in supercapacitors. A special category of electrolytes namely 'gel polymer electrolytes' has also attracted much popularity due to their enhanced electrochemical characteristics [56–58]. Gel polymer electrolytes can be used to fabricate supercapacitors with solid-state features. Solid-state supercapacitors manufactured by gel polymer electrolytes are highly advantageous for application in the next-generation flexible and wearable electronic technologies. Hence, a new type of electrolytes is put forward, namely 'ionic liquid-based gel polymer electrolytes' by combining features of both ionic liquids and gel polymer electrolytes [59, 60]. These have all the advantages of their individual counterparts and the demerits are waived off however, still the cost factor remains a major concern. Detailed characteristics of electrolytes are given in Chap. 10.

2.5.3 Electrolyte Membrane

The functions of an electrolyte separator membrane are (i) to allow the passage of electrolyte ions and (ii) to avoid the short circuiting of the supercapacitor electrodes. The electrolyte membrane with good ionic conductivity is preferred for the supercapacitor application. A simple Xerox paper or a commercially available Whatman™ filter paper can serve the purpose. These types of electrolyte membranes are of low cost and hence affordable. Nanostructured electrolyte membranes are developed in the recent past as well [61–63]. One example is Nafion™ membrane. Nafion™ membranes exhibit high ionic conductivities due to their nanoscale properties and is widely used in commercial supercapacitors. However, the cost of Nafion™ is very high, and hence, it is mandatory to develop novel electrolyte membranes with low-cost polymers [64, 65]. The electrolyte membrane should have porous structure to transport the ions from electrolyte to electrodes, as shown in Fig. 2.14. Detailed characteristics of electrolyte membrane separator are given in Chap. 11.

2.5.4 Current Collector

The function of current collector is to collect the electrons (hence current collector) from the electrode-active material and transport them to the external circuit. Metal plates are used for this purpose such as copper and aluminium. Sometimes,

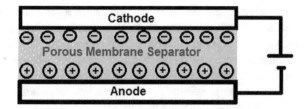

Fig. 2.14 Schematic representation of porous membrane to transport ions to electrodes

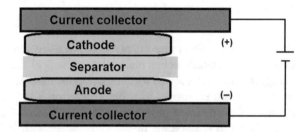

Fig. 2.15 Construction of supercapacitors using current collectors

alloys are also used for the same; one example is steel plate. Generally, two current collectors are used on the surface of cathode and anode of supercapacitors, as shown in Fig. 2.15. Material characteristics of current collector for supercapacitor are discussed in Chap. 12.

2.6   Key Parameters for Estimation of Performance

The various parameters, which used to estimate the performance of the supercapacitor, are specific capacitance, energy density, and power density.

2.6.1   Specific Capacitance

Capacitance values for commercial capacitors are stated as 'rated capacitance C_R'. This is the parameter for which the capacitor has been designed. Typical capacitance of supercapacitor is in Farad (F), three to six orders of magnitude higher than those of conventional capacitors. The capacitance can be calculated from cyclic voltammograms, galvanostatic charge/discharge curves, and from electrochemical impedance spectroscopy curves. Specific capacitance means capacitance with respect to a known entity such as mass, area, and volume. If the capacitance is calculated with respect to the mass of the electrode-active material used, it is known as gravimetric capacitance (i.e., capacitance per gram of the electrode-active material) and represented as F/g. If it is calculated with respect to the area of supercapacitor electrode, it is termed as area-specific capacitance and the unit F/cm^2 is used to represent the same. If the capacitance is calculated with respect to the volume of the supercapacitor itself, it is known as volume-specific capacitance and is denoted by F/cm^3. While calculating the specific capacitance, there usually exists a confusion that whether the specified capacitance is of the electrode or of the supercapacitor itself (which means that of two electrodes, not of the single electrode). Hence, it is better to use the term 'specific capacitance' such that the confusion can be avoided.

2.6.2   Energy Density

Energy density refers to the energy stored in a supercapacitor with respect to either mass of the electrode-active material or volume of the supercapacitor itself. If the energy density is calculated with respect to the mass of the electrode-active material used in the two electrodes of a supercapacitor, it is termed as gravimetric energy density and represented generally by $W h kg^{-1}$ (Watt hour kg^{-1}). If it is calculated with respect to the volume of the supercapacitor, it is known as volume-specific energy density and is represented by $W h cm^{-3}$. But in all cases, the calculation of energy density remains the same as

$$E = \frac{1}{2}CV^2 \qquad (2.9)$$

where C is the specific capacitance (either in $F g^{-1}$ or in $F cm^{-3}$) and V is the operating voltage of the supercapacitor. From (2.9), it is clear that the energy density has a square dependency to the operating voltage; hence, higher energy density can be expected if the high voltage supercapacitor is used and that in turn depends on the type of electrolyte used in the construction of the supercapacitor. As discussed earlier, the maximum operable voltage is limited in the case of aqueous electrolytes, whereas it is high in the case of organic electrolytes. The energy density of supercapacitors is low when compared with batteries but several orders of magnitude greater than that of a conventional capacitor.

2.6.3   Power Density

Power density refers to the energy available from the supercapacitor in a unit time. Similar to the energy density, the power density too can be expressed in terms either of mass of the electrode-active material or volume of the supercapacitor itself. If the power density is calculated with respect to the mass of the electrode-active material used in the two electrodes of a supercapacitor, it is termed as gravimetric power density and it is represented generally by $W kg^{-1}$ (Watt kg^{-1}). If it is calculated with respect to the volume of the supercapacitor, it is known as volume-specific power

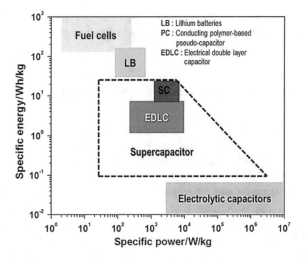

Fig. 2.16 Plots of energy density versus power density of batteries, supercapacitors, and conventional capacitors. Reprinted with permission from [66]

density and is represented by W cm^{-3}. But in all cases, the calculation of power density remains the same as

$$P = \frac{1}{4}\frac{V^2}{R_s} \tag{2.10}$$

where R_s is the equivalent series resistance of the supercapacitor and V is the operable voltage. The equivalent series resistance of a supercapacitor depends on the electrolyte, where aqueous electrolytes offer comparatively less resistance with respect to the organic electrolytes. Hence, the power density will be high if aqueous electrolytes are used in the supercapacitor. Since organic electrolytes offer high resistances, the power density of the supercapacitor will be low. Hence, it can be said that the selection of electrolyte is crucial for determining the power density of the supercapacitor. The power density of supercapacitors is several orders higher than that of batteries. But achieving both high energy and power densities is a major challenge in the case of supercapacitors. Hence, the selection of proper electrode-active materials as well as hierarchical design of the supercapacitor is mandatory for developing high-performance supercapacitors. Energy and power densities of supercapacitors bridge the gap between the batteries and conventional capacitors, as shown in Fig. 2.16.

2.7 Electrochemical Characterizations

The electrochemical performance of supercapacitor electrodes and supercapacitors is examined by various electrochemical techniques such as electrochemical impedance spectroscopy, cyclic voltammetry, and galvanostatic charge/discharge measurements. A brief description on these techniques is given as follows.

2.7.1 Electrochemical Impedance Spectroscopy

Electrochemical impedance is a widely accepted tool used to compute impedance of a charge storage device as a function of frequency on application of alternative voltage superimposed of low amplitude on a steady-state potential [67, 68]. The data are represented graphically in a Bode plot to understand the supercapacitive response between the phase angle and frequency. The other popular method of evaluation of an impedance plot is Nyquist plot, where the imaginary and real parts of the impedances are displayed on a complex plane [69]. From the Nyquist plots, the intersection of the curves at the x-axis represents the bulk electrolyte resistance (R_b) of the supercapacitor electrodes. By using this value, the conductivity (σ) of the supercapacitor electrodes is determined as follows

$$\sigma = \frac{T}{R_b \times A_{ele}} \tag{2.11}$$

where T is the total thickness of the supercapacitor (in cm), R_b is the bulk electrolyte resistance (in Ω), and A_{ele} is the geometrical area of supercapacitor electrodes (in cm^2). The first intercept point of the semi-circle on the real axis of the Nyquist plot is considered as R_b [70–72].

Again, the electrochemical impedance $Z(w)$ is defined as

$$Z(w) = Z^I + jZ^{II} \tag{2.12}$$

where Z^I and Z^{II} are the real and imaginary parts of the impedance, respectively.

The frequency-dependent behavior of a supercapacitor can be described by associating a serial resistance R_s and a capacitance C, as shown in Fig. 2.17. The impedance of the circuit is defined by

$$Z = R_s + \frac{1}{jwC} = \frac{1 + jwR_sC}{jwC} \tag{2.13}$$

Fig. 2.17 Schematic used for describing the frequency-dependent behavior of supercapacitor

Fig. 2.18 Equivalent circuit
of the supercapacitor

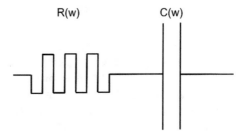

It is possible to describe the supercapacitor by using resistance and capacitance (Fig. 2.18) that are functions of the pulsation w and denoted as $R(w)$ and $C(w)$. The impedance of the equivalent circuit is given by Taberna et al. [67]

$$Z = R(w) + \left[\frac{1}{jwC(w)} \right] = \frac{1}{jKw} \qquad (2.14)$$

where K is the capacitance of the supercapacitor. By solving (2.14), it leads to

$$K = \left[\frac{C(w)}{1 + jwR(w)C(w)} \right] \qquad (2.15)$$

$$K = \left[\frac{C(w)}{1 + w^2 R^2(w)C^2(w)} \right] - \left[\frac{iwR(w)C^2(w)}{1 + w^2 R^2(w)C^2(w)} \right] \qquad (2.16)$$

Equation 2.16 consists of a real part and an imaginary part for the capacitance of the supercapacitor. This (2.16) can be written as

$$K = C^{\mathrm{I}} - jC^{\mathrm{II}} \qquad (2.17)$$

where

$$C^{\mathrm{I}} = \left[\frac{C(w)}{1 + w^2 R^2(w)C^2(w)} \right] \qquad (2.18)$$

$$C^{\mathrm{II}} = \left[\frac{wR(w)C^2(w)}{1 + w^2 R^2(w)C^2(w)} \right] \qquad (2.19)$$

Equation 2.18 gives the expression for the capacitance of the supercapacitor, which is varying with frequency, and (2.19) gives the imaginary part of the capacitance.

2.7.2 Cyclic Voltammetry

Cyclic voltammetry measurement applies a linearly changed electric potential between positive and negative electrodes of a supercapacitor in two-electrode cell configuration [73, 74]. The speed of the potential change in mV s^{-1} is referred as the scan rate, and the range of potential change is known as potential window. The instantaneous current during the anodic and cathodic sweeps is recorded, which is used to characterize the electrochemical reactions involved between the electro-active material and the electrolyte ions. Apart from the two-electrode cell arrangement, a three-electrode cell arrangement was also used for examining the electrode charac-teristics of the SC electrodes. The three electrodes used are the working, counter, and reference electrodes. A counter electrode of platinum and a reference electrode of saturated Ag/AgCl are generally used. In both the cases, the cyclic voltammetry data is plotted as current, I (in A), versus potential, E (in V). The performance parameters of supercapacitor can be calculated from the cyclic voltammetry curves. The overall capacitance value is calculated from the cyclic voltammogram by using (2.20) [75]

$$C = \frac{dq}{dE} = \frac{dq}{dt}\frac{dt}{dE} = i\frac{dt}{dE} = \frac{i}{dE/dt} = \frac{i}{s} \tag{2.20}$$

where 'i' is the instantaneous current in cyclic voltammograms, dE/dt represents the scanning rate, and 's' is the potential sweep rate. The capacitance of the supercapacitor is the average current divided by the scanning rate (2.21) as

$$C = \frac{i_{\text{avg}}}{s} = \frac{1/(E_2 - E_1)\int_{E_1}^{E_2} i(E)dE}{s} \tag{2.21}$$

where E_1 and E_2 are the switching potentials in cyclic voltammetry, $\int_{E_1}^{E_2} i(E)dE$ is the voltammetric charge got by integration of positive or negative sweep in cyclic voltammograms. The upper limit potential is given by (2.22).

$$E_2 = E_1 + st \tag{2.22}$$

where t is the time period of single positive or negative sweep. Therefore,

$$\int_{E_1}^{E_2} i(E)dE \rightarrow s \int_0^t i(t)dt$$

and hence (2.21) becomes

$$C = \frac{\int_0^t i(t)dt}{E_2 - E_1} \tag{2.23}$$

The gravimetric capacitance can be calculated from (2.23) by dividing the capacitance with the mass of electrode-active material as

$$C_s = \frac{C}{m} \qquad (2.24)$$

where m is the total mass of electrode-active material or materials used in the two supercapacitor electrodes.

2.7.3 Galvanostatic Charge/Discharge

Constant current charge/discharge is a versatile and accurate method used to characterize the supercapacitor under direct current. The constant current charge/discharge measurement is conducted by repetitive charging and discharging of the supercapacitor or working electrode at a fixed current level with or without a dwelling period (defined as a time period between charging and discharging while the peak voltage V_o is constant) and normally a plot of E (in V) versus time, t (in s) is the output [69, 76].

2.7.3.1 Calculation of Supercapacitor Parameters

The various supercapacitor performance parameters such as specific capacitance, energy density, and power density can be calculated from the galvanostatic charge/discharge curves, and the equations used for calculating the same are discussed in the next section.

Cell capacitance: The cell capacitance (C_{cell}) of the supercapacitor is calculated by (2.25),

$$C_{cell} = \frac{I t_{dis}}{\Delta E} \qquad (2.25)$$

where I is the charging current, t_{dis} is the discharging time, and ΔE is the operating potential window.

   Areal capacitance: The areal capacitance ($C_{cell,A}$) of the supercapacitor is calculated by (2.26),

$$C_{cell,A} = \frac{C_{cell}}{A_{cell}} \qquad (2.26)$$

where A_{cell} is the total geometric area of two supercapacitor electrodes (i.e., two times the area of single electrode).

Areal energy density: The areal energy density ($E_{cell,A}$) of the supercapacitor is calculated by (2.27),

$$E_{cell,A} = \frac{C_{cell,A} \times (\Delta E)^2}{2 \times 3600}$$ (2.27)

Area-specific capacitance: The area-specific capacitance ($C_{cell,sp,A}$) of the supercapacitor is calculated by (2.28),

$$C_{cell,sp,A} = 4 \times \frac{C_{cell}}{A_{ele}}$$ (2.28)

where A_{ele} is the geometric area of the supercapacitor electrodes.
Area-specific energy density: The area-specific energy density ($E_{cell,sp,A}$) of the supercapacitor is calculated by (2.29),

$$E_{cell,sp,A} = \frac{C_{cell,sp,A} \times (\Delta E)^2}{2 \times 3600}$$ (2.29)

Volumetric capacitance: The volumetric capacitance ($C_{cell,V}$) of the supercapacitor is calculated by the (2.30),

$$C_{cell,V} = \frac{C_{cell}}{V_{cell}}$$ (2.30)

where V_{cell} is the total volume of the supercapacitor.
Volumetric energy density: The volumetric energy density ($E_{cell,V}$) of the supercapacitor is calculated by (2.31),

$$E_{cell,V} = \frac{C_{cell,V} \times (\Delta E)^2}{2 \times 3600}$$ (2.31)

Volumetric power density: The volumetric power density ($P_{cell,V}$) of the supercapacitor is calculated by (2.32)

$$P_{cell,V} = \frac{E_{cell,V} \times 3600}{t_{dis}}$$ (2.32)

Volume-specific capacitance: The volume-specific capacitance ($C_{cell,sp,V}$) of the supercapacitor is calculated by (2.33),

$$C_{cell,sp,V} = 4 \times \frac{C_{cell}}{V_{ele}}$$ (2.33)

where V_{ele} is the total volume of the supercapacitor electrodes.
Volume-specific energy density: The volume-specific energy density ($E_{cell,sp,V}$) of the supercapacitor is calculated by (2.34),

$$E_{cell,sp,V} = \frac{C_{cell,sp,V} \times (\Delta E)^2}{2 \times 3600} \qquad (2.34)$$

Volume-specific power density: The volume-specific power density ($P_{cell,sp,V}$) of the supercapacitor is calculated by (2.35),

$$P_{cell,sp,V} = \frac{E_{cell,sp,V} \times 3600}{t_{dis}} \qquad (2.35)$$

Gravimetric capacitance: The gravimetric capacitance (C_m) of the supercapacitor is calculated by (2.36),

$$C_m = \frac{I \times t_{dis}}{m \times (\Delta E)} = \frac{C_{cell}}{m} \qquad (2.36)$$

where 'm' is the total mass of electrode-active materials used in the supercapacitor (excluding the mass of current collector, separator and electrolyte).
Gravimetric energy density: The gravimetric energy density (E_m) of the supercapacitor is calculated by (2.37),

$$E_m = \frac{C_m \times (\Delta E)^2}{2 \times 3600} \qquad (2.37)$$

Gravimetric power density: The gravimetric power density (P_m) of the supercapacitor is calculated by (2.38),

$$P_m = \frac{E_m \times 3600}{t_{dis}} \qquad (2.38)$$

2.7.3.2 Estimation of Rate Capability of Supercapacitor

The rate capability of the supercapacitor is evaluated by performing the galvanostatic charge/discharge measurement at different current densities. Further, the supercapacitive performance parameters are calculated from the galvanostatic charge/discharge curves. The results obtained at different current densities are compared, and the rate characteristics of the concerned supercapacitor are estimated. Figure 2.19 shows an example of a rate capability plot of a graphene-transition metal-based supercapacitor [77].

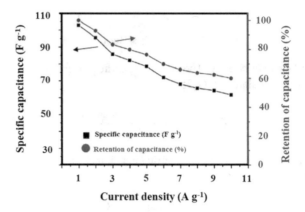

Fig. 2.19 Specific capacitance values of graphene-transition metal-based supercapacitor at different current densities and corresponding rate capability. Redrawn and reprinted with permission from [77]

2.7.3.3 Testing of Flexibility of Supercapacitor

The flexibility of supercapacitor is examined by carrying out the galvanostatic charge/discharge measurement at different supercapacitor bending angles [78]. Further, the galvanostatic charge/discharge measurement is carried out at different bending angles and the $E - t$ response is noted. Finally, the galvanostatic charge/discharge (GCD) curves obtained at different supercapacitor bending are compared. For an example, the GCD curves of a flexible supercapacitor based on graphene oxide and polyaniline at different bending angles [79] are shown in Fig. 2.20a. The stability of the capacitance at different bending cycles is also shown in Fig. 2.20b.

2.7.3.4 Estimation of Cycle Life of Supercapacitor

The life cycle of supercapacitor was determined by the cyclic voltammetry and galvanostatic charge/discharge measurements. In a typical procedure, the charging/discharging or cyclic voltammetry measurement of the SC at a particular current density or scan rate is carried out for 'n' number of cycles. Further, the supercapacitive performance parameters are estimated from the galvanostatic charge/discharge or cyclic voltammetry curves at particular cycle numbers and the results are compared [77, 80] as shown in Fig. 2.21.

2.7.3.5 Two-, Three-, and Four-Electrode Systems

Electrochemical test is used to study the redox reactions. In electrochemical process, electricity is generated by the movement of the electrons from one electrode material to other. These reactions are well known as oxidation-reduction (redox)

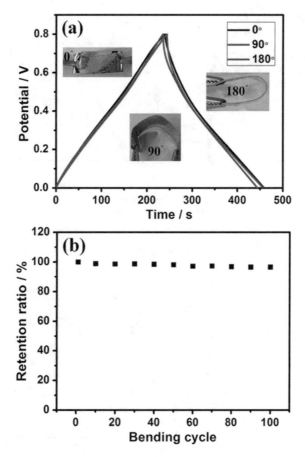

Fig. 2.20 **a** GCD curves of a graphene oxide and polyaniline-based flexible supercapacitors at different bending angles, **b** capacitance retention of the supercapacitor at different bending cycles. Redrawn and reprinted with permission from [79]

reactions. Depending on the application, electrochemical testing can be done either in potentiostatic or in galvanostatic mode or both [81, 82].

The electrodes that are involved during electrochemical testing are discussed as follows:

(a) **Working electrode**

Working electrode is one, which is to be studied, i.e., in this electrode, reaction of interest occurs. Some of the working electrodes such as glassy carbon, metal foam, and carbon cloth, are used to provide conductive pathway for the sample to be tested. For corrosion applications, the material to be investigated is itself used as a working electrode [81, 82].

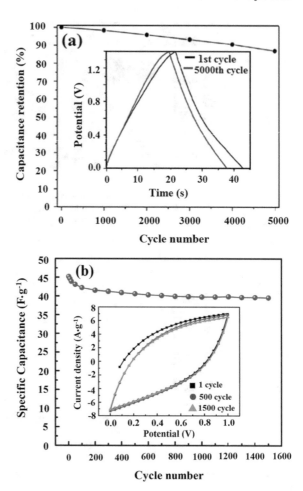

Fig. 2.21 Cyclic stability of supercapacitors measured by **a** galvanostatic charge/discharge and **b** cyclic voltammetry measurements. Redrawn reprinted with permission from [77, 80]

(b) **Counter electrode**

Counter electrode is also known as auxiliary electrode. During electrochemical testing, it is used to close the current circuit [83]. Mainly, inert materials such as platinum, graphite, gold, and glassy carbon are used as the counter electrode. Counter electrode does not participate in electrochemical reactions and acts as current source or sink [81, 82].

(c) **Reference electrode**

Reference electrode is known as the electrode potential, which is used as a reference point for the electrochemical cells. This electrode has high stability and holds constant potential during the entire process of testing. Ideally, current does not flow from the reference electrode, which is again assured by using counter electrode. Some commonly used reference electrodes are mercury/mercury (mercurous) oxide, silver/silver chloride, saturated calomel, mercury/mercury sulfate, copper/copper sulfate, etc. which are displayed along with their reference voltage in Fig. 2.22 [81–83].

Fig. 2.22 Potential scale of commonly used reference electrodes. Redrawn and reprinted with permission from [81]

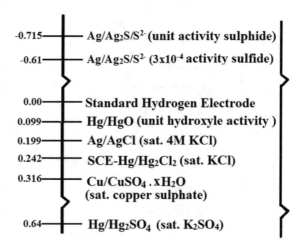

Fig. 2.22 Potential scale of commonly used reference electrodes. Redrawn and reprinted with permission from [81]

Along with working, counter, and reference electrodes, an electrochemical station has fourth connector as sense and fifth connector as a ground connector, which can be used to connect external devices to the same ground of electrochemical station [81, 82].

There are two modes in which electrochemical observations are carried out-

(a) **Potentiostatic mode**: In potentiostatic mode, the potential of working electrode and reference electrode is well defined due to which the potential of counter against reference electrode is controlled [81].

(b) **Galvanostatic mode**: In galvanostatic mode, the flow of current between counter electrode and working electrode and potential difference between reference electrode and working electrode are monitored. The current between working and counter electrode is controlled [81].

The three commonly used electrochemical cell setups, i.e., two-electrode, three-electrode, and four-electrode, are discussed here:

(a) **Two -electrode setup**

In two-electrode setup, working electrode and sense are sorted to form one connector. For the other connector, counter and reference electrodes are sorted. Figure 2.23a shows the schematic diagram of two-electrode setup [81, 82]. This type of electrode setup is used to measure the electrochemistry of the entire cell. It measures the contributions from electrode to electrolyte interface. It offers ultra-fast dynamics of electrode electrochemistry and electrochemical impedance measurements at high frequencies [83].

In two-electrode system, it is difficult to keep steady counter electrode potential (e_c) during passage of current. The voltage drop (iR_s) is not compensated across the solution leading to poor control of working electrode potential (e_w) in two-electrode system [83]. The schematic representation of potential gradients across two electrodes is presented in Fig. 2.23b [81, 82]. This setup is

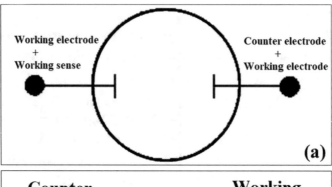

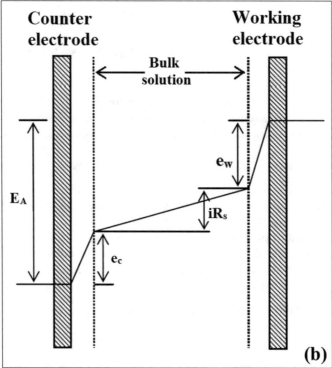

Fig. 2.23 **a** Simple schematic of two-electrode configuration (redrawn and reprinted with permission from [82]) and **b** schematic representation of potential gradients in a two-electrode system while current is flowing. Redrawn and reprinted with permission from [83]

commonly used to measure the cell voltage of electrochemical energy storage and conversion devices such as supercapacitor, batteries, photovoltaic panels, and fuel cells. For supercapacitor application, two-electrode setup is used to measure the performance of the whole assembly of electrodes, electrolyte, and separator in device form [83].

(b) **Three-electrode setup**

Three-electrode setup is widely used electrochemical setup to study the electrochemistry of the electrode material. It consists of working, counter, and reference electrodes. Reference electrode acts as a potential reference for measuring and controlling the potential of the working electrode [81, 82]. Figure 2.24a shows the schematic representation of three-electrode setup. The current flows between counter and working electrode, whereas reference electrode and sense measure potential between them [82]. Figure 2.24b shows the schematic representation of potential gradient in three-electrode system when current is flowing. Also, reference electrode passes negligible current so, iR drop (iR_u) between working and reference electrode is very small [83]. Counter electrode passes the current, which is required to balance the current obtained from the working electrode. Three-electrode setup shows superior control over working electrode potential. For energy storage and conversion devices, it offers preliminary studies to understand the behavior of electrode material with the electrolyte [81–84].

(c) **Four-electrode setup**

In four-electrode setup, the potential between reference electrode and sense is measured during the flow of current between working and counter electrode. This setup is generally used to measure the junction potentials across the membrane or between two non-miscible phases. This allows to analyze the accurate measurement of the resistance of the membrane or liquid interface. Figure 2.25 shows the schematic representation of four-electrode setup [81, 82].

Four–electrode system is also used in zero resistance ammeter experiment, where working and counter electrodes are sorted to maintain zero voltage drop. Here, reference electrode acts as a spectator electrode to working-counter coupling. This experiment is used to study electrochemical noise and galvanic corrosion [81, 82].

2.8   Concluding Remarks

In this chapter, a complete description of supercapacitors including the mechanism of charge, types, advantages, and demerits is included. Supercapacitors are classified into three classes, namely electric double-layer capacitors, redox capacitors, and hybrid capacitors. Electric double-layer capacitors possess non-Faradaic energy storage, and hence, no electrode damage happens for a large number of cycles. But in the case of redox capacitors, as they experience pseudo-Faradaic reactions, there chemical change happens to the electrode-active materials, and hence, they may lose their performance after few thousands of cycles. In order to avoid this, hybrid capacitors have been developed so that the disadvantages of electric double-layer

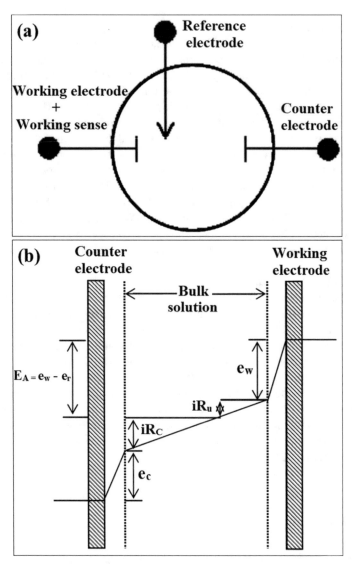

Fig. 2.24 **a** Simple schematic of three-electrode configuration (redrawn and reprinted with permission from [82]) and **b** schematic representation of potential gradients in a three-electrode system while current is flowing. Redrawn and reprinted with permission from [83]

capacitors and redox capacitors have been waived off and all the merits are added up in these advanced capacitors. Supercapacitors exhibit high power densities when compared with batteries, whereas their energy density is lower than that of the batteries. The future supercapacitors should be equipped with high energy density to

Fig. 2.25 Simple schematic of four-electrode configuration. Redrawn and reprinted with permission from [82]

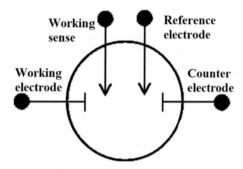

fulfill the requirements. The main challenge of supercapacitor industry is the cost of production, which is making them 'not-reachable' to the society.

Acknowledgements The authors acknowledge the financial support provided by Department of Science and Technology, India (DST/TMD/MES/2K16/37(G)) for carrying out this research work.

References

1. R.B. Marichi, V. Sahu, R.K. Sharma, G. Singh, *Handbook of Ecomaterials*, pp 1–26
2. X. Zhang, L. Ji, O. Toprakci, Y. Liang, M. Alcoutlabi, Polym. Rev. **51**, 239 (2011)
3. M. Lu, F. Beguin, E. Frackowiak, in *Supercapacitors: Materials, Systems, and Applications*
4. M. Winter, R.J. Brodd, Chem. Rev. **104**, 4245 (2004)
5. R.R. Gaddam, N.A. Kumar, R. Narayan, K.V.S.N. Raju, X.S. Zhao, in *Nanomaterials Synthesis Design, Fabrication and Applications* (2019), pp 385–418
6. H.I. Becker, (Google Patents, 1957)
7. D.L. Boos, (Google Patents, 1970)
8. https://www.kanthal.com/en/. Accessed 10 Feb 2019
9. P.M. Biesheuvel, J.E. Dykstra, The difference between Faradaic and Nonfaradaic processes in Electrochemistry
10. A. Pandolfo, A. Hollenkamp, J. Power Sour. **157**, 11 (2006)
11. B.E. Conway, J. Electrochem. Soc. **138**, 1539 (1991)
12. B.E. Conway, *Electrochemical Supercapacitors: Scientific Fundamentals and Technological Applications* (Springer Science & Business Media, 2013)
13. K.K. Kar, S. Rana, J. Pandey, *Handbook of Polymer Nanocomposites Processing, Performance and Application* (Springer, 2015)
14. A. Burke, J. Power Sour. **91**, 37 (2000)
15. Y. Zhang, H. Feng, X. Wu, L. Wang, A. Zhang, T. Xia, H. Dong, X. Li, L. Zhang, Int. J. Hydrogen Energy **34**, 4889 (2009)
16. C. Du, N. Pan, Nanotechnology **17**, 5314 (2006)
17. J. Cherusseri, K.K. Kar, RSC Adv. **5**, 34335 (2015)
18. J. Cherusseri, R. Sharma, K.K. Kar, Carbon **105**, 113 (2016)
19. J. Cherusseri, K.K. Kar, J. Mater. Chem. A **3**, 21586 (2015)
20. Y.B. Tan, J.-M. Lee, J. Mater. Chem. A **1**, 14814 (2013)
21. C. Zhong, Y. Deng, W. Hu, D. Sun, X. Han, J. Qiao, J. Zhang, in *Electrolytes for Electrochemical Supercapacitors* (CRC Press, 2016), p. 347
22. Y. Lin, H. Zhao, F. Yu, J. Yang, Sustainability **10**, 3630 (2018)

23. G. Ma, M. Dong, K. Sun, E. Feng, H. Peng, Z. Lei, J. Mater. Chem. A **3**, 4035 (2015)
24. S. Senthilkumar, R.K. Selvan, J. Melo, J. Mater. Chem. A **1**, 12386 (2013)
25. B.E. Conway, in *Electrochemical Supercapacitors Scientific Fundamentals and Technological Applications* (1999), p. 698
26. J. Cherusseri, K.K. Kar, J. Mater. Chem. A **4**, 9910 (2016)
27. J. Cherusseri, K.K. Kar, PCCP **18**, 8587 (2016)
28. J. Cherusseri, K.K. Kar, RSC Adv. **6**, 60454 (2016)
29. E. Lim, C. Jo, J. Lee, Nanoscale **8**, 7827 (2016)
30. E. Herrero, L.J. Buller, H.D. Abruña, Chem. Rev. **101**, 1897 (2001)
31. S. Trasatti, G. Buzzanca, J. Electroanal. Chem. Interfacial Electrochem. **29**, A1 (1971)
32. B. Senthilkumar, Z. Khan, S. Park, K. Kim, H. Ko, Y. Kim, J. Mater. Chem. A **3**, 21553 (2015)
33. N. Xu, X. Sun, X. Zhang, K. Wang, Y. Ma, RSC Adv. **5**, 94361 (2015)
34. Z. Zhao, S. Hao, P. Hao, Y. Sang, A. Manivannan, N. Wu, H. Liu, J. Mater. Chem. A **3**, 15049 (2015)
35. www.inventlab.ch. Accessed 10 Feb 2019
36. S. Bose, T. Kuila, A.K. Mishra, R. Rajasekar, N.H. Kim, J.H. Lee, J. Mater. Chem. **22**, 767 (2012)
37. K.K. Kar, *Composite Materials: Processing, Applications, Characterizations* (Springer, 2016)
38. Z. Lu, Y. Chao, Y. Ge, J. Foroughi, Y. Zhao, C. Wang, H. Long, G.G. Wallace, Nanoscale **9**, 5063 (2017)
39. L. Jiang, L. Sheng, X. Chen, T. Wei, Z. Fan, J. Mater. Chem. A **4**, 11388 (2016)
40. S.R. Ede, S. Anantharaj, K. Kumaran, S. Mishra, S. Kundu, RSC Adv. **7**, 5898 (2017)
41. P.H. Jampani, O. Velikokhatnyi, K. Kadakia, D.H. Hong, S.S. Damle, J.A. Poston, A. Manivannan, P.N. Kumta, J. Mater. Chem. A **3**, 8413 (2015)
42. L. Li, R. Li, S. Gai, F. He, P. Yang, J. Mater. Chem. A **2**, 8758 (2014)
43. D.-Q. Liu, S.-H. Yu, S.-W. Son, S.-K. Joo, ECS Trans. **16**, 103 (2008)
44. R. Liu, J. Duay, T. Lane, S.B. Lee, PCCP **12**, 4309 (2010)
45. M. Huang, F. Li, F. Dong, Y.X. Zhang, L.L. Zhang, J. Mater. Chem. A **3**, 21380 (2015)
46. M. Qorbani, T.-C. Chou, Y.-H. Lee, S. Samireddi, N. Naseri, A. Ganguly, A. Esfandiar, C.-H. Wang, L.-C. Chen, K.-H. Chen, J. Mater. Chem. A **5**, 12569 (2017)
47. L.M. Santino, S. Acharya, J.M. D'Arcy, J. Mater. Chem. A **5**, 11772 (2017)
48. L. Yuan, B. Yao, B. Hu, K. Huo, W. Chen, J. Zhou, Energy Environ. Sci. **6**, 470 (2013)
49. C.O. Baker, X. Huang, W. Nelson, R.B. Kaner, Chem. Soc. Rev. **46**, 1510 (2017)
50. M. Umashankar, S. Palaniappan, RSC Adv. **5**, 70675 (2015)
51. M. Rajesh, C.J. Raj, B.C. Kim, R. Manikandan, S.-J. Kim, S.Y. Park, K. Lee, K.H. Yu, RSC Adv. **6**, 110433 (2016)
52. Z. Su, C. Yang, C. Xu, H. Wu, Z. Zhang, T. Liu, C. Zhang, Q. Yang, B. Li, F. Kang, J. Mater. Chem. A **1**, 12432 (2013)
53. R. Ramya, R. Sivasubramanian, M. Sangaranarayanan, Electrochim. Acta **101**, 109 (2013)
54. M. Deschamps, E. Gilbert, P. Azais, E. Raymundo-Piñero, M.R. Ammar, P. Simon, D. Massiot, F. Béguin, Nat. Mater. **12**, 351 (2013)
55. C. Zhong, Y. Deng, W. Hu, J. Qiao, L. Zhang, J. Zhang, Chem. Soc. Rev. **44**, 7484 (2015)
56. S. Hashmi, R. Latham, R. Linford, W. Schlindwein, Polym. Int. **47**, 28 (1998)
57. X. Yang, F. Zhang, L. Zhang, T. Zhang, Y. Huang, Y. Chen, Adv. Funct. Mater. **23**, 3353 (2013)
58. H. Yu, J. Wu, L. Fan, Y. Lin, K. Xu, Z. Tang, C. Cheng, S. Tang, J. Lin, M. Huang, J. Power Sour. **198**, 402 (2012)
59. M. Armand, F. Endres, D.R. MacFarlane, H. Ohno, B. Scrosati, in *Materials for Sustainable Energy: A Collection of Peer-Reviewed Research and Review Articles from Nature Publishing Group* (World Scientific, 2011), p. 129
60. Y.J. Kang, H. Chung, C.-H. Han, W. Kim, Nanotechnology **23**, 065401 (2012)
61. H. Dai, H. Zhang, H. Zhong, H. Jin, X. Li, S. Xiao, Z. Mai, Fuel Cells **10**, 754 (2010)
62. Z. Mahmud, N. Zaki, R. Subban, A. Ali, M. Yahya, in *2012 IEEE Colloquium on Humanities, Science and Engineering (CHUSER)* (IEEE, 2012), p. 621

63. M. Rosi, M. P. Ekaputra, M. Abdullah, Khairurrijal, in *AIP Conference Proceedings* (AIP, 2010), p. 55
64. S. Banerjee, K.K. Kar, Recent Pat. Mater. Sci. **7**, 131 (2014)
65. S. Banerjee, K.K. Kar, M.K. Das, Recent Pat. Mater. Sci. **7**, 173 (2014)
66. I. Shown, A. Ganguly, L.-C. Chen, K.-H. Chen, Energy Sci. Eng. **3**, 2 (2015)
67. P. Taberna, P. Simon, J.-F. Fauvarque, J. Electrochem. Soc. **150**, A292 (2003)
68. G.-Q. Zhang, Y.-Q. Zhao, F. Tao, H.-L. Li, J. Power Sour. **161**, 723 (2006)
69. S. Zhang, N. Pan, Adv. Energy Mater. **5**, 1401401 (2015)
70. Y.S. Lim, H.N. Lim, S.P. Lim, N.M. Huang, RSC Adv. **4**, 56445 (2014)
71. A. Singh, A.J. Roberts, R.C. Slade, A. Chandra, J. Mater. Chem. A **2**, 16723 (2014)
72. D. Singh, K. Shahi, K.K. Kar, Solid State Ionics **287**, 89 (2016)
73. K.H. An, K.K. Jeon, J.K. Heo, S.C. Lim, D.J. Bae, Y.H. Lee, J. Electrochem. Soc. **149**, A1058 (2002)
74. Q. Wang, Z. Wen, J. Li, Adv. Funct. Mater. **16**, 2141 (2006)
75. H. Li, J. Wang, Q. Chu, Z. Wang, F. Zhang, S. Wang, J. Power Sour. **190**, 578 (2009)
76. D.P. Singh, K. Shahi, K.K. Kar, Solid State Ionics **231**, 102 (2013)
77. B. De, T. Kuila, N.H. Kim, J.H. Lee, Carbon **122**, 247 (2017)
78. X. Xiang, W. Zhang, Z. Yang, Y. Zhang, H. Zhang, H. Zhang, H. Guo, X. Zhang, Q. Li, RSC Adv. **6**, 24946 (2016)
79. D. Li, Y. Li, Y. Feng, W. Hu, W. Feng, J. Mater. Chem. A **3**, 2135 (2015)
80. W.-W. Liu, X.-B. Yan, J.-W. Lang, C. Peng, Q.-J. Xue, J. Mater. Chem. **22**, 17245 (2012)
81. Autolab Application Note EC08 Basic overview of the working principle of a potentiostat/galvanostat (PGSTAT)—Electrochemical cell setup https://www.ecochemie.nl/download/Applicationnotes/Autolab_Application_Note_EC08.pdf. Accessed 22 Nov 2019
82. Two, Three and Four Electrode Experiments. https://www.gamry.com/application-notes/instrumentation/two-three-and-four-electrode-experiments/. Accessed 22 Nov 2019
83. Potentiostat. https://www3.nd.edu/~kamatlab/documents/facilities/potentiostat.pdf. Accessed 22 Nov 2019
84. R. Kumar, S. Sahoo, E. Joanni, R.K. Singh, W.K. Tan, K.K. Kar, A. Matsuda, Prog. Energy Combust. Sci. **75**, 100786 (2019)

Chapter 3
Characteristics of Transition Metal Oxides

Alekha Tyagi, Soma Banerjee, Jayesh Cherusseri, and Kamal K. Kar

Abstract Transition metal oxides (TMOs) are the oxides of d-block elements in the periodic table with partially filled d-sub-shell. They have attracted the research community with their unique and fabulous properties such as magnetic, optical and electrochemical. The novel properties have envisaged them in many practical applications such as energy storage (e.g., supercapacitors, lithium-ion batteries, etc.), nonvolatile memory devices, sensors, solar cells and infrared detectors. The ability to modulate the physical as well as chemical properties helps in designing novel devices with tunable properties and hence enhances the industrial importances. This chapter mainly focusses on discussing the characteristics of TMOs including physical, chemical, surface, electronic, magnetic, optical, thermoelectric, electrochemical, etc. Synthesis techniques for TMOs are discussed in detail. The environmental impact of these materials along with their potential applications is also included.

A. Tyagi · S. Banerjee · J. Cherusseri · K. K. Kar (✉)
Advanced Nanoengineering Materials Laboratory, Materials Science Programme, Indian Institute of Technology Kanpur, Kanpur 208016, India
e-mail: kamalkk@iitk.ac.in

A. Tyagi
e-mail: alekhatyagi12@gmail.com

S. Banerjee
e-mail: somabanerjee27@gmail.com

J. Cherusseri
e-mail: jayesh@iitk.ac.in

K. K. Kar
Advanced Nanoengineering Materials Laboratory, Department of Mechanical Engineering, Indian Institute of Technology Kanpur, Kanpur 208016, India

© Springer Nature Switzerland AG 2020
K. K. Kar (ed.), *Handbook of Nanocomposite Supercapacitor Materials I*,
Springer Series in Materials Science 300,
https://doi.org/10.1007/978-3-030-43009-2_3

3.1 Introduction

According to the International Union of Pure and Applied Chemistry (IUPAC), a transition metal is the one, whose atom either comprises a partly occupied d-sub-shell or ionizes to give cations with an incomplete d orbital. Hence, transition metals constitute d-block of the modern periodic table. The d-block elements are shown in Table 3.1. Zinc (Zn), cadmium (Cd) and mercury (Hg) are not considered transition elements as the d-sub-shells are completely filled in the atomic as well as have stable cationic state. The oxides formed by these transition metals are termed as transition metal oxides (TMOs). TMOs constitute a fascinating set of solids with a wide diversity in electrical and magnetic properties, which makes them suitable for variety of applications including catalysis and energy storage [1]. In addition, mixed transition metal oxides (MTMOs) are also gaining popularity as better materials for applications like microwave absorption, supercapacitor, etc. MTMOs are ternary metal oxides consisting of two different transition metal species [2].

Table 3.1 d-block elements in the periodic table

Period → Group ↓	3	4	5	6	7	8	9	10	11	12
4	$_{21}$Sc $3d^14s$ 2	$_{22}$Ti $3d^24s$ 2	$_{23}$V $3d^34s$ 2	$_{24}$Cr $3d^54s$ 1	$_{25}$Mn $3d^54s$ 2	$_{26}$Fe $3d^64s$ 2	$_{27}$Co $3d^74s$ 2	$_{28}$Ni $3d^84s$ 2	$_{29}$Cu $3d^{10}4s$ 1	$_{30}$Zn $3d^{10}4s$ 2
5	$_{39}$Y $4d^15s$ 2	$_{40}$Zr $4d^25s$ 2	$_{41}$Nb $4d^35s$ 2	$_{42}$Mo $4d^55s$ 1	$_{43}$Tc $4d^55s$ 2	$_{44}$Ru $4d^65s$ 2	$_{45}$Rh $4d^75s$ 2	$_{46}$Pd $4d^85s$ 2	$_{47}$Ag $4d^{10}5s$ 1	$_{48}$Cd $4d^{10}5s$ 2
6	$_{57}$La $5d^16s$ 2	$_{72}$Hf $5d^26s$ 2	$_{73}$Ta $5d^36s$ 2	$_{74}$W $5d^56s$ 1	$_{75}$Re $5d^56s$ 2	$_{76}$Os $5d^66s$ 2	$_{77}$Ir $5d^76s$ 2	$_{78}$Pt $5d^86s$ 2	$_{79}$Au $5d^{10}6s$ 1	$_{80}$Hg $5d^{10}6s$ 2
7	$_{89}$Ac $6d^17s$ 2	$_{104}$Rf $6d^27s$ 2	$_{105}$Db $6d^37s$ 2	$_{106}$Sg $6d^47s$ 2	$_{107}$Bh $6d^57s$ 2	$_{108}$Hs $6d^67s$ 2	$_{109}$Mt $6d^77s$ 2	$_{110}$Ds $6d^87s$ 2	$_{111}$Rg $6d^97s^2$	$_{112}$Cn $6d^{10}7s$ 2

3.2 Importance of Transition Metal Oxides

Selection of a particular TMO or MTMO system is based on the application under consideration. All the TMOs may or may not function well in all applications. Hence, it is very crucial to look into individual set of properties before moving towards their utilization in practical applications. Recent furtherance in nanodomains has motivated the development of novel TMO-based nanostructures. The nanostructuring of TMOs has assisted in achieving unique properties at nanoscale regime. The materials exhibit novel properties when compared with their bulk state [3]. TMO nanomaterials are used for the manufacture of high energy density batteries and supercapacitors, machinable ceramics, gas filtration membranes, highly sensitive gas sensors, high-end electronic devices, nano-sized catalytic species, durable medical implants, etc.

3.3 Transition Metal Oxide Nanostructures

In this section, preparation strategies for different TMOs are discussed. This provides an insight into how the synthesis techniques influence the growth mechanisms and morphologies of TMOs, which in turn determine the potential applications of TMO nanostructures.

3.3.1 Synthesis of Transition Metal Oxide Nanostructures

3.3.1.1 Ruthenium Oxide

Ruthenium oxide is an example for mixed electronic–protonic conductor [4]. Both the crystalline and amorphous forms of ruthenium oxides are of great importance owing to their good electrical conductivity, high stability, good redox activity, etc. Synthesis of nanostructured ruthenium oxide morphologies for various applications including catalysis and energy storage devices has been extensively explored by the research communities. Various nanostructures as thin films and nanotubes have been synthesized. These are widely accepted due to the astounding nanoscale features, which result in novel characteristics. Electrochemical deposition of ruthenium oxides on substrates like indium tin oxide (ITO), stainless steel and carbon cloth are well studied [5–8]. Gujar et al. have reported the electrochemical deposition of tetragonal ruthenium oxide nanograins on ITO substrate [5]. This method allows much freedom to alter the nanograin size by varying the synthesis parameters such as current, deposition time and employed electrochemical techniques. It is observed that different electrochemical techniques impart different morphological features in RuO_2 [7]. Electrochemical deposition of such oxides is an environment-friendly technique, which can be used for making electrodes for commercial, low-cost supercapacitors. In certain applications, in addition to the nanostructured morphologies, the presence of high surface area is also desired. This can be attained by depositing the nanostructured oxide on some high surface area materials such as carbon nanomaterials [9–11]. This kind of deposition is benefitted from the usage of maximum surface area of the oxide nanostructures for electrode applications. Hsieh et al. have reported the cyclic voltammetric deposition of ruthenium oxide on MWCNTs [12]. Initially, MWCNTs have been grown on titanium substrate and then the oxide is deposited on both the titanium and MWCNTs/titanium substrates by using ruthenium tri-chloride aqueous solution as the deposition bath. The oxide has been uniformly deposited as a layer on the Ti and MWCNTs/Ti substrates. The CVs are scanned at 50 mV/s in the potential window ranging from −0.1 to 1.0 V, at a deposition temperature of 50 °C. It has been observed that an equal amount of oxide has been deposited onto two different substrates. Ruthenium oxide nanoneedles have been synthesized by using a template-assisted electrochemical method [13]. Porous anodic aluminum oxide (AAO) templates are used for the growth of nanoneedles with controlled dimensions using potentiostatic electrodeposition (Fig. 3.1). The micrograph reveals bundles of ruthenium oxide nanoneedles, where the diameter of each nanoneedle is ca. 100 nm.

3.3.1.2   Manganese Oxide

Manganese oxide is a low-cost, naturally abundant material, which has received much attention due to its environment-friendly nature. The photocatalytic and pseudocapacitive properties of manganese oxides are manipulated in water purification and electrochemical devices including supercapacitors, batteries and fuel cells [3, 14]. Rancieite-type oxides prepared by the reduction of permanganates in acidic aqueous medium have been reported [15]. The reduction of $KMnO_4$ or $NaMnO_4$, followed by ion exchange reactions, leads to the formation of nanostructured manganese oxide. Reactions involving proton and alkali ion exchange are commonly employed. The surface area of the oxides is further increased by nitric acid treatment after the first proton exchange reaction. Alkali ions are also removed from the oxides after acid treatment. The scanning electron microscopy (SEM) images of various manganese oxides are shown in Fig. 3.2. The number in sample designations indicates the preparation temperature. Qu et al. have reported the synthesis of manganese oxide nanorods by solution precipitation method using $MnSO_4$ as the precursor [16]. Mixing the solutions of $MnSO_4$ and $K_2S_2O_8$ followed by drop-wise addition of NaOH leads to the precipitation of a dark brown material. The precipitate is then dried after several times of washing with deionized water to finally get manganese oxide powder. A BET surface area of 135 m^2/g has been reported for the powdered sample. It has been reported that manganese oxides are deposited onto the silica surface in order to form a core–shell nanostructure, with SiO_2 as the core and manganese oxide as the shell [17]. A precipitation method is opted for the preparation of core–shell structure with the help of NH_4OH and NaOH as the precipitating agent. Different morphologies of shells are observed when the concentration of the precipitating agent

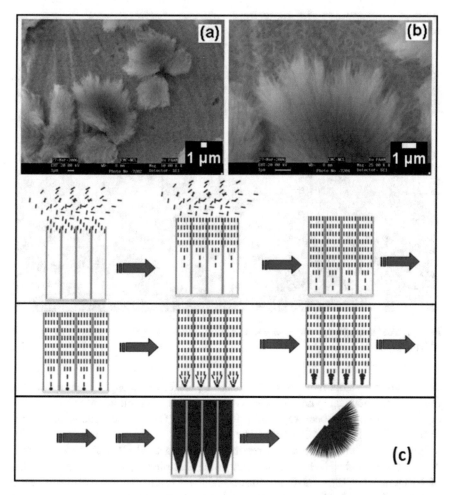

Fig. 3.1 a, b SEM micrographs of RuO_2 nanoneedles prepared by the potentiostatic method using AAO as template, **c** schematic for the mechanism of the RuO_2 nanoneedle formation. Redrawn and reprinted with permission [13]

is increased. These include particles, thin layers and urchin-like morphologies. The particles formed are found to be of Mn_3O_4, thin layers are of mixed Mn_2O_3 and Mn_3O_4, and urchin-like morphologies are of α-MnO_2 phases.

3.3.1.3 Nickel Oxide

Nickel oxide is a low-cost material with distinct redox activity. The various processing methods available for the preparation of nanostructured nickel oxide morphologies have increased their demand in diverse application areas [18]. Electrochemical

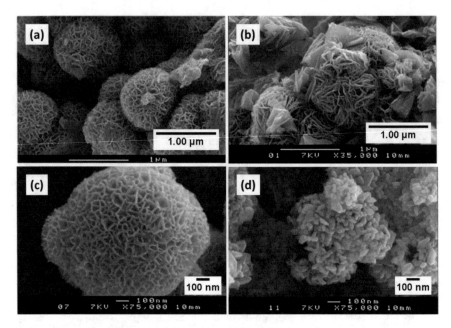

Fig. 3.2 SEM micrographs of samples **a** K60, **b** Na60, **c** Na20 and **d** K100. Redrawn and reprinted with permission [15]

deposition is a viable method for synthesizing nickel oxide on different substrates [19]. Electrochemical synthesis of α-Ni(OH)$_2$ films on Ni foil in Ni(NO$_3$)$_2$ · 6H$_2$O aqueous solution has been reported [20]. The α-Ni(OH)$_2$ nanoparticles are found to have a loosely packed structure. The synthesis of nickel oxide nanoparticles by a facile precipitation approach is reported by Xiang et al. [21]. A nickel source, NiCl$_2$, is used, and NH$_4$HCO$_3$ is selected as the precipitating agent. The precipitates of Ni(OH)$_2$ · NiCO$_3$ · xH$_2$O are further calcined in air to form the nickel oxide nanoparticles of 10–15-nm diameter. Nickel oxide nanoparticles are also prepared by sol-gel method [22, 23]. The nickel (II) nitrate hexahydrate is used as a nickel source, and agarose polysaccharide gel is utilized for the synthesis [22]. Initially, the gel is prepared followed by calcination leading to the formation of nickel oxide nanoparticles. Lu et al. have reported a straightforward and scalable preparation of nickel oxide/Ni composites by using nickel nanoparticles as the starting material [24]. Nickel nanoparticles have been synthesized by a modified polyol process. These nanoparticles have undergone mechanical compaction leading to the preparation of monolithic and stable nickel oxide/Ni composite. This way of preparing composite has a benefit of reducing the resistivity; hence, they can be used in various applications such as electrodes for energy storage devices.

3.3.1.4 Molybdenum Oxide

Molybdenum oxides are of much importance among other TMOs due to high stability. Hence, they have achieved a significant role as cathode materials in lithium batteries [25, 26]. Molybdite (α-MoO_3) is a two-dimensional electro-active material. The mesoporous molybdenum oxide films have been prepared by an evaporation-induced self-assembly process [27]. Initially, a dip-coating is done on a polar substrate with a sol-gel precursor in ethanol and structure-directing agent. After being waited for the complete solvent evaporation, an inorganic/organic composite with mesoporous structure has been formed by the system co-assembly process. Finally, with the help of thermal treatment, a unique mesoporous solid architecture is developed. The transmission electron microscopy (TEM) image of self-organized molybdenum oxide thin film is shown in Fig. 3.3a. The inset is the magnified TEM micrograph. Figure 3.3b shows the tapping mode atomic force microscopy (AFM) map of the molybdenum oxide thin film. The AFM map reveals that a homogeneous oxide film is formed with a flat-top surface, in which surface pores remain open. The inset is a three-dimensional AFM map taken at a tilt of 50°. Molybdenum oxide nanowires have been prepared by selecting a biomolecule-assisted hydrothermal method [28]. A hydrothermal reaction has been carried out using $(NH_4)_6Mo_7O_{24}$ $\cdot 4H_2O$ and L-aspartic acid in HNO_3 solution. This leads to the development of ultra-long molybdenum oxide wires of length greater than 60 μm. The molybdenum oxide nanoparticles synthesized using a hot-wire CVD technique yield high crystallinity [29, 30]. Molybdenum filament is used as precursor for the preparation of oxide nanoparticles. The as-synthesized bulk powder consists of molybdenum oxide nanospheroids with diameters ranging from 5 to 20 nm. Various other morphologies of molybdenum oxide obtained by various synthesis routes have also been reported,

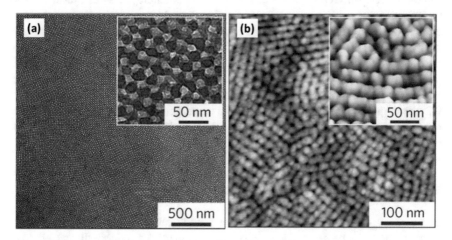

Fig. 3.3 **a** Low- and high-magnification bright-field TEM micrographs and **b** AFM image of the top surface of molybdenum oxide thin film. Redrawn and reprinted with permission [27]

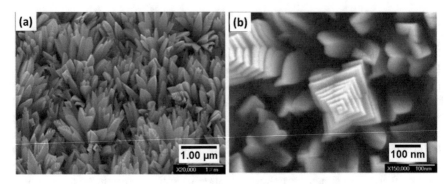

Fig. 3.4 Field emission SEM micrographs of iridium oxide nanorods. Redrawn and reprinted with permission [34]

such as four-armed molybdenum oxide [31], thin films [32] and molybdenum oxide nanotubes [33].

3.3.1.5 Iridium Oxide

Iridium oxide is an electrically conductive TMO having a tetragonal rutile structure. The iridium oxide single crystals possess lowest room temperature resistivity as low as 32 $\mu\Omega$. Due to the conductive nature, iridium oxide finds extensive use in ferroelectric capacitors as thin-film electrodes [43]. Chen et al. have reported the preparation of iridium oxide nanorods and thin films by cold-wall CVD method [34]. A source reagent, (5-methylcyclopentadienyl) (1,5-cyclooctadiene) iridium ((MeCp)Ir(COD)), is used, and Si (100) is employed as the substrate for the growth process. Various parameters such as deposition pressure and temperature influence the growth of iridium oxide nanostructures. A stable thin-film phase has been observed when the growth temperature is 400 °C, and vertically aligned iridium oxide nanorods are formed when the pressure is 30 torr. Figure 3.4 shows the SEM micrographs of the vertically aligned iridium oxide nanorods. The average radius and length of iridium oxide single crystals are found to be 150 ± 50 nm and 3.5 ± 0.2 μm, respectively. A recent review on the preparation techniques of iridium oxide nanocrystals discusses the factors affecting the morphology and structure of iridium oxide nanostructures [35].

3.3.1.6 Iron Oxide

Iron oxides are very abundant, environment-friendly and cheap oxides [36]. Various nanostructures are being synthesized, such as nanoparticles and thin films. Cathodic electrodeposition is a widely used method for the fabrication of iron oxide thin films. Electrodeposition is carried out in $FeCl_2$ solution with chitosan as binder [37]. A

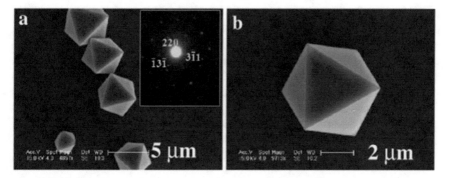

Fig. 3.5 FESEM micrographs of Fe_3O_4 octahedra: **a** low and **b** high magnification. Reprinted with permission from [40]

good adhesion is obtained in the presence of chitosan. The addition of binder also has helped in reducing the crack inside the deposited iron oxide thin films. Kulal et al. have reported the chemical synthesis of iron oxide thin films by successive ionic layer adsorption and reaction (SILAR) method [38]. The anionic and cationic precursors used for the synthesis are 0.1 M sodium hydroxide and 0.05 M ferrous sulfate, respectively. The thin films are uniformly adhered to the substrate surface. Xie et al. have prepared highly ordered iron oxide nanotube arrays [39]. Anodization technique has been used for the synthesis. The nanotube arrays have unique nanostructures with high surface area. Zhao et al. have reported the perfect octahedral crystals of Fe_3O_4 of 4 μm edge size, utilizing polyol synthesis method assisted by polyethylene glycol p-(1,1,3,3-tetramethylbutyl)-phenyl ether commonly known as Triton X-100 (Fig. 3.5) [40]. The microcrystals show low coercivities and high saturation magnetization.

3.3.1.7 Cobalt Oxide

Electrodeposition is used for the preparation of thin-film cobalt oxides in eutectic solvent of glyceline [41]. This kind of nano-architectures can also be prepared by other techniques such as solvothermal, electrochemical precipitation and spray pyrolysis [42–44]. Flowerlike cobalt hydroxide hierarchical microspheres have been synthesized by Yuan et al. [45]. Hydrothermal strategy has been opted for the preparation. The microspheres are self-assembled in the form of one-dimensional nanobelts with a mesoporous structure.

3.3.1.8 Vanadium Oxide

Synthesis of amorphous vanadium oxide has been reported by Lee and Goodenough, by heating the quenched vanadium oxide fine powders [46]. A dark brown precipitate has been obtained in this procedure. Passerini et al. have reported the vanadium oxide aerogels prepared by a sol-gel process [47]. The as-prepared aerogels are highly porous structures with pore diameters ranging from 10 to 20 nm. Thin films of aerogels are coated on conducting substrates and possess large specific surface area up to 40 m^2/g. Dong et al. have synthesized vanadium oxide aerogels and used sticky carbon for immobilizing these aerogels [48]. Kim et al. have synthesized thin films of amorphous and hydrous vanadium oxide onto three-dimensional CNT substrates [49]. Electrochemical deposition has also been carried out for depositing the oxides on CNT substrates, and thickness of the as-deposited thin films is found to be ~6 nm. Various other nanostructured morphologies of vanadium oxides are also reported, such as nanorods [50], nanofibers [51], nanotubes and nanowires [52–54].

3.3.1.9 Tin Oxide

Wu et al. have synthesized nanocrystalline tin oxide by a sol-gel process. Nanocrystallinity has been developed by inhibiting the crystallite growth through replacement of functional groups [55]. The fully crystallized tin oxide is formed with a mean crystallite size of 20 Å. Tin oxide nanoparticles have also been prepared by using $SnCl_4 \cdot 5H_2O$ and ethylene glycol as precursor [56]. The tin oxide nanoparticles so prepared exhibit spherical morphology having a size ~12 nm. The tin oxide diskettes have been prepared by thermal evaporation method [57]. Diskettes with different morphologies, namely the solid wheel and spiral step-like are obtained. Various groups have prepared unique morphologies of tin oxide nanostructures, such as nanotubes, sandwiched nanoribbons and nanowires by implementing processes like elevated temperature heating and hydrothermal synthesis [58, 59].

3.3.1.10 Copper Oxide

Cuprous oxide nanoparticles have been prepared by thermal decomposition procedures using bis(salicylidiminato) Cu(II) and bis(salicylaldoximato)Cu(II) complexes [60]. It is found that at the initial stages, copper nanoparticles are forming followed by oxidation to form cuprous oxide nanoparticles. The as-synthesized nanoparticles exhibit an average diameter of 10 nm. A shape-controlled synthesis of cuprous oxide nanocubes has been reported by Bai et al. [61]. For the synthesis, polyvinylpyrrolidone and ascorbic acid have been used as surfactant and reducing agent, respectively. The microstructures of the as-synthesized cuprous oxides are shown in Fig. 3.6. Atomic layer deposition has been used for depositing cuprous oxide nanoparticles on CNT substrate [62]. The precursors used for the synthesis are (n-butylphosphane)copper(I)acetylacetonate (($^nBu_3P)_2Cu(acac)$) and wet oxygen. It has been observed that the morphological features of the nanoparticles vary in accordance with the defect sites introduced on CNTs.

Fig. 3.6 **a** SEM and **b** TEM images of cuprous oxide nanocubes. Redrawn and reprinted with permission [61]

3.3.1.11 Titanium Oxide

Various nanostructures of titanium oxide (titania) such as thin films, nanotubes and nanoparticles have been synthesized. Titania nanoparticles have been synthesized by low-pressure spray pyrolysis method [63]. Diameters of the titania nanoparticles prepared from titanium isopropoxide (TTIP) and water-soluble transition metal sources are found to be ~10 nm. Jha et al. have reported the microorganism-mediated synthesis of titania nanoparticles [64]. Low-cost microbes such as Lactobacillus species and *Saccharomyces cerevisiae* are used for the preparation. The sizes of the nanoparticles vary from 8 to 35 nm. Various other synthesis techniques have also been reported for the synthesis of nanoparticles such as premixed stagnation swirl flames [65]. Titania anatase nanowires have been synthesized by microwave heating method [66, 67]. The diameter and length of the as-prepared anatase nanowires are found to be in the range of 5–10 nm and 500 nm–2 μm, respectively. Rao et al. have prepared titania thin films by spray deposition technique [68]. Titania nanotube arrays have been prepared by anodic oxidation of titanium foils [69]. The as-prepared titania nanotubes have length and diameter of 1 μm and 90 nm, respectively. Anatase titania and zinc-doped titania nanotubes prepared using electrochemical anodization are examined for photocatalytic activity by Benjwal et al. [70]. Hydrothermal methods are also extensively used for the preparation of nanotubes [71–74]. Kwon et al. have described a unique titanium oxide nanostructure comprising anatase nanobranches covered with rutile phase nanorods (Fig. 3.7).

3.3.1.12 Tungsten Oxide

Tungsten oxide (WO_x) where x lies in between 2 and 3 has emerged as an attractive material for applications including photocatalysis [75], sensors [76] and electrochromic devices [77]. Fabrication of nanostructures of variable dimensions is possible with well-defined synthesis strategies. This gives rise to morphologies with

Fig. 3.7 FESEM micro-
graphs of titania
nanostructure:
a cross-sectional view and
b top view. Redrawn and
reprinted with permission
[73]

hierarchical porous structures with specifically exposed facets. Synthesis of 0-D
WO_3 quantum dots of 1 nm size is reported using a super-microporous silica template [78]. Stoycheva et al. have reported tungsten oxide nanoneedles prepared from
tungsten hexaphenoxide using an aerosol-assisted chemical vapor deposition technique (Fig. 3.8) [76]. The nonaligned nanoneedles have most sensitive response
towards gas sensing. Hierarchical nanostructures derived from 0- and 1-D structures
have the ability to show improved performance than the constituent units. A simple
technique of thermal evaporation is used to obtain $WO_{3-\delta}$ nanowire networks for
application in field emission, gaschromics and electrochemistry [79].

3.3.1.13 Zinc Oxide

Zero-dimensional hollow ZnO nanostructures are widely explored in the recent literature owing to successful applications in catalysis, sensing and biomolecular systems

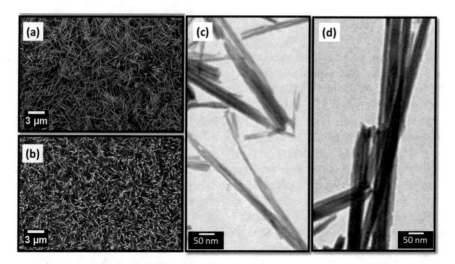

Fig. 3.8 **a, b** FESEM and **c, d** TEM micrograph of tungsten oxide nonaligned and quasi-aligned nanoneedles. Redrawn and reprinted with permission from [76]

[80–82]. A facile hydrothermal followed by pyrolysis strategy is opted for the synthesis of zinc oxide hollow spheres using Streptococcus thermophilus as a biological template [82]. Wang et al. have described the synthesis of aligned ZnO nanorods for sensor and optoelectronic arrays using a self-assembly over alumina substrate [83].

3.4 Characteristics of Transition Metal Oxides

As the name implies, TMOs are oxides of transition metals. Hence, they are conductors of electricity; however, the conductivity range remains much lower when compared with their metallic counterparts. High density and high melting and boiling points are the common characteristics of the transition metals. The reason behind these properties relies on the presence of delocalized d-electron-mediated metallic bonding leading to cohesion. This in turn increases when the numbers of shared electrons are large. However, the group 12 elements exhibit significantly lower melting and boiling points and the reason is the presence of completely filled d-sub-shells, which prevents the d–d bonding. For example, mercury (Hg) possesses a melting point of −38.83 °C.

3.4.1   Size

Size of any material is an important factor since many properties show variation in response to changes in atomic sizes. Many of the properties of TMOs are not adequate for industrial applications. When the size changes from micrometers to nanometers, their properties are also altered significantly. That means at lower sizes, they exhibit novel properties, which are not observed in their bulk state. The advancement in the nanotechnology helps in synthesizing new materials called nanomaterials having at least one dimension in nanoregime (1–100 nm). Recent progresses in the research dealing with TMOs show that the importance of nanostructured and nano-sized TMOs has achieved much importance when compared with their bulk counterparts. One-, two- and three-dimensional nanostructures are possible to synthesize, and each nanostructure can be of a particular size, which can be varied with the help of some experimental parameters. The size of nanostructured TMOs can be controlled accurately by using well-defined experimental procedures [84, 85]. Kumar et al. have discussed the effect of solvents on size of TMOs nanoparticles of ZnO, CuO, Co_3O_4 and Fe_3O_4 [84].

3.4.2   Shape

The shapes of TMOs are easily tunable by opting specific synthesis pathways. Spherical, cylindrical and planar structures with tunable size can be easily synthesized. Some examples of the morphologies include nanodisks and nanoneedles (Figs. 3.9a, b) [86], nanospheres (Fig. 3.9c) [87], dendritic microstructures (Fig. 3.9d) [88], tubular core–shell nanostructures (Fig. 3.9e) [89], nanoflowers (Fig. 3.9f) [90], etc. The mixed three-dimensional TMO nanostructures are also possible to prepare. Jana et al. have reported a simple reproducible strategy to control the size and size distribution of magnetic nanocrystals [91]. This involves the heat treatment of metal containing fatty acid in non-coordinating solvents to get nanocrystals of well-defined and controlled dimensions and shapes in nanoregime.

3.4.3   Surface Area

TMOs in their bulk form exhibit lesser surface area. However, nanostructured TMOs possess huge surface area in comparison. As these materials are in the nanometer dimension, their surface area-to-volume ratio is extremely high. The surface area can be increased by constructing three-dimensional nano-architectures. Not only the surface area, but the pore sizes, and pore volume distributions are also important parameters to be controlled to effectively use metal oxide nanostructures for specific applications. The pictorial representation of hierarchical porous channel is given

Fig. 3.9 a, b SEM micrographs of ZnO nanodisks and nanoneedles(redrawn and reprinted with permission [86]); **c** TEM micrograph of hollow nanospheres of Co_3O_4 (redrawn and reprinted with permission [87]); **d** SEM image of Fe_3O_4 dendrites (redrawn and reprinted with permission [88]); **e** SEM image of CuO/CoO tubular heterostructures (redrawn and reprinted with permission [89]); and **f** SEM images of Co_3O_4 nanoflower (redrawn and reprinted with permission [90])

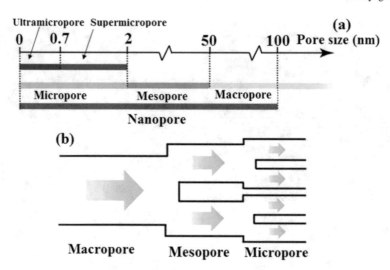

Fig. 3.10 Schematic diagram of the hierarchical pore structure. Redrawn and reprinted with permission [92]

in Fig. 3.10 [92]. The pores with size less than 2 nm are classified as micropores that with pores in the range of 2–50 nm are mesopores and macropores that have pore size higher than 50 nm. The surface area and pore size distribution of the nanostructured TMOs can be tuned by altering the morphology. In favor of this argument, Sun et al. have prepared ultra-thin two-dimensional nanosheets of TiO_2, ZnO, Co_3O_4, MnO_2, WO_3 and Fe_3O_4 by a self-assembly bottom-up approach [93]. The nanosheets exhibit manifold increment in specific surface areas in comparison with conventional transition metal nanoparticles as shown in Fig. 3.11. A simple strategy of fixing the amount of nitric acid during sol-gel synthesis is applied to control the pore size distribution in TMO aerogels reported by Suh et al. [94].

3.4.4 Adhesion

For particular application, such as in electrodes, the individual units of active materials should have good adhesion between them in order to function well. Any discontinuity will certainly decline the performance of the device. TMOs coated on glass substrate are very popular for numerous applications. Benjamin et al. have discussed the importance of intermediate metal oxide film used to ensure proper adhesion over the glass substrate [95]. Further, to achieve good adhesion between the TMO units, some polymeric binder, e.g., polytetrafluoroethylene (PTFE), is generally used. Indium oxide thin films are deposited on glass substrate, and the procedure for fabricating the same is shown in Fig. 3.12 [96]. This includes the micro-patterning of TMO without etching using a green approach.

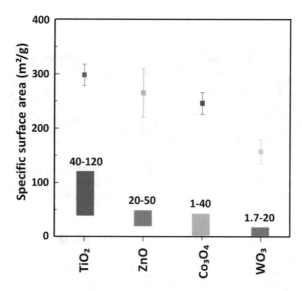

Fig. 3.11 Specific surface area of 2-D nanosheets in comparison with conventional nanoparticles. Redrawn and reprinted with permission [93]

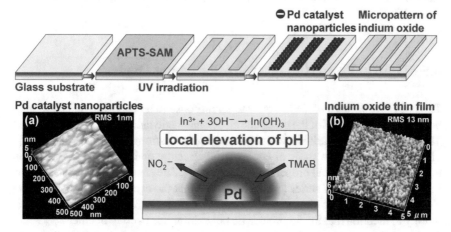

Fig. 3.12 Schematic illustrating the deposition of indium oxide micropattern over glass substrate through a green approach. AFM 3-D maps of **a** palladium (Pd) nanoparticles and **b** indium oxide micropattern. Reprinted with permission from [96]

3.4.5 Color

Transition metal compounds exist in nature in vibrant colors owing to electronic transitions between *d*-orbitals (*d–d* transitions). In addition, the charge transfer (CT) transition is also a contributing phenomenon behind the colors of transition metal

Fig. 3.13 Colored aqueous solutions of transition metal salts: cobalt nitrate (red), potassium dichromate (orange), potassium chromate (yellow), nickel chloride (turquoise blue), copper sulfate (blue), potassium permanganate (purple). Reprinted with permission from [97]

salts. CT transitions are of two types: ligand-to-metal charge transfer (LMCT) transition and metal-to-ligand charge transfer (MLCT) transition. Transition metal in higher oxidation state generally undergoes LMCT transitions. This involves movement of electrons from a ligand-dominated orbital to a metal-dominated orbital. Lower oxidation states lead to a MLCT transition. For instance, LMCT transitions are responsible for the color in chromate, dichromate and permanganate ions. The color of some of the aqueous solutions of transition metal compounds is shown in Fig. 3.13.

3.5 Optical Properties of Transition Metal Oxides

The optical properties of TMOs refer to its response to an exposure to electromagnetic radiation. In most of the cases, the electromagnetic radiation is the visible light. The ability to show variable valence states leads to strong absorption in UV and visible range [98]. The study of energy band structure and energy states assists in understanding the optical properties of TMOs [99]. The energy band analysis of various TMOs of early $3d$ series (Sc, Ti, V) shows that the overlapping of nonbonding d-orbitals of adjacent cations results in the formation of a $3d$ conduction band. It is worthy to mention that the $3d$ orbitals do not exhibit this overlap in other $3d$ TMOs. This leads to occupation of isolated energy states by their $3d$ electrons. Moreover, the electron transport between neighboring cations occurs through electron exchange process. This process particularly necessitates activation energy. The conduction in TMOs is attributed to the movement of highly mobile holes in $2p$ orbital of the oxygen and lesser mobile electron and holes in the $3d$ orbitals of the transition metal cation lattice [100]. Deb et al. have studied the optical and photoelectric characteristics of single crystal and thin-film molybdenum trioxide. Study reveals that the electronic structure of the oxide has strong influence on the optical properties [101]. Optical spectroscopy in combination with first-principle studies comes very handy to understand the fundamental dynamics of nanostructured TMOs [102]. Recently,

Lin et al. have reported in a theoretical study that confined structures like quantum dots and wells allow a fine control over the electronic and hence optical properties of the material [103].

3.6 Surface Properties of Transition Metal Oxides

TMOs are known to exist in unique morphologies. It is possible to fabricate TMOs in a variety of shapes and size by changing the precursors, synthesis procedure and conditions. These structural features affect their surface energies, which in turn affect the chemical properties. The structural defects and features regulate the acidity and basicity of the surface atoms and hence play a significant role in determining catalytic activities. Various characterization techniques such as infrared spectroscopy are commonly employed to understand the details about the presence of acidic and basic sites on the surface of TMOs. Batzill et al. have reviewed the influence of surface crystallographic orientation of TMOs on the photocatalytic activity [104]. TMOs also undergo photo-assisted adsorption and desorption [105]. In order to examine the surface structure, it is important to consider that the oxides are ideal crystals in which the arrangement of atoms is included up to the outer surface plane. In the case of bulk structure, the surfaces are created by cleavages along particular planes of the bulk structure. As a result of cleavage, the position of surface atoms will be different as compared to their bulk structure. The freshly created surfaces will have high surface energy and try to minimize this excess energy through reconstruction mechanisms. Hence, a thermodynamically stable surface is created. The stability of newly constructed surface structures is examined by performing studies on the surface polarity, defect sites and intensity, and the degree of coordinative unsaturation. The surface properties of TMOs are of key interest in many established and emerging technologies. Borghetti et al. have studied the surface properties of TiO_2 on polar faces of ZnO. The phase and surface orientation are characterized by X-ray photoemission spectroscopy (Fig. 3.14) [106]. The study proclaims that the Ti–ZnO interaction depends upon the two polar ZnO surfaces. They have concluded that on Zn–ZnO, reduction of ZnO is carried out by Ti to form titania whereas on O–ZnO, titanium deposition leads to a bimetallic compound (Ti, Zn, O). Further, they have also carried out the structural study of titanium dioxide surfaces by using photoemission spectroscopy [107]. The titanium valence decay involving L and M sub-shells marks Ti $4sp$–O $2p$ hybridizations that are the characteristics of different phases of titania and respective orientations. The Ti LM valence templates for rutile (110), anatase (101) and (100) single crystals can assist in assessment of titania nanopowders, which envisaged to bridge the gap between standard and complex titania surfaces (Fig. 3.15). Brodskii et al. have studied the electro-surface characteristics of TMOs and transition metal hydroxides in aqueous solutions [108]. An infrared study of the surface properties of metal oxides and the interaction of ammonia with the surface of Fe_2O_3, ZnO, MoO_3 and V_2O_5 have been reported [109]. Also, the TMOs are the best candidates for

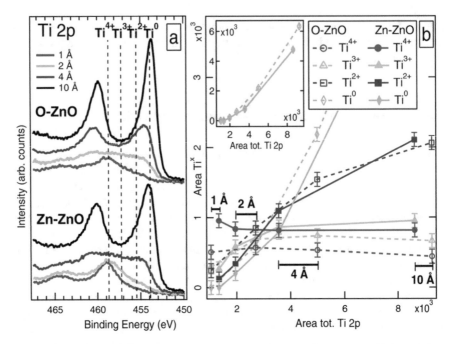

Fig. 3.14 XPS studies of the titanium deposited on the polar faces of zinc oxide: **a** high resolution (HR)-XPS spectra of Ti 2*p* on O–ZnO Zn–ZnO polar faces and **b** areas of component titanium ions obtained from deconvoluted HR-XPS spectra versus total Ti 2*p* area of O–ZnO and Zn–ZnO films. The horizontal bars mark the corresponding Ti thickness. Inset represents the full range of Ti$_0$ area versus the total area. Reprinted with permission from [106]

electrochemical energy devices since they can provide maximum available surface area for the process.

3.7 Electronic Properties of Transition Metal Oxides

3.7.1 Electronic Configuration of Transition Metals

The electronic configuration refers to the arrangement of electrons of an atom around the nucleus in atomic orbitals. It describes that each electron is orbiting the nucleus in an orbital under the influence of an average field due to electrons in other orbitals. Slater determinants are used to mathematically describe the electronic configurations. Electronic configuration of an element helps in describing the chemical bonding that may be exhibited in compounds. The electronic configurations of all the transition metal elements are presented in Table 3.1.

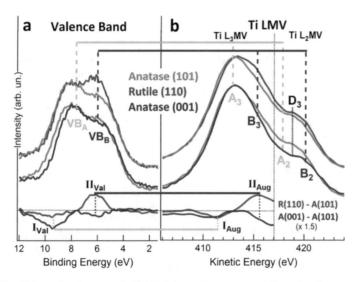

Fig. 3.15 **a** Valence band spectra; **b** Ti LMV Auger spectra comparison of rutile, anatase (101) and (001). Corresponding difference spectra: R(110)–A(101) and A(001)–A(101) are presented at the bottom of both the spectra. Reprinted with permission from [107]

3.7.2 *Energy Band and Orbital Interactions*

In recent years, there has been a great interest in understanding the 3*d* transition metal oxides in terms of their electrical properties. An electron linked with an atom in a solid has three characteristics, i.e., the charge, spin states and orbital [110]. Orbital refers to the geometry of electron cloud around the atomic nucleus. The TMOs possess an anisotropic *d*-electron cloud, and the Coulombic interaction is of utmost importance to understand their metal–insulator transitions and resulting in exciting phenomenon of colossal magnetoresistance and high-temperature superconductivity [111]. Mott and Hubbard have tried to explain the electrical and magnetic inactivity of some *d*-block element oxides in spite of possessing partially filled 3*d* band. A strong *d–d* Coulombic interaction is responsible for the observed behavior. 3*d* bandwidth and the *d–d* Coulombic interactions are the two opposing forces acting in determination of conductivity behavior. If the bandwidth dominates over the Coulombic interaction, it imparts metallic character to the compound. On the other hand, overpowering *d–d* Coulombic interaction leads to the localized 3*d* electrons forming an insulator with local magnetic moment. The single-band Hubbard model is used to understand this Mott-Hubbard-type metal–insulator transition [112–114].

3.7.3 Electronic Conductivity

The TMOs cover the entire range of electronic conductivity spectrum, i.e., from metals to semiconductors to insulators. Adler et al. have studied the formation of equilibrium lattice of a solid from individual atoms at large atomic separations and thereby the development on electronic structure in the process [115]. TMO also undergoes semiconductor–metal transitions. The pure band approach is applicable to metallic compounds. In order to use it for nonmetallic solids, it is required to consider serious modification [116, 117]. A transition from metallic at high temperatures to poor semiconducting below crystallographic transition point is observed in vanadium and titanium oxides. This is accounted by the direct cation–cation interaction and presents antiferromagnetic ordering [118]. Agreement with the optical and electrical data can be obtained by considering the complete localization of d-electrons. Among the variety of TMOs, most of them are semiconductors with wide band gap. For practical applications, the electronic conductivity of these materials should be tailored by fabricating composites with electrically conductive elements such as carbon nanotubes and graphene [119, 120].

3.8 Magnetic Properties of Transition Metal Oxides

The different types of magnetism exhibited by TMOs are shown in Fig. 3.16. Arrows in the schematic mark the spin direction of electrons. Compounds of d-block elements are mostly paramagnetic owing to the spin moment of one or more unpaired electrons in d-sub-shell. Elements with four to seven d-electrons readily form high-spin as well as low-spin octahedral complexes [121]. However, in case of tetrahedral

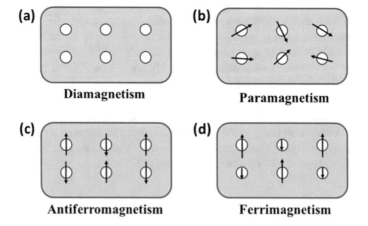

Fig. 3.16 Different kinds of magnetism in TMOs

transition metal complexes, generally high-spin complexes are formed because of the low crystal field splitting energy (CFSE) than the pairing energy [122, 123]. The relativistic nephelauxetic effect in transition metal complexes has been studied in the literature [124].

Matsumoto et al. have studied the room-temperature ferromagnetism in transparent metal-doped titania [125]. When doped with cobalt atoms, titania exhibits ferromagnetism above the room temperature. Titania shows a magnetic moment of 0.32 Bohr magnetons/Co atom and a 60% positive magnetoresistance at 2 K. The ferromagnetic behavior of ZnO-based diluted magnetic semiconductors (DMSs) has been studied and reported using first-principle calculations [126]. The ferromagnetic states in Mn-doped ZnO-based DMSs are shown in Fig. 3.17 as a function of the carrier concentration. The ferromagnetic states are found to be more stable owing to positive energy difference as compared to the spin-glass state. Antiferromagnetic behavior in doped TMOs arises from the individual spin alignment in the solid state. Shen and Wang have come up with electronic models explaining the antiferromagnetism and phase separation in doped TMOs [127]. A density functional theory (DFT)-based solution to the gap problem of antiferromagnetic TMOs and parent compounds of high-Tc superconductors are explained by Fritsche et al. [128].

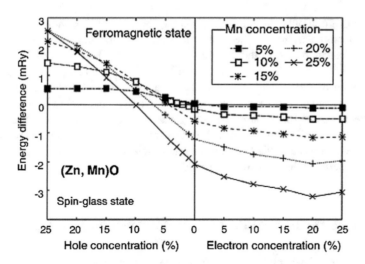

Fig. 3.17 Energy difference between the ferromagnetic and the spin-glass state in Mn-doped zinc oxide-based DMSs versus carrier (hole and electron) concentration. The y-axis represents the stability of ferromagnetic states. Redrawn and reprinted with permission [126]

3.9 Chemical Properties of Transition Metal Oxides

3.9.1 Chemical Stability

TMOs are highly stable in various chemical environments. These are significantly resistant to degradation by the chemical agents. This in turn increases the life of device. Campbell has studied the stability of TMO thin films and proposed that the energy of adhesion at the interface of thin films leads to additional stability over the bulk oxide by an exponential factor depending on the surface energy of oxide, the thickness of oxide film and the oxygen concentration in the bulk oxide [129]. Ataca et al. have investigated the stability of single-layer TMOs by studying in-plane stiffness and ab initio molecular dynamic (MD) calculations [130]. This report mentions the unique physical and chemical properties of honeycomb-structured TMOs, which are far superior to other widely studied 2-D structures such as graphene.

3.9.2 Catalytic Properties

The TMOs are popular in industrial processes because of their homogeneous and heterogeneous catalytic activities. The effectiveness of transition metals as catalysts further enhances owing to the variable oxidation states. Examples of TMOs with good catalytic activity are vanadium oxide (VO_x) in the contact process, vanadium and niobium oxide cluster ions for catalysis of oxygen evolution reaction (OER) [131], cobalt and copper oxides for carbon monoxide oxidation [132], tungsten and molybdenum oxide for oxidative dehydrogenation of alkanes [133], etc. Nano-sized catalyst particles allow more reactant molecules to interact in comparison with the bulk. They can weaken the bonds in reacting molecules and reduce the activation energy to facilitate the formation of product.

3.10 Electrochemical Properties of Transition Metal Oxides

TMOs are potential candidates with enhanced electrochemical properties, which can be successfully utilized in electrochemical energy conversion and storage devices as electrodes and catalysts/catalyst support. The most commonly used TMOs for electrochemical applications include ruthenium oxide, iron oxide, manganese oxide, vanadium oxide, molybdenum oxide, iridium oxide, nickel oxide, etc. The electrodes with high surface area are an essential prerequisite for electrochemical application. Nanotechnology has helped the researchers to design electro-active materials with high surface area. The research and development in nano-sized TMOs with various interesting morphologies have reformed them into excellent candidates for electrode applications.

3.11 Thermoelectric Properties of Transition Metal Oxides

Thermoelectric materials convert the waste heat directly into electrical energy under the influence of Seebeck effect [134]. TMOs are gaining popularity as thermoelectric materials owing to a wide range of thermal stability and ability to modify electronic and photonic transport properties by structural changes [135]. Titania and manganese oxide in stoichiometric, non-stoichiometric and doped forms are widely explored for thermoelectric properties [136–140]. Thermoelectric TMOs find applications in areas including cooling and refrigeration, energy harvesting from waste heat, thermoelectric generators and sensors. Yang et al. have reported a single Sb-doped ZnO microbelts exhibiting a Seebeck coefficient of ~ -350 μV/K, useful for nanogenerator application [141].

3.12 Environmental Impact of Transition Metal Oxides

Most of the TMOs are environmentally safe to use. Not much safety issues are reported for these materials. Hence, these materials are invariably used as electrodes in batteries, e.g., lithium-ion batteries, supercapacitors, fuel cells, etc. These materials have increased the device safety to a large extent. However, the recycling and disposal of some of the oxide material are still major challenges to overcome.

3.13 Post-transition Metal Oxides

The post-transition metals, also known as the poor metals, constitute elements from group 13 and 14 of the periodic table. They reside to the right of the transition metals. The group 12 elements are sometimes considered owing to their strikingly different properties compared to the rest of transition metals because of filled d-orbitals. They normally have higher electronegativity and lower melting points than the transition metals. They are generally softer than other transition metals. They include lead and tin in addition to group 13 elements (Table 3.2).

Table 3.2 Examples for the post-transition metals in the periodic table

Group/Period	12	13	14	15
3		$_{13}$Al		
4	$_{30}$Zn	$_{31}$Ga		
5	$_{48}$Cd	$_{49}$In	$_{50}$Sn	
6	$_{80}$Hg	$_{81}$Tl	$_{82}$Pb	$_{83}$Bi

3.14 Cost of Transition Metal Oxides

The successful marketing of any product lies on the cost of end product. Although there are various factors that influence the cost of product, one of them is the type of the functional material, which is used for manufacturing the product. In this scenario, TMOs are the best candidates as these are available in bulk at a relatively low cost. Although certain metal oxides such as ruthenium oxide and vanadium oxide are costly, however a majority of oxides are of low cost. The cost of the material is directly proportional to its availability. The advancements in the field of nanoscience and nanotechnology help to synthesize nanostructured TMOs in bulk quantities at a relatively low cost.

3.15 Applications of Transition Metal Oxides

TMOs are suitable for diverse range of applications like flexible electronics, catalysis and biomedicine owing to the unique electronic, surface and biocompatible properties, respectively. Some domains, where TMOs find applications, are discussed in this section (Fig. 3.18).

3.15.1 Nanoelectronics

The exfoliated two-dimensional nanosheet-like structures are emerging as suitable materials for miniaturized digital electronic devices, e.g., field-effect transistors (FET), logic gates and photo-conducting diodes owing to the novel electronic properties [142]. The quantum confinement in these structures will assist in modulation of electronic properties by controlling the electron transport mechanisms. High mobility and controllable carrier concentration make ionic amorphous oxides (e.g., In_2O_3–Ga_2O_3–ZnO) suitable for thin-film flexible transparent transistors [143].

3.15.2 Photonic and Optoelectronic Devices

Designing of photonic and optoelectronic devices requires development of porous and high specific surface area thin films of semiconducting TMOs like TiO_2 and ZnO. TiO_2 is the most widely explored material for solar cells because of high conversion efficiency for white light and chemical stability [144]. Recent explorations of transparent and flexible thin-film transistors (TFTs) result in realization of liquid crystal display (LCD) and see-through organic light-emitting diode (OLED) [145].

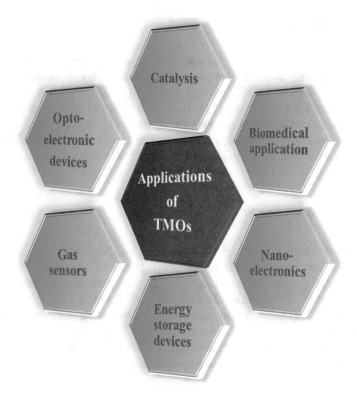

Fig. 3.18 Applications of TMOs

3.15.3   Gas Sensors

Semiconducting TMOs have been widely implemented for gas sensing applications. It is worth noticing that the gas sensing ability of a metal oxide depends upon its sensitivity towards the surface reactions. The gas detection ability of TMOs is dependent on the conductive responses [146]. Factors controlling the gas sensing capability of a material include morphology, temperature and humidity. Room-temperature gas sensing ability of various TMOs nanostructures is explored in the recent literature [147]. SnO_2-decorated V_2O_5 nanowires are reported for alcohol (ethanol) sensing with a sensitivity of 100 ppm [148]. Bao et al. have reported a plate-like nickel oxide/tungsten oxide nanocomposite for NO_2 sensing [149, 150]. Nanoplates are sensitive to concentrations as small as 30 ppm with a response time of 2.5 s. SnO_2 nanocrystalline tubes synthesized using electrospinning are also reported for NO_2 sensing. Nanotubes are sensitive to a concentration of 9.7 ppb with 20 s response time [151]. Sandeep et al. have designed ZnO nanocapsules/nitrogen-doped graphene nanohybrid for ammonia sensing [152]. Hence, an effective structural tuning in TMOs makes it possible to fabricate a highly sensitive gas sensing device.

3.15.4   Catalysis

TMOs are also suitable for catalytic applications owing to the high specific surface area and hence better availability of active sites. Vanadium and niobium oxide cluster ions are recognized as suitable catalyst for OER [131]. Cobalt and copper metal oxides are used for CO oxidation [132]. Cobalt and nickel oxides are widely used for methane to synthesis gas (syngas) oxidation [153], etc. Magnetic oxides are used as catalysts in Fischer–Tropsch hydrocarbon synthesis, alcohol oxidation and high-temperature carbon monoxide–hydrogen shift reactions.

3.15.5   Biomedicine

Zinc oxide is a suitable material for biomedical imaging and diagnosis owing to the low-cost and biocompatible nature [154]. It has an edge over organic dyes and other chalcogenide quantum dots owing to better photoluminescence and non-toxicity. In addition, magnetite (Fe_3O_4) and maghemite (γ-Fe_2O_3) are the most widely utilized TMOs for in-vitro and in-vivo biomedical applications. These find usage in contrast agents for magnetic resonance imaging (MRI), targeted drug delivery and therapeutic agents for cancer treatment [155]. The surfaces of these magnetic nanoparticles are modified with biocompatible polymers to ensure binding with the bioactive entities [156].

3.15.6   Electrodes in Supercapacitors and Lithium-Ion Batteries

Supercapacitors and batteries have attracted the interest of research communities in the field of energy devices owing to their capability to alter the energy scenario. These are suitable for portable electronics and power vehicles. Nanostructured TMO-based SC electrodes utilize reversible faradaic redox reactions on their surface to store energy, and hence the specific capacitances are 10–100 times that of electric double layer capacitors (EDLCs) [157]. These are versatile materials for pseudocapacitor application due to the good redox properties, high specific surface area, tunability in the size and shape and chemical and environmental stability [158]. The research and development in TMO nanoparticles have made them as excellent electro-active materials for SC electrodes. The charge transfer reactions happening at the electrode/electrolyte interface determine the amount of charge that can be stored in the particular electrode. Electrochemical tuning of surface properties of electrode materials results in the improved performance of TMO-based SCs. The commonly used TMOs are ruthenium oxide (RuO_2), manganese oxide (MnO_x), titanium oxide (TiO_2), nickel oxide (NiO), iron oxide (FeO_x), vanadium oxide (VO_2), molybdenum oxide

(MbO_x), iridium oxide (IrO_2), cobalt oxide (Co_3O_4), etc. In this category, ruthenium oxide is the most widely used [159]. It is excellent in its electrochemical performance; however, the only disadvantage is the high cost [160]. Hence, the research is going in a direction to make use of other low-cost TMOs for the supercapacitor application. It is possible to tune the surface area of these nanostructures during the synthesis itself. The pore size can also be modified by altering the morphology of the electrode. For electrode applications, the nanostructured metal oxides should have mesoporous architecture, where an enhanced pseudo-faradaic reaction can be expected. However, the microporous and macroporous architectures are less preferred for charge storage applications as the charge storage capability is limited for such architectures. This increases the probability of electrolyte ions to interact with the electro-active species, thereby increasing the specific capacitance and charge storage. The nature of substrates and their morphology on which the TMOs are being deposited also influence the specific capacitance of electrode materials. The power and energy densities of TMO electrode-based supercapacitors are very high because of rapid electron transfer through the current collectors and large surface area for the faradaic reactions. The cell lives of these supercapacitors are very long since the electrodes undergo completely reversible oxidation–reduction reactions without any morphological change in the electrode. The self-discharge rates of these supercapacitors are also very small.

3.16 Concluding Remarks

This chapter deals with the synthesis and properties of TMOs. The ability to control the atoms at nanoscale leads to the synthesis of interesting morphologies in TMOs. This gives rise to interesting novel properties due to unique combination of properties like higher surface areas, redox properties and semiconducting. The nanostructured TMOs are gaining popularity in the field of catalysis, electronics and energy storage. For a better insight, an overview on the applications of TMOs in the domains of nanoelectronics, energy and catalysis is also presented.

Acknowledgements The authors acknowledge the financial support provided by the Department of Science and Technology, India (DST/TMD/MES/2K16/37(G)), for carrying out this research work. The authors are thankful to Ms Tanvi Pal for drafting a few figures.

References

1. C.N.R. Rao, Ann. Rev. Phys. Chem. **40**, 291 (1989)
2. Y. Zhao, X. Li, B. Yan, D. Xiong, D. Li, S. Lawes, X. Sun, Adv. Energy Mater. **6**, 1502175 (2016)
3. M.M. Najafpour, M. Hołyńska, S. Salimi, Coord. Chem. Rev. **285**, 65 (2015)
4. D.R. Rolison, P.L. Hagans, K.E. Swider, J.W. Long, Langmuir **15**, 774 (1999)

5. T.P. Gujar, W.-Y. Kim, I. Puspitasari, K.-D. Jung, O.-S. Joo, Int. J. Electrochem. Sci. **2**, 8 (2007)
6. J.M. Sieben, E. Morallón, D. Cazorla-Amorós, Energy **58**, 519 (2013)
7. R.K.V. Prataap, R. Arunachalam, R. Pavul Raj, S. Mohan, L. Peter, Curr. Appl. Phys. **18**, 1143 (2018)
8. K.K. Kar, *Composite Materials: Processing, Applications, Characterizations* (Springer, Berlin, 2016)
9. Z. Lv, D. Xie, X. Yue, C. Feng, C. Wei, J. Power Sources **210**, 26 (2012)
10. M.-C. Tsai, T.-K. Yeh, C.-H. Tsai, Electrochem. Comm. **8**, 1445 (2006)
11. K.K. Kar, S. Rana, J. Pandey, *Handbook of Polymer Nanocomposites Processing, Performance and Application* (Springer, Berlin, 2015)
12. T.-F. Hsieh, C.-C. Chuang, W.-J. Chen, J.-H. Huang, W.-T. Chen, C.-M. Shu, Carbon **50**, 1740 (2012)
13. M. Subhramannia, B.K. Balan, B.R. Sathe, I.S. Mulla, V.K. Pillai, J. Phys. Chem. C **111**, 16593 (2007)
14. P. Benjwal, K.K. Kar, RSC Adv. **5**, 98166 (2015)
15. E. Beaudrouet, A. Le Gal La, D.Guyomard Salle, Electrochim. Acta **54**, 1240 (2009)
16. Q. Qu, P. Zhang, B. Wang, Y. Chen, S. Tian, Y. Wu, R. Holze, J. Phys. Chem. C **113**, 14020 (2009)
17. S.-H. Ryu, S.-G. Hwang, S.-R. Yun, K.-K. Cho, K.-W. Kim, K.-S. Ryu, Bull. Korean Chem. Soc. **32**, 2683 (2011)
18. K. Fominykh, J.M. Feckl, J. Sicklinger, M. Döblinger, S. Böcklein, J. Ziegler, L. Peter, J. Rathousky, E.-W. Scheidt, T. Bein, D. Fattakhova-Rohlfing, Adv. Funct. Mater. **24**, 3123 (2014)
19. M.-S. Wu, Y.-A. Huang, J.-J. Jow, W.-D. Yang, C.-Y. Hsieh, H.-M. Tsai, Int. J. Hydrogen Energy **33**, 2921 (2008)
20. G. Fu, Z. Hu, L. Xie, X. Jin, Y. Xie, Y. Wang, Z. Zhang, Y. Yang, H. Wu, Int. J. Electrochem. Sci. **4**, 11 (2009)
21. L. Xiang, X.Y. Deng, Y. Jin, Scr. Mater. **47**, 219 (2002)
22. M. Alagiri, S. Ponnusamy, C. Muthamizhchelvan, J. Mater. Sci.: Mater. Electron. **23**, 728 (2012)
23. A.S. Danial, M.M. Saleh, S.A. Salih, M.I. Awad, J. Power Sources **293**, 101 (2015)
24. Q. Lu, M.W. Lattanzi, Y. Chen, X. Kou, W. Li, X. Fan, K.M. Unruh, J.G. Chen, J.Q. Xiao, Angew. Chem. Int. Ed. **50**, 6847 (2011)
25. S.H. Choi, Y.C. Kang, Chemsuschem **7**, 523 (2014)
26. P. Meduri, E. Clark, J.H. Kim, E. Dayalan, G.U. Sumanasekera, M.K. Sunkara, Nano Lett. **12**, 1784 (2012)
27. T. Brezesinski, J. Wang, S.H. Tolbert, B. Dunn, Nature Mater. **9**, 146 (2010)
28. R. Liang, H. Cao, D. Qian, Chem. Commun. **47**, 10305 (2011)
29. L.A. Riley, S.-H. Lee, L. Gedvilias, A.C. Dillon, J. Power Sources **195**, 588 (2010)
30. S. Mitra, K. Sridharan, J. Unnam, K. Ghosh, Thin Solid Films **516**, 798 (2008)
31. X.W. Lou, H.C. Zeng, J. Am. Chem. Soc. **125**, 2697 (2003)
32. M. Mattinen, P.J. King, L. Khriachtchev, M.J. Heikkilä, B. Fleming, S. Rushworth, K. Mizohata, K. Meinander, J. Räisänen, M. Ritala, M. Leskelä, Mater. Today Chem. **9**, 17 (2018)
33. B. Shen, H. Xie, L. Gu, X. Chen, Y. Bai, Z. Zhu, F. Wei, Adv. Mater. **30**, 1803368 (2018)
34. R.-S. Chen, Y.-S. Huang, Y.-M. Liang, D.-S. Tsai, Y. Chi, J.-J. Kai, J. Mater. Chem. **13**, 2525 (2003)
35. I. Ali, K. AlGhamdi, F.T. Al-Wadaani, J. Mol. Liq. **280**, 274 (2019)
36. S. Banerjee, P. Benjwal, M. Singh, K.K. Kar, Appl. Surf. Sci. **439**, 560 (2018)
37. N. Nagarajan, I. Zhitomirsky, J. Appl. Electrochem. **36**, 1399 (2006)
38. P.M. Kulal, D.P. Dubal, C.D. Lokhande, V.J. Fulari, J. Alloys Compd. **509**, 2567 (2011)
39. K. Xie, J. Li, Y. Lai, W. Lu, Z. Zhang, Y. Liu, L. Zhou, H. Huang, Electrochem. Commun. **13**, 657 (2011)

40. L. Zhao, H. Zhang, J. Tang, S. Song, F. Cao, Mater. Lett. **63**, 307 (2009)
41. A.M.P. Sakita, R.D. Noce, C.S. Fugivara, A.V. Benedetti, Phys. Chem. Chem. Phys. **18**, 25048 (2016)
42. A. Agiral, H.S. Soo, H. Frei, Chem. Mater. **25**, 2264 (2013)
43. V. Srinivasan, J.W. Weidner, J. Power Sources **108**, 15 (2002)
44. V.R. Shinde, S.B. Mahadik, T.P. Gujar, C.D. Lokhande, Appl. Surf. Sci. **252**, 7487 (2006)
45. C. Yuan, L. Yang, L. Hou, D. Li, L. Shen, F. Zhang, X. Zhang, J. Solid State Electrochem. **16**, 1519 (2012)
46. H.Y. Lee, J.B. Goodenough, J. Solid State Chem. **148**, 81 (1999)
47. S. Passerini, J.J. Ressler, D.B. Le, B.B. Owens, W.H. Smyrl, Electrochim. Acta **44**, 2209 (1999)
48. W. Dong, D.R. Rolison, B. Dunn, Electrochem. Solid-State Lett. **3**, 457 (2000)
49. I.-H. Kim, J.-H. Kim, B.-W. Cho, Y.-H. Lee, K.-B. Kim, J. Electrochem. Soc. **153**, A989 (2006)
50. S. Pavasupree, Y. Suzuki, A. Kitiyanan, S. Pivsa-Art, S. Yoshikawa, J. Solid State Chem. **178**, 2152 (2005)
51. D. Yu, C. Chen, S. Xie, Y. Liu, K. Park, X. Zhou, Q. Zhang, J. Li, G. Cao, Energy Environ. Sci. **4**, 858 (2011)
52. M.E. Spahr, P. Bitterli, R. Nesper, M. Müller, F. Krumeich, H.U. Nissen, Angew. Chem. Int. Ed. **37**, 1263 (1998)
53. H.-J. Muhr, F. Krumeich, U.P. Schönholzer, F. Bieri, M. Niederberger, L.J. Gauckler, R. Nesper, Adv. Mater. **12**, 231 (2000)
54. C. Xiong, A.E. Aliev, B. Gnade, K.J. Balkus, ACS Nano **2**, 293 (2008)
55. N. Wu, Science **285**, 1375 (1999)
56. G. Zhang, M. Liu, J. Mater. Sci. **34**, 3213 (1999)
57. Z.R. Dai, Z.W. Pan, Z.L. Wang, J. Am. Chem. Soc. **124**, 8673 (2002)
58. Z.R. Dai, J.L. Gole, J.D. Stout, Z.L. Wang, J. Phys. Chem. B **106**, 1274 (2002)
59. P. Bhattacharya, J.H. Lee, K.K. Kar, H.S. Park, Chem. Eng. J. **369**, 422 (2019)
60. M. Salavati-Niasari, F. Davar, Mater. Lett. **63**, 441 (2009)
61. Y. Bai, T. Yang, Q. Gu, G. Cheng, R. Zheng, Powder Technol. **227**, 35 (2012)
62. M. Melzer, *Atomic Layer Deposition and Microanalysis of Ultrathin Layers* (Chemnitz University of Technology, 2011)
63. W.-N. Wang, I.W. Lenggoro, Y. Terashi, T.O. Kim, K. Okuyama, Mater. Sci. Eng., B **123**, 194 (2005)
64. A.K. Jha, K. Prasad, A.R. Kulkarni, Colloids Surf., B **71**, 226 (2009)
65. J. Wang, S. Li, W. Yan, S.D. Tse, Q. Yao, Proc. Combust. Inst. **33**, 1925 (2011)
66. L. Li, X. Qin, G. Wang, L. Qi, G. Du, Z. Hu, Appl. Surf. Sci. **257**, 8006 (2011)
67. S.K. Singh, M.J. Akhtar, K.K. Kar, Compos. Part B: Eng. **167**, 135 (2019)
68. A. Ranga Rao, V. Dutta, Sol. Energy Mater. Sol. Cells **91**, 1075 (2007)
69. J. Zhao, X. Wang, R. Chen, L. Li, Solid State Commun. **134**, 705 (2005)
70. P. Benjwal, K.K. Kar, Mater. Chem. Phys. **160**, 279 (2015)
71. N.M. Makwana, C.J. Tighe, R.I. Gruar, P.F. McMillan, J.A. Darr, Mater. Sci. Semicond. Process. **42**, 131 (2016)
72. P.M. Tsimbouri, L. Fisher, N. Holloway, T. Sjostrom, A.H. Nobbs, R.M.D. Meek, B. Su, M.J. Dalby, Sci. Rep. **6**, 36587 (2016)
73. S.J. Kwon, H.B. Im, J.E. Nam, J.K. Kang, T.S. Hwang, K.B. Yi, Appl. Surf. Sci. **320**, 487 (2014)
74. P. Benjwal, B. De, K.K. Kar, Appl. Surf. Sci. **427**, 262 (2018)
75. Y. Hou, F. Zuo, A.P. Dagg, J. Liu, P. Feng, Adv. Mater. **26**, 5043 (2014)
76. T. Stoycheva, F.E. Annanouch, I. Gràcia, E. Llobet, C. Blackman, X. Correig, S. Vallejos, Sens. Actuators, B **198**, 210 (2014)
77. C.Y. Ng, K. Abdul Razak, Z. Lockman, J. Alloys Compd. **588**, 585 (2014)
78. H. Watanabe, K. Fujikata, Y. Oaki, H. Imai, Chem. Commun. **49**, 8477 (2013)
79. J. Zhou, Y. Ding, S.Z. Deng, L. Gong, N.S. Xu, Z.L. Wang, Adv. Mater. **17**, 2107 (2005)

80. B. Liu, H.C. Zeng, Chem. Mater. **19**, 5824 (2007)
81. Z. Deng, M. Chen, G. Gu, L. Wu, J. Phys. Chem. B **112**, 16 (2008)
82. H. Zhou, T. Fan, D. Zhang, Microporous Mesoporous Mater. **100**, 322 (2007)
83. X. Wang, C.J. Summers, Z.L. Wang, Nano Lett. **4**, 423 (2004)
84. R.V. Kumar, Y. Diamant, A. Gedanken, Chem. Mater. **12**, 2301 (2000)
85. M.T. Reetz, W. Helbig, J. Am. Chem. Soc. **116**, 7401 (1994)
86. J.H. Zeng, B.B. Jin, Y.F. Wang, Chem. Phys. Lett. **472**, 90 (2009)
87. J. Park, X. Shen, G. Wang, Sens. Actuators, B **136**, 494 (2009)
88. G. Sun, B. Dong, M. Cao, B. Wei, C. Hu, Chem. Mater. **23**, 1587 (2011)
89. J. Wang, Q. Zhang, X. Li, B. Zhang, L. Mai, K. Zhang, Nano Energy **12**, 437 (2015)
90. Q. Yan, X. Li, J. Hazard. Mater. **209–210**, 385 (2012)
91. N.R. Jana, Y. Chen, X. Peng, Chem. Mater. **16**, 3931 (2004)
92. T. Liu, F. Zhang, Y. Song, Y. Li, J. Mater. Chem. **5**, 17705 (2017)
93. Z. Sun, T. Liao, Y. Dou, S.M. Hwang, M.-S. Park, L. Jiang, J.H. Kim, S.X. Dou, Nat. Commun. **5**, 3813 (2014)
94. D.J. Suh, T.-J. Park, Chem. Mater. **8**, 509 (1996)
95. P. Benjamin, C. Weaver, N.F. Mott, Proc. R. Soc. London, Ser. A **261**, 516 (1962)
96. Y. Masuda, M. Kondo, K. Koumoto, Cryst. Growth Des. **9**, 555 (2009)
97. L. learning, Lumen Chem. (2019)
98. C. Sanchez, *Sol-Gel Optics* (International Society for Optics and Photonics, 1990), pp. 40–51
99. S.K. Singh, D.R. Chauhan, Int. J. Eng. Sci. **3**, 37 (2014)
100. F.J. Morin, Bell System Tech. J **37**, 1047 (1958)
101. S.K. Deb, Proc. R. Soc. London, Ser. A **304**, 211 (1968)
102. J.L. Musfeldt, in *Functional Metal Oxide Nanostructures*, ed. by J. Wu, J. Cao, W.-Q. Han, A. Janotti, H.-C. Kim (Springer New York, NY, 2012), pp. 87–126
103. C. Lin, A. Posadas, M. Choi, A.A. Demkov, J. Appl. Phys. **117**, 034304 (2015)
104. M. Batzill, Energy Environ. Sci. **4**, 3275 (2011)
105. K. Kase, M. Yamaguchi, T. Suzuki, K. Kaneko, J. Phys. Chem. **99**, 13307 (1995)
106. P. Borghetti, Y. Mouchaal, Z. Dai, G. Cabailh, S. Chenot, R. Lazzari, J. Jupille, Phys. Chem. Chem. Phys. **19**, 10350 (2017)
107. P. Borghetti, E. Meriggio, G. Rousse, G. Cabailh, R. Lazzari, J. Jupille, J. Phys. Chem. Lett. **7**, 3223 (2016)
108. V.A. Brodskii, A.F. Gubin, A.V. Kolesnikov, N.A. Makarov, Glass Ceram. **72**, 220 (2015)
109. YuV Belokopytov, K.M. Kholyavenko, S.V. Gerei, J. Catal. **60**, 1 (1979)
110. Y. Tokura, N. Nagaosa, Science **288**, 462 (2000)
111. S.-W. Cheong, Nat. Mater. **6**, 927 (2007)
112. T. Herrmann, W. Nolting, J. Magn. Magn. Mater. **170**, 253 (1997)
113. P.A. Igoshev, M.A. Timirgazin, V.F. Gilmutdinov, A.K. Arzhnikov, V.Y. Irkhin, J. Phys, Condens. Matter **27**, 446002 (2015)
114. S.M. Griffin, P. Staar, T.C. Schulthess, M. Troyer, N.A. Spaldin, Phys. Rev. B **93** (2016)
115. D. Adler, J. Solid State Chem. **12**, 332 (1975)
116. K. Hübner, G. Leonhardt, Phys. Status Solidi B **68**, K175 (1975)
117. D. Adler, Radiat. Eff. **4**, 123 (1970)
118. G.H. Jonker, S. van Houten, in *Halbleiterprobleme* (Springer, Berlin, 1961), pp. 118–151
119. K.K. Kar, *Developments in Nanocomposites* (Research Publishing, 2014)
120. K.K. Kar, *Carbon Nanotubes: Synthesis, Characterization and Applications* (Research Publishing Service, 2011)
121. W. Moffitt, W. Thorson, Phys. Rev. **108**, 1251 (1957)
122. J. Krzystek, A. Ozarowski, J. Telser, Coord. Chem. Rev. **250**, 2308 (2006)
123. T. Martin, C.J. Vala, R. Ballhausen, S.L. Dingle, Holt. Mol. Phys. **23**, 217 (1972)
124. F. Neese, E.I. Solomon, Inorg. Chem. **37**, 6568 (1998)
125. Y. Matsumoto, Science **291**, 854 (2001)
126. K. Sato, H. Katayama-Yoshida, Phys. B **308**, 904 (2001)
127. S.-Q. Shen, Z.D. Wang, Phys. Rev. B **58**, R8877 (1998)

128. L. Fritsche, A.J. Pérez-Jiménez, T. Reinert, Int. J. Quantum Chem. **91**, 216 (2003)
129. C.T. Campbell, Phys. Rev. Lett. **96**, 066106 (2006)
130. C. Ataca, H. Sahin, S. Ciraci, J. Phys. Chem. C **116**, 8983 (2012)
131. K.A. Zemski, D.R. Justes, A.W. Castleman, J. Phys. Chem. B **106**, 6136 (2002)
132. S. Royer, D. Duprez, ChemCatChem **3**, 24 (2011)
133. K. Chen, A.T. Bell, E. Iglesia, J. Catal. **209**, 35 (2002)
134. C. Gayner, K.K. Kar, Prog. Mater Sci. **83**, 330 (2016)
135. S. Walia, S. Balendhran, H. Nili, S. Zhuiykov, G. Rosengarten, Q.H. Wang, M. Bhaskaran, S. Sriram, M.S. Strano, K. Kalantar-zadeh, Prog. Mater Sci. **58**, 1443 (2013)
136. M.K. Nowotny, T. Bak, J. Nowotny, J. Phys. Chem. B **110**, 16283 (2006)
137. M.K. Nowotny, T. Bak, J. Nowotny, C.C. Sorrell, Phys. Status Solidi B **242**, R88 (2005)
138. L. Su, Y.X. Gan, Compos. Part B **43**, 170 (2012)
139. F. Song, L. Wu, S. Liang, Nanotechnology **23**, 085401 (2012)
140. M.F. Hundley, J.J. Neumeier, Phys. Rev. B **55**, 11511 (1997)
141. Y. Yang, K.C. Pradel, Q. Jing, J.M. Wu, F. Zhang, Y. Zhou, Y. Zhang, Z.L. Wang, ACS Nano **6**, 6984 (2012)
142. M. Osada, T. Sasaki, J. Mater. Chem. **19**, 2503 (2009)
143. H. Hosono, J. Non-Cryst, Solids **352**, 851 (2006)
144. K. Kalyanasundaram, M. Grätzel, Coord. Chem. Rev. **177**, 347 (1998)
145. X. Yu, T.J. Marks, A. Facchetti, Nature Mater. **15**, 383 (2016)
146. C. Wang, L. Yin, L. Zhang, D. Xiang, R. Gao, Sensors **10**, 2088 (2010)
147. N. Joshi, T. Hayasaka, Y. Liu, H. Liu, O.N. Oliveira, L. Lin, Microchim. Acta **185**, 213 (2018)
148. R. Wang, S. Yang, R. Deng, W. Chen, Y. Liu, H. Zhang, G.S. Zakharova, RSC Adv. **5**, 41050 (2015)
149. M. Bao, Y. Chen, F. Li, J. Ma, T. Lv, Y. Tang, L. Chen, Z. Xu, T. Wang, Nanoscale **6**, 4063 (2014)
150. K.K. Kar, A. Hodzic, *Carbon Nanotube Based Nanocomposites: Recent Development* (Research Publishing, Singapore, 2011)
151. C. Jiang, G. Zhang, Y. Wu, L. Li, K. Shi, CrystEngComm **14**, 2739 (2012)
152. S.K. Singh, N.K. Tiwari, A.K. Yadav, M.J. Akhtar, K.K. Kar, IEEE Sens. J. **1** (2019)
153. B. Christian Enger, R. Lødeng, A. Holmen, Appl. Catal., A **346**, 1 (2008)
154. H.-M. Xiong, Adv. Mater. **25**, 5329 (2013)
155. A.S. Teja, P.-Y. Koh, Prog. Cryst. Growth Charact. Mater. **55**, 22 (2009)
156. A. Akbarzadeh, M. Samiei, S. Davaran, Nanoscale Res. Lett. **7**, 144 (2012)
157. G. Wang, L. Zhang, J. Zhang, Chem. Soc. Rev. **41**, 797 (2012)
158. C. Yuan, H.B. Wu, Y. Xie, X.W.D. Lou, Angew. Chem. Int. Ed. **53**, 1488 (2014)
159. S.-M. Chen, R. Ramachandran, V. Mani, R. Saraswathi, Int. J. Electrochem. Sci. **9**, 14 (2014)
160. Y. Wang, J. Guo, T. Wang, J. Shao, D. Wang, Y.-W. Yang, Nanomater. **5**, 1667 (2015)

Chapter 4
Characteristics of Activated Carbon

Prerna Sinha, Soma Banerjee, and Kamal K. Kar

Abstract The wise use of natural resources is the necessity of today's growing demand for energy for human civilization. The utilization of sustainable natural resources such as biomass and biomass waste has been noticed in the recent trend for the societal benefit and need. In this context, activated carbon has been used for decades in applications such as environmental remediation, energy storage and resource recovery. Biomass-derived activated carbon has attracted great attention because of natural porosity, hierarchical structure and inherent heteroatom doping characteristics. In this review, different synthesis mechanisms to derive biochar from a variety of biomass along with property dependence of activating agents have been discussed. Furthermore, morphology, physiochemical property, structural and elemental characterization have also been discussed in detail. Various applications of activated carbon derived from natural and waste biomass have been included to have a broad overview of this fascinating form of carbon.

4.1 Introduction

Activated carbon is one of the important forms of carbon, which displays excellent physicochemical properties that enable it to be used in various applications. Application of activated carbon is very diverse from the field of catalysis, gas purification, chemical storage and purification to electrode material for energy storage systems

P. Sinha · K. K. Kar (✉)
Advanced Nanoengineering Materials Laboratory, Materials Science Programme, Indian Institute of Technology Kanpur, Kanpur 208016, India
e-mail: kamalkk@iitk.ac.in

P. Sinha
e-mail: findingprerna09@gmail.com

S. Banerjee
Advanced Nanoengineering Materials Laboratory, Department of Mechanical Engineering, Indian Institute of Technology Kanpur, Kanpur 208016, India
e-mail: somabanerjee27@gmail.com

© Springer Nature Switzerland AG 2020
K. K. Kar (ed.), *Handbook of Nanocomposite Supercapacitor Materials I*,
Springer Series in Materials Science 300,
https://doi.org/10.1007/978-3-030-43009-2_4

[1]. The properties exhibited by activated carbon include high surface area, variable pore size and volume, chemical inertness and stability [1–4]. Earlier, activated carbon has been derived from petroleum coke and coal, which are non-renewable. Also, the synthesis condition is harsh since the processing temperature used to be above 1000 °C for the preparation of activated carbon [1, 5].

From the past few years, research has been promoted towards the use of agricultural and municipal waste products for the synthesis of activated carbon [4]. These biomass products are cheap, sustainable and rich sources of carbon. This property makes them the appropriate raw material for the preparation of activated carbon [5–7]. With respect to environmental concern, these agricultural and municipal wastes will get decomposed eventually with time, but most of them degrade slowly until which it will stay as lumps and create different types of pollution [8]. The best way to address these issues is to use them as valuable resource material. Along with having a good source of carbon, these biomaterial wastes contain other elements and compounds, which can be exploited wisely according to the application [1, 9].

The conversion of various biomasses to activated carbon requires a simple process, which is environment-friendly as well [2]. Carbonization or pyrolysis is an energy recovery process that generates biochar, oil and gaseous products. The application of thermal energy removes moisture and volatile material from the biomass, resulting in solid biochar. The obtained char is used as a by-product for the preparation of activated carbon [6, 10]. Generally, activated carbon is prepared by mixing dehydrating or oxidizing agents like KOH, $ZnCl_2$ and H_3PO_4 or reactive gases such as CO_2 and steam, which react with the carbon framework to produce activated carbon [6, 11]. Depending on the application of activated carbon, synthesis techniques can be tailored during carbonization and activation process. Another common technique to alter the properties of activated carbon is surface modification [3, 12]. In this technique, some functional groups are purposely attached at the surface to enhance reactivity [7]. For example, high surface area and porous structure of activated carbon are required for the application in absorbent and gas storage [13–15]. Energy storage applications of activated carbon require high surface area and capability to dope with heteroatoms along with good conductivity [16–23]. For catalysis, the addition of foreign functional groups at the surface enhances the reactivity of activated carbon. Hence, the properties of activated carbon can be tailored easily that makes it an excellent candidate to meet various application demands [24, 25].

Properties exhibited by activated carbon include high specific surface area and tunable pore size distribution. Activated carbon derived from various biomasses shows different morphologies, i.e. from sheet-like to hierarchical porous structure of auricularia-derived activated carbon to three-dimensional cheese-like carbon nano-architecture of corn waste-derived activated carbon [26, 27], and many more interesting structures are present in the literature. Hence, different types of textural morphology can be obtained depending on the precursor. Also, the presence of inherent heteroatom doping such as oxygen, nitrogen and sulphur functionalities promotes surface wettability and also acts as electroactive sites [1, 9, 28–30]. Various types of biomasses used to derive activated carbon are present in the literature [20–22, 25,

26, 28–43], and their synthesis condition, properties and application are discussed in detail in the succeeding sections of this chapter.

4.2   Preparation of Activated Carbon

Various techniques have been employed to synthesize activated carbon. Generally, at the initial stage of preparation, solid carbon blocks are extracted from the precursor by thermal heating called carbonization or pyrolysis. In the carbonization process, the non-carbon species are eliminated from the biomasses by thermal decomposition, resulting in the enrichment of carbon content in the carbonaceous materials. At the initial stage, the porosity of the char remains low which is further improved in the later stage by utilizing the activation process. Hence, a proper choice of carbonization parameters is essential to improve the properties of the final product. The second alternative involves the extraction of char by wet biomass conversion also called hydrothermal carbonization. It is similar to the natural process by which coal is formed. Carbonization helps in the generation of various morphological features in activated carbon. In this regard, a comparative study of '*Auricularia*'-derived activated carbon from direct carbonization with $ZnCl_2$ activation and with hydrothermal alkali carbonization is carried out, which leads to the formation of highly porous layered morphology. In another work, Jiang et al. have prepared ultra-small carbon nanosheets-like structure via two-step carbonization and $ZnCl_2$ activation of auricularia. Activation results in a layered porous structure implying that $ZnCl_2$ can act as a spacer for preventing adjacent cell walls from agglomeration during activation and carbonization [26]. On contrary to this, Long et al. have derived porous layered stacked activated carbon from auricularia using hydrothermal synthesis. Hydrothermal treatment in the alkali environment hydrolyzes the biomolecules, resulting in an increase in porosity and formation of three-dimensional interconnected porous framework [42]. After extraction of biochar by carbonization technique, the second step involves the development of porous structure and enhancement in the surface area via chemical and physical routes to prepare activated carbon. Figure 4.1 shows

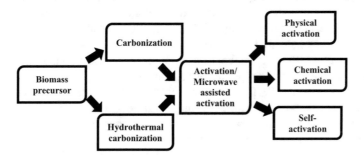

Fig. 4.1   Schematic representation of the preparation of activated carbon from biomass precursor

the schematic representation of different techniques involved to prepare activated carbon from biomass.

4.2.1 Factors Deciding the Properties of Activated Carbon

In order to prepare high performance activated carbon, it is important to maintain a suitable control over the following parameters during preparation stage.

(i)   **Raw materials**: Many organic materials with high carbon contents are used as the precursor for the synthesis of the activated carbon. The production of porous carbon structured activated carbon is dependent on a number of parameters such as high carbon content, low inorganic or ash content, high density, sufficient volatile content, stability and low degradation upon storage and inexpensive.

(ii) **Surface area and pore size distribution**: Temperature and choice of activating agent are the main parameters that determine the physicochemical properties of activated carbon. High temperature leads to the decomposition of carbonaceous species and promotes crystallinity. However, extreme crystallinity leads to a decrease in surface area.

(iii) **Extent of graphitization**: High-temperature annealing promotes graphitization by self-healing of graphite lattice, which leads to the decomposition of carbon network around heteroatom. Although graphitization results in high conductivity among carbon matrix, this affects the depletion of heteroatom content, reducing the content of carrier concentration.

(iv) **Content of heteroatom**: The presence of heteroatom favours the adsorption of ions at the surface of the carbon matrix. However, over doping of heteroatoms leads to a steric effect, lowering the conductivity within the carbon network.

(v)   **Surface concentration of heteroatoms**: The involvement of surface-active species improves the wettability of the carbon matrix. A sufficient amount of heteroatoms is needed to get the desired wettability of the carbon matrix.

(vi) **Type of functional groups**: Heteroatom functionalities either improve the reactivity or enhance the conductivity along with the carbon matrix. For example, the presence of pyridinic-N acts as Lewis base, whereas graphitic-N promotes conductivity [46, 47].

(vii) **Activation time**: Activation time has a profound effect on the carbonization and hence the final properties of the activated carbon. In an increase in activation time, the BET surface area enhances while the percentage yield of activated carbon decreases. This may be attributed to the volatilization of the organic substances during carbonization process.

(viii) **Activation temperature**: In addition to activation time, activation temperature also plays a dominating role in determining the BET surface area and yield of carbonized products. The BET surface area is found to be improved with an increment in activation temperature. This is due to the fact that an increase in temperature leads to an advancement of new pores and broadening of the existing pores due to the release of volatile matter. Again, the increase in activation temperature results in the reduction of the yield of activated carbon due to the release of a high volume of volatile matters.

The choice of a precursor determines the overall properties of the synthesized activated carbon (Fig. 4.2). In accordance with this, Yin et al. have utilized '*coconut fibers*' as carbon source to prepare 3D hierarchical porous activated carbon. The multi-tubular hollow structure of coconut fibers is retained with the creation of micro- and mesoporous after carbonization and activation. Synthesis of activated carbon derived from coconut shell consists of a two-step process as shown in Fig. 4.2c. The formation of multi-tubular hollow structure shortens the ion diffusion path and

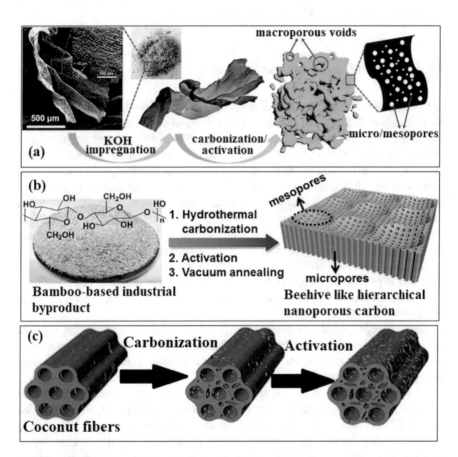

Fig. 4.2 Schematic representation for the synthesis process of **a** Enteromorpha-derived carbons [22], **b** bamboo-based industrial by-product-derived beehive-like hierarchical nanoporous structures [44] and **c** three-dimensional highly porous activated carbon derived from coconut fibers [36]. Redrawn and reprinted with permission from [22, 36, 44]

increases the electrical conductivity, which is beneficial for electrochemical application [36]. Rana et al. have synthesized high surface area N-doped amorphous carbon using '*soya*' (protein-rich source). High preparation temperature results in increment in the surface area along with nitrogen heteroatom doping. The presence of nitrogen functionalities obtained from protein content acts as active sites for Faradic reaction, adsorption and improves the conductivity along with carbon skeleton [46]. Hierarchical porous carbon has been obtained via carbonization and activation of '*cattle bone*'. Generally, bones are a hierarchically structured natural material with lamellar microstructure, which acts as a template for the formation of pores during activation [33]. Subsequently, similar work has been carried out by He et al. to obtain hierarchical porosity from '*pig bone*'. The ordered structure of precursor leads to the formation of a mesopore dominant hierarchical porous network with nitrogen and oxygen doping. The introduction of hierarchical porosity among the carbon networks facilitates adsorption and ion transportation [48].

4.2.2 Carbonization and Activation

The performance of an activated carbon depends on the production conditions. In the first step of the synthesis procedure, pyrolysis of carbonization of the biomass has been carried out leading to the formation of char. During the entire process of carbonization or pyrolysis, volatile matters and moistures have been removed. From this char, activated carbon can be prepared by using three different processes, physical activation, chemical activation and physicochemical (combination of physical and chemical) activation. A process of physical activation involves gas activators such as steam and CO_2, whereas chemical activation proceeds through the use of chemical agents such as metal oxides, alkaline metals and acids. The activation process leads to the formation of a high porous structure, large surface area and high pore volume [49]. Hence, activated carbon can be prepared by two-step processes, where the precursor is pyrolyzed in a controlled or inert atmosphere normally at 400–900 °C to produce biochar followed by activation using gases or chemical compounds at around 450–900 °C [1].

The main objective of an activation process is to increase the pore volume, enhance the pore diameter and improve the pore structures and porosity of the activated carbon. As mentioned earlier, during the first stage of an activation process, activating agents form the microporous structures. In the later stage of activation, the existing pores are widened due to the formation of large pores as a result of burning off the wall between the small pores. This leads to the generation of microporosity in the activated carbon with an eventual decrement in the volume of micropores. Hence, it is worth to mention at this point that, the extent of burning of carbon material or in other words extent of activation is a decisive parameter to produce high-quality activated carbon.

4.2.2.1 Physical Activation

Physical activation is carried out in different oxidizing environments such as air, oxygen, steam, carbon dioxide or their mixtures [50]. During physical activation, carbon precursor is pyrolyzed in an inert atmosphere to remove volatile compounds, followed by gasification using oxidizing gases at around 350–1000 °C. The oxygen species present in the activators are used to burn away the tar products, which are trapped within the pores, and left during pyrolysis. This results in the opening of pores. More pores are developed as activators burn the reactive areas of carbon matrix, resulting in CO and CO_2 emission. The extent of formation of porosity depends upon the nature of gas used and the temperature of activation. The properties of activated carbon strongly depend on precursor, choice of activating agent, temperature and degree of activation. These properties define the porous structure of activated carbon ranging from high to moderate porosity. In general, higher the activation temperature and time, the larger will be pore formation. However, the development of high porosity leads to the broadening of pore sizes [50]. Feng et al. have studied the variation of pore structure using air and carbon dioxide as a gasification agents [51]. The precursor is pyrolyzed by air gasification followed by carbon dioxide activation. During air gasification, the surface area initially increases and then gradually slows down. No significant change in the surface area and pore structure has been observed after carbon conversion during air activation. For carbon dioxide gasification, the surface area and volume of small micropores increase with the progress in gasification, resulting in high surface area and a wide variety of pore sizes. Gasification in air undergoes a fast exothermic reaction of carbon with air or oxygen, making it difficult to control the process and leads to the rapid enlargement of micropores into mesopores [50]. This results in excessive burning of carbon leading to a reduction in the yield of activated carbon. However, high reactivity of oxygen makes this process of low energy and cost-effective and therefore it requires low activation energy compared to steam or carbon dioxide [52, 53]. Steam and CO_2 are preferred over the air as an activating agent due to the provisions of high surface area and increment in pore size distribution. Gasification during CO_2 activation can be easily controlled due to the endothermic reaction between carbon and CO_2 as follows:

$$C + CO_2 \leftrightarrow 2CO, \quad \text{where } \Delta H = 163 \text{ kJ mol}^{-1} \tag{4.1}$$

The surface area, pore volume and average pore size distribution of activated carbon increase with activation time and can be controlled by changing synthesis conditions [54–56].

4.2.2.2 Chemical Activation

Chemical activation holds few advantages over the physical activation processes such as (i) higher carbon yield, (ii) high surface area and (iii) development of microporosity that can be tailored as per need of desired application [50]. Some of the activating agents commonly used are $ZnCl_2$, H_3PO_4, NaOH and KOH. $ZnCl_2$, and H_3PO_4 act as dehydrating agent, whereas KOH and NaOH act as oxidizing agent [57, 58]. H_3PO_4 promotes dehydration at a lower temperature with the evolution of CO and CO_2. During heat treatment, activators produce dehydrating effects on the carbon matrix, which results in a cross-linking reaction (formation of cyclic compounds and condensation process). Dehydration of carbon reduces particle dimension and creates microporosity. Also, high temperature promotes the reaction between H_3PO_4 and organic species forming phosphate and polyphosphate bridges. For $ZnCl_2$, the small size of the compound leads to small and uniform pore distribution [58]. In contrary, NaOH and KOH act as oxidizing agent, where redox reaction takes place as [50]:

$$6\,KOH + 2C \leftrightarrow 2\,K + 3H_2 + 2K_2CO_3$$

KOH etches the carbon matrix in order to generate pores by oxidizing carbon into carbonate ions followed by intercalation of remaining potassium species. Also, at a high temperature of about 700 °C, decomposition of K_2CO_3 into carbon dioxide further leads to the formation of pores through carbon gasification [59–62].

Chemical activation can be done in one-step process, where activating agents are mixed with the precursor and heated in an inert atmosphere over the temperature range of (450–900 °C). KOH activation has been utilized in many research works to date. Yu et al. have synthesized heteroatom-doped porous carbon by direct carbonization and KOH activation process of algae named 'Enteromorpha'. Figure 4.2a shows the schematic representation of the preparation of activated carbon from Enteromorpha. SEM micrograph of derived activated carbon reveals that the presence of KOH etching agent leads to the formation of macropores on the carbon surface. The presence of amorphous phase with a partially ordered graphitic structure provides active sites for adsorption and intercalation. The existence of nitrogen and oxygen doping leads to rapid ion transportation and adsorption of ions at the surface [22]. The second approach is a two-step process, where the precursor is pyrolyzed at first (i.e., carbonization) followed by mixing of activating agent with the char and activation at high temperature. Both the steps yield activated carbon of high surface area as compared to the physical activation process.

One-step carbonization activation and two-step carbonization followed by activation lead to changes in morphological characteristics. Wang et al. have prepared three-dimensional micrometre pore sizes of activated carbon by one-step carbonization and KOH activation of 'willow catkins'. Before heating, willow catkins show a hollow tube-like structure. After activation, SEM image shows three-dimensional porous network structures with large pores of micrometre size. At higher magnification, thickness of the wall looks around <100 nm as shown in Fig. 4.3. TEM images reveal sheet like morphology. High-resolution TEM shows that sheet-like carbon structure rich in micropores of 1 nm [63]. On the other hand, preparation of activated carbon by low-temperature carbonization followed by KOH activation, retains biomass morphology. SEM images reveal the retention of hollow structure after activation as shown in Fig. 4.3. Although a small amount of samples collapse

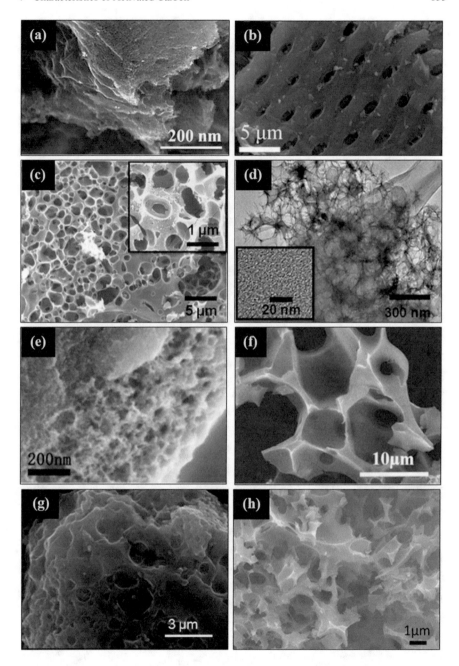

◀**Fig. 4.3** SEM images of **a** kelp-derived activated carbon [17], **b** cotton stalk-derived activated carbon [37], **c** corn fiber-derived activated carbon (inset: magnified FESEM images), **d** HRTEM images of corn fiber-derived activated carbon (inset: magnified HRTEM images showing the pores) [27], **e** rice husk-derived activated carbon [64], **f** hemp hurd-derived activated carbon [29], **g** soya-derived activated carbon [46], **h** willow catkins-derived activated carbon [63] and **i–j** [19], **k** paulownia flower-derived activated carbon showing irregularly arranged ribbon structure [21], **l** tremella-derived activated carbon showing irregular and rough surface [18], **m** SEM and **n** TEM images of cattle bone exhibiting 3D porous networks having numerous mesopores and macropores [48]. Redrawn and reprinted with permission from [17–19, 21, 27, 29, 37, 46, 63]

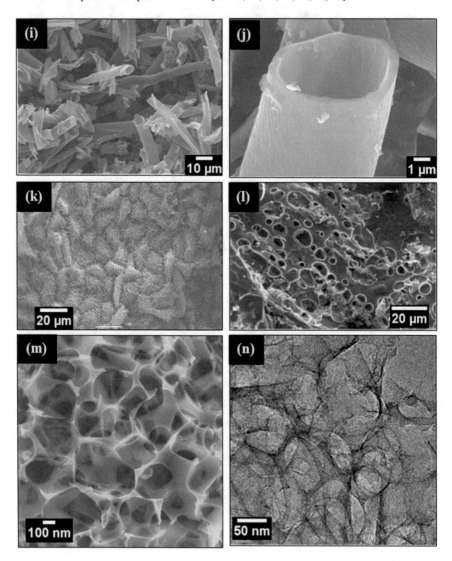

Fig. 4.3 (continued)

into fragments, surface shows a distinct stripe texture with fewer granules [19]. The above studies suggest that two-step synthesis technique retains the original feature and morphology of biomass, whereas one-step synthesis destroys the biomass texture. This can be due to the simultaneous process occurring during high-temperature heating. The decomposition organic molecules into small fragments of carbon and simultaneous etching of KOH in the nascent carbon matrix. This leads to the distortion in biomass structure, resulting in a hierarchical porous texture. During activation, the development of porous structure starts with microporosity. With an increase in the amount of activating agent, pore size distribution becomes heterogeneous. Tian et al. have obtained flute-type micropore activated carbon via KOH activation derived from 'cotton stalk'. The change in amount of KOH leads to variation in pore size distribution. As the amount of KOH increases, the average distance between the pore decreases. This can be attributed to the reaction between KOH and carbon matrix accelerated due to the presence of excess KOH. Figure 4.3 shows the SEM micrograph of flute-type microporous activated carbon. The morphological features can be attributed to the expansion of carbon lattices during activation [37]. Another study on 'Tremella' leads to the formation of numerous incomplete holes expose to the surface. During activation, these holes provide active sites to undergo activation. After activation, irregular and rough surfaces are obtained suggesting that KOH activation leads to the formation of porous texture as shown in Fig. 4.3. Also, along with the reaction between KOH and carbon, some volatile substances decompose in high temperatures, which results in the modification in surface morphology [18]. KOH activation also leads to the widening of microporosity to heterogeneous micropores, whereas $ZnCl_2$ activation produces wide micropores and low mesopores. Gao et al. have derived porous carbon from 'rice husk' by KOH activation. The result shows that small pore size distribution with amorphous carbon matrix has been developed at low activation temperature, whereas the activation temperature increases more ordered carbon with broader pore size distribution is developed. This shows that the increase in activation temperature leads to enhancement in ordered structure with an increase in surface area as shown in Fig. 4.3 [64].

It is seen that the increase in activation temperature induces the broadening of pore size distribution as well. Literature reports that two facile steps of carbonization and high-temperature KOH activation lead to the high surface area, heteroatom-doped activated carbon. Chang et al. have used 'Paulownia flower' to synthesize activated carbon using carbonization followed by KOH activation. Carbonization leads to the conversion of organic molecules into carbon fragments. SEM images of the carbonized sample show thick sheets with dense profile and rough surface. This may be attributed to the shrinkage of sheets as a result of evaporation of water, volatile oils and decomposed gaseous substances. The SEM micrograph of as-prepared activated carbon, as shown in Fig. 4.3, displays a crumpled appearance with a rough surface, which is composed of numerous irregularly arranged ribbons. TEM micrograph reveals ragged and entangled ribbons resulting from many mesopores and macropores. Also, the SAED pattern reveals moderate graphitization. The presence of heteroatoms leads to super-hydrophilicity, which enhances the wettability of ions at the carbon surface [21]. Rana et al. have synthesized heteroatom-doped activated carbon from '*soya*' through NaOH activation. SEM micrograph in Fig. 4.3 depicts the porous surface indicating the etching of carbon matrix during NaOH activation [46]. Gopiraman et al. have prepared 3-dimensional cheese-like carbon nanoarchitecture from '*corn*' residues by NaOH activation. SEM micrograph depicts cheese-like morphology with abundant pores as shown in Fig. 4.3 [27].

H_3PO_4 activation develops large mesopores and macropores as reported [58]. Activation temperature also influences the development of pores. It is found that with an increase in temperature porosity rises to a maximum and then starts decreasing with further increase in temperature owing to shrinkage and collapse of the structure [57–59, 62, 65, 66]. Subramanian et al. have derived activated carbon from '*banana fibers*' by KOH and $ZnCl_2$ activation. SEM morphology of $ZnCl_2$ activation reveals largely aligned fibrous stacks of carbon, whereas loose, disjointed structure has been evidenced by KOH activation. Hence, change in activator has a pronounced effect on deciding the morphology of activated carbon [45].

The co-activation approach shows high porosity and well-developed mesoporous structure [30]. Eutectic NaOH and KOH melt have been used as activating agent. For KOH activation, temperature strongly affects the pore size distribution [59, 67–72]. This can be explained as the formation of micropores is due to the decomposition of volatile compounds, which occurs at the temperature range of 500–700 °C. Above 700 °C, the release of CO_2 from intermediate compounds such as K_2CO_3 has promoted carbon gasification, which results in opening up of closed pores and pore widening of the existing micropores. Also, metallic potassium compounds intercalate within carbon laminar structure that widens the spaces between carbon atomic layers and thereby increases the pore volume [59, 73]. A combination of chemical activation with $ZnCl_2$ and H_3PO_4 leads to weight loss and shrinkage in carbon structure, when temperature is above 500 °C. This can be related to the thermal stability of $ZnCl_2$ and H_3PO_4 that undergo rearrangement and breakdown of carbonaceous aggregates and shrinkage of pores [57, 59, 66]. $ZnCl_2$ and H_3PO_4 activation leads to widening of pores more than 1 nm [59, 65, 66, 74].

4.2.2.3 Combination of Physical and Chemical Activation Processes

In order to tune the pore size distribution and to further enhance the porosity, physical and chemical activations are combined to produce activated carbon as per requirement. It involves two-step activation process consisting of chemical activation normally with $ZnCl_2$ and H_3PO_4 followed by physical activation with carbon dioxide [67, 68]. A small amount of chemical activating agent allows the generation of narrow microporosity without affecting the bulk density. The presence of gases during physical activation permits the appropriate development of pore structure formed during chemical activation.

4.2.2.4 Self-activation

Self-activation is a process, where gases are emitted during pyrolysis of biomass used for activation. This approach delivers environment-friendly synthesis of activated carbon, where no foreign chemicals are involved [75]. This is a chemical-free one-step process, where carbonization and activation have been combined in a single step. This process has an added advantage of environment-friendliness at the low cost of production due to the saving of costs for activating agents.

4.2.3 Hydrothermal Carbonization and Activation

Hydrothermal carbonization is a thermochemical technique of conversion that utilizes water for the conversion of the biomasses to the carbonaceous materials through fractionations. Carbonization temperature depends on the type of starting materials and their decomposition temperature (150–350 °C is generally employed). In this process, water plays the multifunctional role of acts both as a catalyst and solvent to promote hydrolysis. The high ionization constant of water at high temperature leads to hydrolysis of biomasses that are catalyzed further using different acids and bases. The hydrothermal carbonization also generates different organic acids such as acetic, formic and lactic acids, which reduce the pH of the system. This acid helps in hydrolysis followed by the decomposition of the oligomers and monomers into smaller fragments. Hence, the hydrothermal process promotes hydrolysis and dehydration of biomasses and produces the hydrochar of high oxygen functional contents [76–79].

Schematic, as shown in Fig. 4.2b, represents the synthesis method of bamboo-based industrial by-product, combined with hydrothermal carbonization, activation and vacuum annealing. The hydrothermal treatment undergoes carbonization, self-assembly into nano-fragments and generation of closed macro-/mesopore. Derived activated carbon shows the presence of pores with interconnected carbon nanosheet structure under FESEM. At higher magnification, beehive-like hierarchical nanoporous morphology having slits or cylindrical micropores has been observed [44]. Jiang et al. have prepared uniform carbon fiber-like activated carbon from '*bamboo chopsticks*' by hydrothermal treatment in alkali medium. The study suggests that reaction time and concentration of alkali are the key parameters for the formation of carbon fibers [80]. Zhang et al. have studied hydrothermal activation of various '*pollens*' in order to derive activated carbon. After hydrothermal step, macroscale porous structures are retained. The presence of porous structure promotes even mixing and carries reactions with KOH during activation. Alkali activation after hydrothermal carbonization leads to the formation of an irregular structure of carbon having a lateral size of 10–20 μm. This shows that regardless of carbon precursor, after activation, similar irregular structures are observed [81]. Sun et al. have also studied two-step hydrothermal processing and chemical activation of '*hemp*' for the preparation of activated carbon. The morphology depends on the type of solution used.

It can be clearly seen that hydrothermal reaction develops microscale structure of hemp bast. The precursor in a neutral environment maintains fiber-like structure, whereas in the acidic medium biochar is broken up to irregular chips of smaller sizes. The decomposition of hemp fibers gets accelerated in acidic environment during hydrothermal reaction. SEM micrograph in Fig. 4.3f of activated carbon derived from hemp bast displays 3D structure of a porous thin sheet covering a large area. The sheet shows brittle behavior and can be easily broken into smaller chips. The difference in morphologies of biochar and activated samples indicates that the activation process consists of complicated reactions rather than decomposition and separation. Also, the study concludes that activated carbon derived from different environments exhibits similar morphologies for the same activation conditions [29].

4.2.3.1 Microwave-Assisted Activation

Unlike conventional heating, where thermal energy is produced by a conventional heating system, microwave is an energy-saving approach due to its rapid heating, low energy consumption and fast process [82, 83]. During this process, thermal energy is supplied by the conversion of microwave energy. Microwave heating holds an advantage over conventional heating. Thermal processes take time to reach the desired level of activation [83, 84]. In addition to this, surface heating cannot maintain uniform temperature inside the sample. To address these shortcomings, microwave-assisted heating can act as an alternative and fast heating approach. In microwave heating, energy is directly supplied by dipole rotation and ionic conduction inside the particles [85, 86]. The activation time in microwave heating has a pronounced effect on surface area and porosity of activated carbon. He et al. have studied the preparation of activated carbon by microwave-assisted KOH activation of '*petroleum coke*' with varying activation time. As-synthesized activated carbon shows high surface area with micropores [87]. The result shows that surface area and porosity increase as the activation time increases [1, 6]. Deng et al. have used microwaves heating techniques to prepare activated carbon from '*cotton stalk*' by KOH and K_2CO_3 activation. Microwave-assisted KOH activation leads to a gasification reaction, where KOH is reduced to metallic potassium. K_2CO_3 gets reduced to K_2O, CO_2, CO and K. Above a boiling temperature of potassium, the potassium diffuses into carbon matrix, resulting in the further development of porous structure. The results show that KOH activation leads to the formation of a large quantity of micropore and mesopores, whereas K_2CO_3 leads to the generation of a large number of mesopores [85, 88]. Foo et al. have carried out similar study and optimized parameters for the preparation of activated carbon from '*mangosteen peel*' by microwave-assisted K_2CO_3 activation. Potassium compound formed during K_2CO_3 activation widens the existing pores and promotes porosity. However, excess of K_2CO_3/char ratio results in the blocking of pores and decreases in surface area. In addition to this, high radiation power of say >800 W leads to a detrimental impact causing the burning of carbon and destruction of pore. Microwave radiation time is another key factor, which decreases carbon yield by promoting the side reactions between carbon and potassium by-products [89].

4.2.4 Surface Modification and Heteroatom Doping

Textural and surface chemistry determine the properties of activated carbon. The types and concentration of functional groups may be altered by appropriate thermal and chemical post-treatments. Ismanto et al. have prepared activated carbon from '*cassava peel*' by KOH chemical activation followed by CO_2 physical activation. Additionally, modification of functional groups at the surface is carried out using oxidative chemical agents like sulphuric acid, nitric acid and hydrogen peroxide characterized as Boehm titration. Oxygen-containing groups, e.g. carbonyl, hydroxyl, carboxyl and quinine, are present on the surface of the material, which enhances the properties of the activated carbon by increasing polarity and hydrophilicity [31, 90–92]. A one-step process of direct carbonization of '*Kelp*' in ammonia develops three-dimensional, highly porous, high surface area activated carbon with oxygen and nitrogen-enriched functional groups. The SEM image displayed in Fig. 4.3a of nitrogen-doped porous carbon consists of multiple-oriented porous carbon piles exhibiting 3D architecture, whereas HRTEM reveals disordered mesoporous carbon [17]. Zhu et al. have synthesized nitrogen-doped porous carbon from '*black liquor*' by two-step hydrothermal processes and KOH activation. During alkali activation, melamine is added as a nitrogen source and pore modifier. Addition of melamine leads to an increase in surface area and promotes the formation of nitrogen bonds in carbon matrix [93]. For many applications, such as energy storage systems and catalysis, the presence of heteroatoms at surface greatly affects the electrochemical performance of the activated carbon.

4.3 Characterizations of Activated Carbon

4.3.1 Physiochemical Study (BET Surface Area and Pore Size Distribution)

The most important property exhibited by activated carbon is a high specific surface area and tunable pore size distribution. Till now, many biomasses derived precursors have been used to prepare high surface area porous activated carbon. Specific surface area is calculated through BET nitrogen adsorption–desorption isotherm. However, activation leads to the formation of porous structure in carbon material, resulting in an increase in surface area. In general, nitrogen sorption isotherm of activated carbon shows Type-I profile suggesting microporous structure, whereas some activated carbon shows Type-IV profile, suggesting a mesoporous structure. Sometimes a combination of Type-I and Type-IV isotherm also obtained indicating the coexistence of both mesopores and micropores [19]. Tian et al. have derived activated carbon from '*cotton stalk*' showing Type-I/IV isotherm indicating that the majority of adsorption takes place at low pressure less than 0.1. This shows the dominance of micropores in the as-prepared activated carbon [37]. Wang et al. have also obtained a combination of Type-I and Type-IV characteristics in '*willow catkins*' derived activated carbon. Sorption isotherm shows gas uptake at low pressure and flat plateau at high pressure indicating the dominance of microporous structure and a small amount of mesoporous structure [19]. Although Rufford et al. have studied the textural properties of as-prepared activated carbon derived from '*sugarcane bagasse*' by N_2 and CO_2 sorption analyses, $ZnCl_2$ activation shows Type-I isotherm for both gases showing characteristics of microporous material. The study shows that on increasing the activation temperature, specific surface area decreases due to the volatilization of $ZnCl_2$ at the above boiling point of $ZnCl_2$. In the absence of $ZnCl_2$ particles, the pore shrinkage has been evidenced in carbon structure during heat treatment [94]. Gopiraman et al. have obtained Type-I isotherm of activated carbon derived from '*corn residues*' by NaOH activation. The surface area of activated carbon increases with an increase in NaOH concentration till 1:2. A further increase in NaOH leads to distortion of pores due to severe oxidation [27]. Yu et al. have obtained a combination of Type-I/IV isotherms for '*Enteromorpha*' derived carbon samples. The adsorption at low pressure ($P/P_o < 0.01$) and broadening of knee in medium pressure ($0.01 < P/P_o < 0.4$) reveal hierarchical micropores and mesopores network. Since KOH activation leads to the reaction between KOH and C, at a higher temperature of say >700 °C, the formation of CO_2 enhances the decomposition of K_2CO_3. This results in the production of porous structure through carbon gasification [95]. Lists of textural characteristics are provided in Table 4.1. The study shows that BET specific surface area at first decreases with KOH concentration, from 2073 $m^2\ g^{-1}$ for equal impregnation ratio to 1532 $m^2\ g^{-1}$ for 2:1 KOH: carbonized ratio and then increases further to 1879 $m^2\ g^{-1}$. Pore size distribution shows that the amount of pore increases with an increase in KOH concentration [22].

Parameters that affect the development of porosity are (i) activating agent, (ii) activation temperature and time, (iii) gas flow rate, (iv) type of gases used during pyrolysis and activation, (v) mixing procedure (solution and mechanical mixing) and (vi) heating rate [59]. The role of temperature during activation has been studied by Li et al. on the formation of activated carbon from '*sunflower seed shell*' by impregnation-activation and carbonization-activation methods. BET surface area and pore volume are found to be improved with the temperature. This is due to the fast activating reaction at higher temperature. Moreover, at the same temperature, high concentration of KOH also leads to large BET surface area, indicating the importance of amount of activator on the properties of activated carbon. Among these two synthesis processes, carbonization-activation process generates maximum pore structure leading to an increase in surface area [32]. Table 4.1 shows the BET specific surface area and various pore structures of activated carbon derived from different biomasses. The surface area and pore volume have a direct correlation to the amount of activating agent [2].

Table 4.1 BET specific surface area and pore structure properties of activated carbon derived from various biomasses

AC sample (precursor)	SBET (m² g−1)	Smic (m² g−1)	Smeso (m² g−1)	Vt (cm³ g−1)	Vmic (cm³ g−1)	Vmeso (cm³ g−1)	Avg. pore size (nm)	References
Kelp	677.5	–	–	–	–	–	3.42	[17]
N-doped Kelp	1015.4	–	–	–	–	–	3.87	
Cotton stalk	2525.36	–	–	1.23	–	–	–	[37]
Corn cob	3475.2	–	–	1.83	–	–	2.11	[65]
Coconut fibers	2898	1937	961	1.59	1.11	0.48	–	[36]
Animal bone	2157	–	–	2.26	0.77	–	4.18	[33]
Cattle bone	2520	689	1832	3.183	0.308	2.682	–	[48]
Algae	1879	–	–	0.99	10.70	46.22	–	[22]
Sugarcane bagasse	1788	–	–	1.74	0.19	1.55	–	[94]
Waste coffee beans	1019	–	–	0.48	0.21	–	–	[97]
Willow catkins	1107	997	–	0.51	0.46	–	1.8	[19]
Lotus-HA	3037	–	–	2.27	0.41	1.86	–	[81]
Peony-HA	2673	–	–	1.50	0.48	1.02	–	
Rape-HA	2765	–	–	2.20	0.41	1.79	–	
Camellia-HA	2819	–	–	2.06	0.45	1.62	–	
Waste tea leaves	2841	–	–	1.366	–	–	2.67	[41]
Auricularia	1607	327	1280	1.51	0.13	1.38	–	[26]
Rice husk	1768	–	–	1.07	0.75	–	2.42	[40]
Sunflower seed shell	2585	1760	825	1.41	0.83	0.58	2.1	[32]
Black liquor	2646	1416	1230	1.285	0.580	–	–	[93]
Hemp	2879	1609	244	1.16	0.86	0.30		[29]
Rice husk	3145	–	–	2.68	–	–	7.23	[64]
Human hair	1306	103	1203	0.9	0.28	0.62	2.05	[98]
Carton box	2731	–	–	1.68	0.89	–	2.0	[30]
Chicken feather	2426	2096	–	0.870	0.856	–	1.196	[99]
Willow catkins	1586	1296	290	0.78	0.60	–	1.06	[63]

The porous structure of carbonized coconut fiber (CFC) and activated coconut fiber (HPAC) are analysed by nitrogen adsorption-desorption isotherm as shown in Fig. 4.4a. The CFC and HPAC show Type-I isotherm, having sharp nitrogen adsorption at relatively low pressure below 0.1, which indicates microporous structure [96]. Result shows that CFC has a surface area of 33 m^2 g^{-1} and pore volume of 0.08 cm^3 g^{-1}, as shown in Fig. 4.4b, whereas after KOH activation the surface area increases. This remains true until the KOH/CFC ratio reaches 5 while the surface area gets lowered as the ratio has been increased further. The detail textural analyses are shown in Table 4.1. Moreover, pore size distribution has been calculated by the NLBFT model as shown in Fig. 4.4b. In accordance with the surface area, pore volume significantly increases for activated samples [36]. Huang et al. have studied the physicochemical characteristics of carbonized and KOH activated sample derived from 'cattle bone'. Carbonization of animal bone gives Type-IV isotherm according to IUPAC classification. At low pressure, the steep increase in nitrogen uptake suggests the formation of micropores, due to the dissociation of organic molecules. The continuous

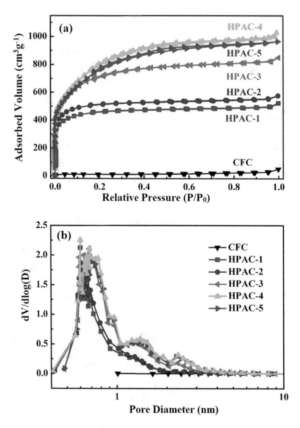

Fig. 4.4 **a** Nitrogen adsorption/desorption isotherm **b** pore size distribution of activated carbon derived from coconut fibers. Redrawn and reprinted with permission from [36]

acceleration of adsorption from 0.4 to 1 indicates the development of macroporosity. This shows that during carbonization, porosity is formed by the presence of natural minerals. KOH activation leads to an increase in surface area with the development of hierarchical porous structure with micropores, mesopores and macropores [33]. Zhang et al. have used various '*pollens*'-derived activated samples to obtain large BET surface area >2673 m^2 g^{-1}. Among the entire samples activated carbon derived from Lotus pollen shows 3037 m^2 g^{-1}, which is much higher than commercial activated carbon, RP20 (1677 m^2 g^{-1}). This can be due to the porous structure, which promotes reaction between KOH and carbon. The details of the various properties are given in Table 4.1. Lotus-HA shows a huge pore volume of 2.27 cm^3 g^{-1}, where 82% of volume corresponds to mesopores, which is much higher than commercial activated carbon, RP20 of 0.64 cm^3 g^{-1} [81].

4.3.2   Study of Functional Groups of Heteroatom

The types and amount of functional groups play a dominating role to decide the capacitance and hence the energy density of a supercapacitor device. The presence of functional groups promotes pseudocapacitance in porous carbons. The generation of functional groups will be determined by the types of precursor used. For example, the oxygen content in biomass is quite higher compared to that in petroleum-based precursors. Hence, biomasses as precursors are capable to promote oxygen functionalities in the final activated carbon. It is worthy to mention that the presence of oxygen functional groups in activated carbon is advantageous to improve the pseudo-capacitance, however, have a detrimental effect on the electrical conductivity of the material. However, the introduction of nitrogen functional groups leads to introduce the pseudocapacitance along with an improvement in electrical conductivity of the final product.

4.3.2.1   Structural Characterization

Structural studies of activated carbon derived from biomass show a wide range of properties irrespective of synthesis condition. Activated carbon exhibits amorphous behavior under X-ray diffraction (XRD). Many literatures are found to get two distinct broad peaks located at around 24° and 43.6° [17, 21, 27, 33–35, 39, 43, 44, 63, 64, 100]. Figure 4.5a shows '*Jute*'-derived activated carbon exhibiting amorphous characteristics showing broad peaks at 24° and 43.6°.

The peaks at 24° and 43.6° correspond to (002) and (100) or (100/101) facets of graphitic carbon [98] in XRD pattern of activated carbon. These peaks belong to the disordered phase of carbon materials [17]. Increasing the amount of KOH leads to reduction and broadening of the intensity of (002) at diffraction angle 22° [64]. The increase in peak intensity at low angle with activation temperature supports high-density micropores as shown in Fig. 4.5b [98]. Low intensity and broad peak imply

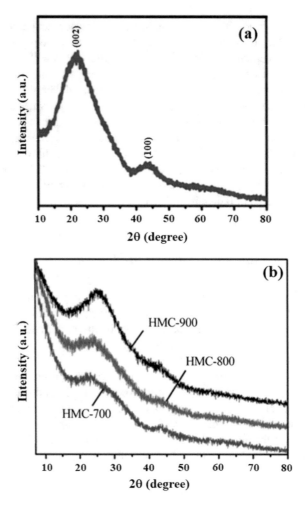

Fig. 4.5 XRD pattern of activated carbon derived from **a** recycled jute [39] and **b** human hair [98]. Redrawn and reprinted with permission from [39, 98]

disordered turbostratic porous carbon structure. The interlayer distance of graphitic layer d_{002} is calculated to be 0.36–0.37 nm. Thickness and average width of graphitic domain L_c and L_a are calculated by the Scherrer equation. The value corresponds to 0.68–0.85 and 0.92–1.02 nm, suggesting that activated carbon is composed of 2–3 graphitic layers [22]. The peak at 43.7° confirms the formation of a higher degree of interlayer condensation of graphite; hence, this approves the presence of graphitic phase of carbon [39].

Raman spectroscopy of activated carbon displays two characteristics peaks at 1349 cm^{-1} assigned as A$_{1g}$ symmetry (D-band) and 1597 cm^{-1} assigned as sp^2

C–C (G-band) [101–104]. The ratio of intensity of these two peaks at D- and G-bands defines the degree of graphitization [105]. Higher degree of graphitization leads to high conductivity in the samples but lacks porosity, whereas high disorder and defects lead to poor conductivity [17, 22, 27, 37, 39, 44, 81, 106]. Qian et al. have derived activated carbon from '*human hair*' (HMC), which shows two carbon peaks at 1320 cm^{-1} corresponding to D-band of defects and disorder and 1590 cm^{-1} corresponds to G-band graphitic peaks. The D/G ratio indicates the degree of structural order. The results explain that high carbonization temperature leads to advanced structural alignment as shown in Fig. 4.6a [98]. Similar Raman spectra are obtained for '*waste paper basket*'-derived activated carbon as shown in Fig. 4.6b [107]. The effect of temperature on '*soya*'-derived activated carbon reveals the existence of graphitic nature in Fig. 4.6c. The ratio of D- to G-band is found to be the highest for SC-800 and lowest for SC-1000 as shown in Fig. 4.6d [46].

FT-IR provides information on the functional groups present in the activated carbon. The broad hump at a higher wavenumber of 3420 cm^{-1} corresponds to hydrogen-bonded hydroxyl groups and the N–H symmetric stretching vibration. The broad

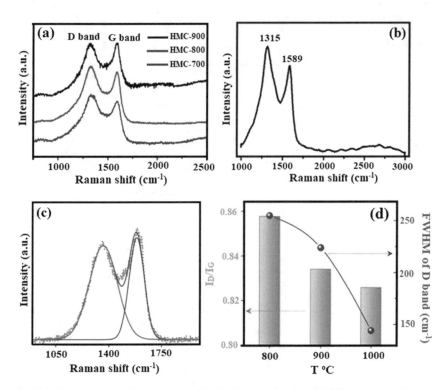

Fig. 4.6 Raman spectra of **a** human hair-derived activated carbon [98], **b** waste paper basket-derived activated carbon [107], **c** soya-derived activated carbon, SC-800, SC-900 and SC-1000; and **d** variation in intensity ratio of D- to G-bands and FWHM of D-band (shown as blue dots) with respect to pyrolysis temperature [46]. Redrawn and reprinted with permission from [46, 98, 107]

humps around 2925 and 2854 cm^{-1} are due to C–H vibrations and stretching. The peaks at 1460 and 1385 cm^{-1} are credited as C–H deformation vibration, whereas the presence of hump at wavenumber 1740 cm^{-1} is due to the C=O stretching vibrations. The peak at 1604 cm^{-1} is due to N=H in-plane deformation vibrations, which might be due to the presence of aromatic rings [108]. The broad peak present at a wavenumber of 1246 cm^{-1} is due to the presence of C–N stretching vibrations. Also, the band at 1162 cm^{-1} is due to the vibration of C–O bond in esters, phenols and ethers [63, 109]. The FT-IR analysis confirms the presence of oxygen- and nitrogen-containing functional groups [110–112]. The effect of activation temperature on the '*willow catkins*' (WCs)-derived activated carbon along with the major FT-IR peak position for rice husk-derived activated carbon have been represented in Fig. 4.7 for example. Study reveals that the bands become broader and weaker as activation temperature increases due to the strong absorption of carbon materials [40, 63].

Fig. 4.7 FT-IR spectra of activated carbon derived from **a** willow catkins [63] and rice husk [40]. Redrawn and reprinted with permission from [40, 63]

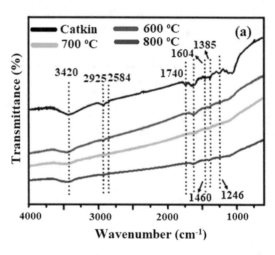

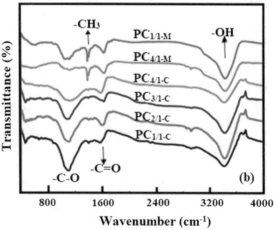

4.3.2.2 Elemental Characterization

XPS is one of the most important techniques to study to binding states of heteroatoms present in the activated carbon matrix. The presence of heteroatoms like N, O and S enhances the wettability of the prepared activated carbon. Also, the presence of functional groups promotes electroreactive sites in the matrix. The presence of peaks at binding energies of around 285.8 401.3, 532.8 eV corresponds to carbon, nitrogen and oxygen functional groups in activated carbon. Carbon mainly shows functionalities at around 284.6, 285.5, 286.5, 288.0 and 290.4 eV corresponding to sp^2 C=C and sp^3 C–C, C–O/C–N, C=O/N–C =N and O–C=O bonds as shown in Fig. 4.8a [42]. The presence of carbon at C=C increases conductivity due to the presence of π-bond. Binding of carbon with oxygen induces polarity, which acts as active sites for the promotion of redox reaction [18, 22, 27, 37, 41, 113]. Further, the high-resolution N1s spectra, Fig. 4.8b, can be fitted into four components, which can be attributed as

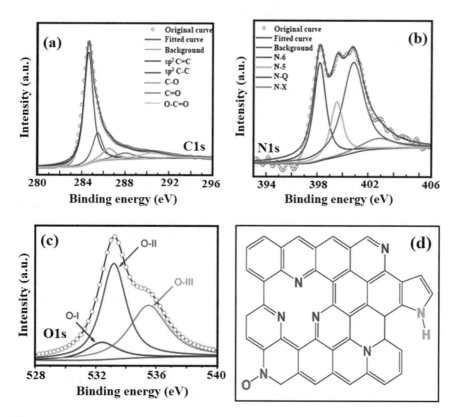

Fig. 4.8 High-resolution XPS spectra of **a** C1s, **b** N1s activated carbon derived from Auricularia [26], **c** O1s activated carbon derived from Tremella [18] and **d** schematic representation of different nitrogen functional groups [26]. Redrawn and reprinted with permission from [18, 26]

pyridinic-N (N-6 around 398.2 eV), pyrrolic-N (N-5, around 399.5 eV), quaternary-N (N–Q, around 400.8 eV) and pyridine-N-oxide (N–X, around 402.8 eV). The presence of N-6 and N-5 is mainly attributed to the enhanced pseudocapacitance by Faradaic reactions, whereas N–Q enhances the electrical conductivity of carbon [18, 22, 37, 41, 113]. The oxygen spectra as shown in Fig. 4.8c can be categorized into three peaks for keto oxygen or quinone group, O-I (532.4 eV), esters or phenolic group, O-II (533.2 eV) and carboxylic group or water, O-III (535.5 eV) [114]. The presence of surface oxygen groups boosts the wettability of activated carbon [18, 26, 37, 41, 42, 64, 113]. The surface functionalities of '*Paulownia flower*'-derived activated carbon have been evidenced by four peaks at a binding energy of 285, 348, 400 and 532 eV, corresponding to C1s, Ca2p, N1s and O1s orbital, respectively. This indicates the presence of C, Ca, N and O elements in activated carbon. The intensity ratio of O1s increases for activated samples as compared to carbonized one indicating the incorporation of more oxygen elemental by the alkali during activation step. The atomic content of N elemental decreases from 3.96 to 1.09%, indicating the alkali activation at high temperature introduces oxygen functional groups. However, this process removes nitrogen elemental from carbonized carbon motif [21]. The abundance of surface functionalities leads to the formation of defects and active sites in activated carbon [17]. The presence of nitrogen- and oxygen-containing groups improves hydrophilicity and fast ion transportation [63]. Heteroatoms promote conductivity and improve the wettability of activated carbon.

4.4 Applications

Application of activated carbon is diverse. It is used in food processing, petroleum pharmaceuticals and automobile industry. This is due to the high surface area with extensive pore sizes, which act as absorptive material. Along with the above-stated application, recent research has promoted the use of activated carbon for energy storage systems in batteries, supercapacitor, fuel cell and solar cell. It is also used as microwave absorber, CO_2 adsorbent, H_2 storage and in wastewater treatment [115]. Basically, the properties of activated carbon can be tuned, from high porosity to high conductivity, depending on the application. Also, the extraction of activated carbon from biomass provides the inherent doping of heteroatoms of mainly nitrogen and oxygen functionalities. These properties enhance the wettability and conductivity of carbon matrix. The presence of heteroatom in activated carbon finds application as electrode material in the energy storage system and also as catalysts. In addition to the above-stated properties, for a particular application, the selection of appropriate biomass provides the required template to obtain the desired morphology.

4.4.1   Energy Storage

4.4.1.1   Electric Double-Layer Capacitor

Activated carbon has attracted much attention due to the abundance, low cost, chemical and thermal stability, ease of processing and ability to tune textural and structural characteristics. The BET surface area of activated carbon is found to be above 800 m^2 g^{-1} [17, 18, 98]. Hence, they are widely used as electrode material for supercapacitor applications. The high surface area provides enough surface for electrolytic ions to get physically adsorbed at the electrode surface, hence forming an electric double layer. The accumulation of charge at electrode-electrolyte interfaces gives the overall capacitance [115]. Along with having high surface area and pore size distribution, chemical inertness helps electrode material to achieve high cycle life during the charge discharge process. Also, many activated carbons contain heteroatom-doped surface functionalities, which promote a redox reaction to deliver pseudocapacitance effects. This increases the overall capacitance of electrode [28].

4.4.1.2   Asymmetrical Supercapacitor

Asymmetrical supercapacitor has two electrodes made of different materials. One is the electrochemical capacitor, which comprised of carbon electrodes and the other is pseudocapacitive material made of metal oxides/nitrides, conductive polymer or composite or composite of carbon/pseudocapacitive material. The assembly of asymmetrical materials combines the energy storage characteristics of pseudocapacitive materials with the benefit of rapid energy delivery and stability. Hence, a combination of both can easily overcome the limitations of the constituent materials [5].

4.4.1.3   Lithium-Ion Capacitors

Lithium ion capacitor is a battery-type capacitor that works on a combination of battery and supercapacitor charge storage mechanism. Li$^+$ enables intercalation–deintercalation in battery-type electrodes. The electric double-layer adsorption and desorption occur at the surface of activated carbon electrode [116, 117]. The rate capability of Li-ion capacitor gets improved with the introduction of activated carbon as compared to battery electrode. Li-ion capacitor enables fast charge capability, excellent cycle life, high energy and power density [5].

4.4.2   Photocatalyst

The tunable pore structure and surface properties make activated carbon as an excellent material for catalysis. Chemical stability and being economical promote its use in the diverse field of applications. Use of activated carbon helps in mineralization of organic pollutants into neutral by-products of H_2O, CO_2 and mineral acids [10, 115].

4.4.3   Gas Adsorbent

Activated carbon made from waste biomass can be used as adsorbents for numerous environmental applications including the elimination of hazardous compounds from industrial waste gases or water. High surface area, large pore volume, good chemical stability and tunable porosity of activated carbon make it an excellent material for hydrogen storage [10].

4.4.4   Microwave Absorber

Activated carbon can be used as an efficient material for absorbing microwaves. Generally, the high surface area of activated carbon enables more interaction of microwaves, resulting in more losses, eliminating high-frequency microwaves. Also, porosity leads to multiple reflections, interface polarization and scattering loss. The presence of solid–air interfaces gives interfacial polarization under oscillating electromagnetic waves. This amplifies the microwave absorption properties [35, 86].

4.4.5   Wastewater Treatment

Activated carbon adsorption can be applied to reduce chlorine, lead, organic chemicals, unpleasant taste and odour from water. Wide porosity range of activated carbon finds application as adsorbent in wastewater treatment plant. The contaminant gets attracted towards the surface of carbon particles, which traps it from the water.

4.5   Understanding the Associated Challenges of Biomass-Derived Activated Carbon

Renewable carbon materials sound promising due to the high energy and power densities compared to commercial carbon-based electrodes and all at a low cost.

However, the correlation among different factors such as surface functional groups, microstructures, chemical composition, wettability with electrodes and surface area are still to be explored to a great extent with the aim to develop activated carbon of high quality. Understanding the essential requirement of biomasses to attain high electrical conductivity along with high specific surface area is a matter to be resolved. The relation between the types of biomasses with the pore size distribution is still to be considered in depth. The scarcity in understanding is mainly due to the complex chemical reactions occurs during activation, heterogeneous morphologies of biomasses, resulting in variation in the diffusion of activating agents, etc. Research to date establishes the fact that the creation of a large amount of micropores leads to the enhancement in surface area, which is beneficial in terms of energy density and capacitance of a supercapacitor. The original morphology of the plant tissue used for the preparation of the activated carbon also plays a dominating role to determine the pore size distribution of the final product.

4.6 Concluding Remarks

Activated carbon shows wide applications in the field of materials science. Their structural and chemical characteristics are continuously evolving to fulfill the required demands of intended applications. Till now, research is focused to control their properties through the development of suitable activation procedures and selection of appropriate biomass precursors for structural and chemical characteristics. A plenty of morphological features can be developed in activated carbon via the use of different carbon precursors. Different types of synthesis techniques along with several approaches to modify surface properties have been discussed in detail. The study of surface area has been carried out through BET analysis of nitrogen adsorption and desorption isotherm. The textural characteristics are observed through electron microscopy, where different morphologies of activated carbon are observed depending on the type of precursor used. XRD and Raman spectroscopy show the structural assembly of carbon matrix. Raman studies explain the defects and disorder in the carbon matrix, which can be directly related to the porosity of the material. FT-IR and XPS analysis confirm the presence of functional groups in the carbon framework. The presence of N, O and S heteroatoms functionalities enhance the wettability of the activated carbon and act as electroactive sites for electrochemical reaction. Further, activated carbon is widely used in different applications such as CO_2 uptake, hazardous gas adsorbent, hydrogen storage and catalysis. It plays an important role as electrode material for supercapacitors and hybrid batteries.

Acknowledgements The authors acknowledge the financial support provided by the Department of Science and Technology, India (DST/TMD/MES/2K16/37(G)), for carrying out this research work. Authors are thankful to Ms Tanvi Pal for drafting few figures.

References

1. A.M. Abioye, F.N. Ani, Renew. Sustain. Energy Rev. **52**, 1282 (2015)
2. J.M. Chem, J. Mater. Chem. A Mater. Energy Sustain. **5**, 2411 (2017)
3. K.K. Kar, J.K. Pandey, S. Rana, *Handbook of Polymer Nanocomposites Processing Performance and Application* (Springer, Berlin, Heidelberg, 2015)
4. P. Benjwal, R. Sharma, K.K. Kar, Mater. Des. **110**, 762 (2016)
5. S. Faraji, F.N. Ani, Renew. Sustain. Energy Rev. **42**, 823 (2015)
6. J. Deng, M. Li, Y. Wang, Green Chem. **18**, 4824 (2016)
7. R. Kumar, S. Sahoo, E. Joanni, R.K. Singh, W.K. Tan, K.K. Kar, A. Matsuda, Prog. Energy Combust. Sci. **75**, 100786 (2019)
8. L. Wei, G. Yushin, Nano Energy **1**, 552 (2012)
9. T.K. Enock, C.K. King'ondu, A. Pogrebnoi, Y.A.C. Jande, Int. J. Electrochem. **2017**, 1 (2017)
10. S. Bagheri, N. Muhd Julkapli, S. Bee Abd Hamid, Int. J. Photoenergy **2015**, 1 (2015)
11. S. Banerjee, R. Sharma, K.K. Kar, in Compos. Mater., S. Banerjee, R. Sharma, and K.K. Kar. Banerjee, R. Sharma, and K.K. Kar (Springer, Berlin, Heidelberg, 2017), pp. 251–280
12. K.K. Kar, *Composite Materials: Processing Applications Characterization* (Springer, Heidelberg, New York, Dordrecht London, Berlin, Heidelberg, 2017)
13. P.J.M. Suhas, M.M.L. Carrott, Ribeiro Carrott. Bioresour. Technol. **98**, 2301 (2007)
14. K. Kadirvelu, C. Namasivayam, Adv. Environ. Res. **7**, 471 (2003)
15. S. Sircar, T.C. Golden, M.B. Rao, Carbon **34**, 1 (1996)
16. J. Cherusseri, K.K. Kar, Phys. Chem. Chem. Phys. **18**, 8587 (2016)
17. J. Li, K. Liu, X. Gao, B. Yao, K. Huo, Y. Cheng, X. Cheng, D. Chen, B. Wang, W. Sun, D. Ding, M. Liu, L. Huang, A.C.S. Appl, Mater. Interfaces **7**, 24622 (2015)
18. N. Guo, M. Li, X. Sun, F. Wang, R. Yang, Mater. Chem. Phys. **201**, 399 (2017)
19. K. Wang, R. Yan, N. Zhao, X. Tian, X. Li, S. Lei, Y. Song, Q. Guo, L. Liu, Mater. Lett. **174**, 249 (2016)
20. C. Chen, Y. Zhang, Y. Li, J. Dai, J. Song, Y. Yao, Y. Gong, I. Kierzewski, J. Xie, L. Hu, Energy Environ. Sci. **10**, 538 (2017)
21. J. Chang, Z. Gao, X. Wang, D. Wu, F. Xu, X. Wang, Y. Guo, K. Jiang, Electrochim. Acta **157**, 290 (2015)
22. W. Yu, H. Wang, S. Liu, N. Mao, X. Liu, J. Shi, W. Liu, S. Chen, X. Wang, J. Mater. Chem. A **4**, 5973 (2016)
23. W. Si, J. Zhou, S. Zhang, S. Li, W. Xing, S. Zhuo, Electrochim. Acta **107**, 397 (2013)
24. P. Serp, J.L. Figueiredo, *Carbon Materials for Catalysis* (Wiley Inc., Hoboken, NJ, USA, 2008)
25. E. Auer, A. Freund, J. Pietsch, T. Tacke, Appl. Catal. A Gen. **173**, 259 (1998)
26. L. Jiang, L. Sheng, X. Chen, T. Wei, Z. Fan, J. Mater. Chem. A **4**, 11388 (2016)
27. M. Gopiraman, D. Deng, B.S. Kim, I.M. Chung, I.S. Kim, Appl. Surf. Sci. **409**, 52 (2017)
28. V.V.N. Obreja, Phys. E Low Dimension. Syst. Nanostruct. **40**, 2596 (2008)
29. W. Sun, S.M. Lipka, C. Swartz, D. Williams, F. Yang, Carbon **103**, 181 (2016)
30. D. Wang, G. Fang, T. Xue, J. Ma, G. Geng, J. Power Sources **307**, 401 (2016)
31. A.E. Ismanto, S. Wang, F.E. Soetaredjo, S. Ismadji, Bioresour. Technol. **101**, 3534 (2010)
32. X. Li, W. Xing, S. Zhuo, J. Zhou, F. Li, S.Z. Qiao, G.Q. Lu, Bioresour. Technol. **102**, 1118 (2011)
33. W. Huang, H. Zhang, Y. Huang, W. Wang, S. Wei, Carbon **49**, 838 (2011)
34. R. Wang, P. Wang, X. Yan, J. Lang, C. Peng, Q. Xue, A.C.S. Appl, Mater. Interfaces **4**, 5800 (2012)
35. S.K. Singh, H. Prakash, M.J. Akhtar, K.K. Kar, A.C.S. Sustain, Chem. Eng. **6**, 5381 (2018)
36. L. Yin, Y. Chen, D. Li, X. Zhao, B. Hou, B. Cao, Mater. Des. **111**, 44 (2016)
37. X. Tian, X. Ma, Z. Li, S. Yan, L. Ma, F. Yu, G. Wang, X. Guo, Y. Ma, C. Wong, J. Power Sources **359**, 88 (2017)

38. H. Wang, Z. Xu, A. Kohandehghan, Z. Li, K. Cui, X. Tan, T.J. Stephenson, C.K. King'Ondu, C.M. B. Holt, B.C. Olsen, J.K. Tak, D. Harfield, A.O. Anyia, D. Mitlin, ACS Nano 7, 5131 (2013)
39. C. Zequine, C.K. Ranaweera, Z. Wang, P.R. Dvornic, P.K. Kahol, S. Singh, P. Tripathi, O.N. Srivastava, S. Singh, B.K. Gupta, G. Gupta, R.K. Gupta, Sci. Rep. 7, 1174 (2017)
40. X. He, P. Ling, M. Yu, X. Wang, X. Zhang, M. Zheng, Electrochim. Acta 105, 635 (2013)
41. C. Peng, X. Bin Yan, R.T. Wang, J.W. Lang, Y.J. Ou, Q.J. Xue, Electrochim. Acta 87, 401 (2013)
42. C. Long, X. Chen, L. Jiang, L. Zhi, Z. Fan, Nano Energy 12, 141 (2015)
43. G. Zhang, H. Chen, W. Liu, D. Wang, Y. Wang, Mater. Lett. 185, 359 (2016)
44. W. Tian, Q. Gao, Y. Tan, K. Yang, L. Zhu, C. Yang, H. Zhang, J. Mater. Chem. A 3, 5656 (2015)
45. V. Subramanian, C. Luo, A.M. Stephan, K.S. Nahm, S. Thomas, B. Wei, J. Phys. Chem. C 111, 7527 (2007)
46. M. Rana, K. Subramani, M. Sathish, U.K. Gautam, Carbon 114, 679 (2017)
47. M. Rana, G. Arora, U.K. Gautam, Sci. Technol. Adv. Mater. 16, 14803 (2015)
48. D. He, J. Niu, M. Dou, J. Ji, Y. Huang, F. Wang, Electrochim. Acta 238, 310 (2017)
49. N. Mohamad Nor, L.C. Lau, K.T. Lee, A.R. Mohamed, J. Environ. Chem. Eng. 1, 658 (2013)
50. M. Sevilla, R. Mokaya, Energy Environ. Sci. 7, 1250 (2014)
51. B. Feng, S.K. Bhatia, Carbon 41, 507 (2003)
52. F. Rodríguez-Reinoso, M. Molina-Sabio, Carbon 30, 1111 (1992)
53. E. Dawson, G.M. Parkes, P. Barnes, M. Chinn, Carbon 41, 571 (2003)
54. S. Román, J.F. González, C.M. González-García, F. Zamora, Fuel Process. Technol. 89, 715 (2008)
55. F.C. Wu, R.L. Tseng, C.C. Hu, C.C. Wang, J. Power Sources 144, 302 (2005)
56. A. Singh, D. Lal, J. Appl. Polym. Sci. 115, 2409 (2010)
57. M. Jagtoyen, F. Derbyshire, Carbon 36, 1085 (1998)
58. M. Molina-Sabio, F. Rodríguez-Reinoso, Colloids Surfaces Physicochem. Eng. Asp. 241, 15 (2004)
59. H. Teng, L.Y. Hsu, Ind. Eng. Chem. Res. 38, 2947 (1999)
60. C. Guan, K. Wang, C. Yang, X.S. Zhao, Microporous Mesoporous Mater. 118, 503 (2009)
61. M.J. Bleda-Martínez, J.A. Maciá-Agulló, D. Lozano-Castelló, E. Morallón, D. Cazorla-Amorós, A. Linares-Solano, Carbon 43, 2677 (2005)
62. M.A. Lillo-Ródenas, J. Juan-Juan, D. Cazorla-Amorós, A. Linares-Solano, Carbon 42, 1365 (2004)
63. K. Wang, N. Zhao, S. Lei, R. Yan, X. Tian, J. Wang, Y. Song, D. Xu, Q. Guo, L. Liu, Electrochim. Acta 166, 1 (2015)
64. Y. Gao, L. Li, Y. Jin, Y. Wang, C. Yuan, Y. Wei, G. Chen, J. Ge, H. Lu, Appl. Energy 153, 41 (2015)
65. W.T. Tsai, C.Y. Chang, S.L. Lee, Bioresour. Technol. 64, 211 (1998)
66. L.-Y. Hsu, H. Teng, Fuel Process. Technol. 64, 155 (2000)
67. M.J. Illán-Gómez, A. García-García, C. Salinas-Martinez De Lecea, A. Linares-Solano, Energy Fuels 10, 1108 (1996)
68. D. Lozano-Castelló, D. Cazorla-Amorós, A. Linares-Solano, D.F. Quinn, Carbon 40, 989 (2002)
69. P.M. Eletskii, V.A. Yakovlev, V.B. Fenelonov, V.N. Parmon, Kinet. Catal. 49, 708 (2008)
70. J.J. Niu, J.N. Wang, Solid State Sci. 10, 1189 (2008)
71. M. Sevilla, A.B. Fuertes, R. Mokaya, Energy Environ. Sci. 4, 1400 (2011)
72. M. Sevilla, A.B. Fuertes, Energy Environ. Sci. 4, 1765 (2011)
73. A. Ahmadpour, D.D. Do, Carbon 35, 1723 (1997)
74. A. Puziy, O. Poddubnaya, A. Martínez-Alonso, F. Suárez-García, J.M. Tascón, Carbon 40, 1493 (2002)
75. C. Xia, S.Q. Shi, Green Chem. 18, 2063 (2016)
76. Z. Liu, F.-S. Zhang, J. Wu, Fuel 89, 510 (2010)

77. M. Sevilla, J.A. Maciá-Agulló, A.B. Fuertes, Biomass Bioenerg. **35**, 3152 (2011)
78. M.-M. Titirici, R.J. White, C. Falco, M. Sevilla, Energy Environ. Sci. **5**, 6796 (2012)
79. A. Jain, R. Balasubramanian, M.P. Srinivasan, Chem. Eng. J. **283**, 789 (2016)
80. J. Jiang, J. Zhu, W. Ai, Z. Fan, X. Shen, C. Zou, J. Liu, H. Zhang, T. Yu, Energy Environ. Sci. **7**, 2670 (2014)
81. L. Zhang, F. Zhang, X. Yang, K. Leng, Y. Huang, Y. Chen, Small **9**, 1342 (2013)
82. K.Y. Foo, B.H. Hameed, Desalination **275**, 302 (2011)
83. K.Y. Foo, B.H. Hameed, Microporous Mesoporous Mater. **148**, 191 (2012)
84. K.Y. Foo, B.H. Hameed, Chem. Eng. J. **187**, 53 (2012)
85. H. Deng, G. Li, H. Yang, J. Tang, J. Tang, Chem. Eng. J. **163**, 373 (2010)
86. S.K. Singh, M.J. Akhtar, K.K. Kar, A.C.S. Appl, Mater. Interfaces **10**, 24816 (2018)
87. X. He, Y. Geng, J. Qiu, M. Zheng, S. Long, X. Zhang, Carbon **48**, 1662 (2010)
88. H. Deng, G. Zhang, X. Xu, G. Tao, J. Dai, J. Hazard. Mater. **182**, 217 (2010)
89. K.Y. Foo, B.H. Hameed, Chem. Eng. J. **180**, 66 (2012)
90. T.A. Centeno, F. Stoeckli, Electrochim. Acta **52**, 560 (2006)
91. R.Q. Wang, M.G. Deng, Adv. Mater. Res. **347–353**, 3456 (2011)
92. W. Shen, Z. Li, Y. Liu, Recent Patents. Chem. Eng. **1**, 27 (2008)
93. L. Zhu, F. Shen, R.L. Smith, L. Yan, L. Li, X. Qi, Chem. Eng. J. **316**, 770 (2017)
94. T.E. Rufford, D. Hulicova-Jurcakova, K. Khosla, Z. Zhu, G.Q. Lu, J. Power Sources **195**, 912 (2010)
95. H. Wang, Q. Gao, J. Hu, J. Am. Chem. Soc. **131**, 7016 (2009)
96. M. Kunowsky, A. Garcia-Gomez, V. Barranco, J.M. Rojo, J. Ibañez, J.D. Carruthers, A. Linares-Solano, Carbon **68**, 553 (2014)
97. T.E. Rufford, D. Hulicova-Jurcakova, Z. Zhu, G.Q. Lu, Electrochem. Commun. **10**, 1594 (2008)
98. W. Qian, F. Sun, Y. Xu, L. Qiu, C. Liu, S. Wang, F. Yan, Energy Environ. Sci. **7**, 379 (2014)
99. Q. Wang, Q. Cao, X. Wang, B. Jing, H. Kuang, L. Zhou, J. Power Sources **225**, 101 (2013)
100. M.J. Ahmed, M.A. Islam, M. Asif, B.H. Hameed, Bioresour. Technol. **243**, 778 (2017)
101. A. Sadezky, H. Muckenhuber, H. Grothe, R. Niessner, U. Poschl, Carbon **43**, 1731 (2005)
102. W. Luo, B. Wang, C.G. Heron, M.J. Allen, J. Morre, C.S. Maier, W.F. Stickle, X. Ji, Nano Lett. **14**, 2225 (2014)
103. E. Marinho-Soriano, P.C. Fonseca, M.A.A. Carneiro, W.S.C. Moreira, Bioresour. Technol. **97**, 2402 (2006)
104. P. Chamoli, M.K. Das, K.K. Kar, J. Phys. Chem. Solids **113**, 17 (2018)
105. J. Cherusseri, K.K. Kar, RSC Adv. **6**, 60454 (2016)
106. X. Liu, D. Chao, Y. Li, J. Hao, X. Liu, J. Zhao, J. Lin, H. Jin Fan, Z. Xiang Shen, Nano Energy **17**, 43 (2015)
107. D. Puthusseri, V. Aravindan, B. Anothumakkool, S. Kurungot, S. Madhavi, S. Ogale, Small **10**, 4395 (2014)
108. T.H. Liou, S.J. Wu, J. Hazard. Mater. **171**, 693 (2009)
109. A. Kaushik, M. Singh, G. Verma, Carbohydr. Polym. **82**, 337 (2010)
110. C.C.O. Alves, A.S. Franca, L.S. Oliveira, Biomed. Res. Int. **2013**, 1 (2013)
111. I. Ghouma, M. Jeguirim, U. Sager, L. Limousy, S. Bennici, E. Däuber, C. Asbach, R. Ligotski, F. Schmidt, A. Ouederni, Energies **10**, 1508 (2017)
112. H.N. Tran, F.-C. Huang, C.-K. Lee, H.-P. Chao, Green Process. Synth. **6**, (2017)
113. Y. Han, N. Shen, S. Zhang, D. Li, X. Li, J. Alloys Compd. **694**, 636 (2017)
114. Y.-Q. Zhao, M. Lu, P.-Y. Tao, Y.-J. Zhang, X.-T. Gong, Z. Yang, G.-Q. Zhang, H.-L. Li, J. Power Sources **307**, 391 (2016)
115. S. Banerjee, P. Benjwal, M. Singh, K.K. Kar, Appl. Surf. Sci. **439**, 560 (2018)
116. B. De, A. Yadav, S. Khan, K.K. Kar, A.C.S. Appl, Mater. Interfaces **9**, 19870 (2017)
117. A. Yadav, B. De, S.K. Singh, P. Sinha, K.K. Kar, A.C.S. Appl, Mater. Interfaces **11**, 7974 (2019)

Chapter 5
Characteristics of Graphene/Reduced Graphene Oxide

Pankaj Chamoli, Soma Banerjee, K. K. Raina and Kamal K. Kar

Abstract Graphene, the thinnest two-dimensional material, is extensively explored by interdisciplinary fields of research communities due to the excellent electronic, mechanical, optical and thermal properties. This chapter provides a comprehensive review of the structure, synthesis, properties and applications of graphene/reduced graphene oxide. Graphene, the one atomic thick layer of sp^2 hybridized carbon atoms arranged in honeycomb lattice, is the building block of all carbon materials such as graphite, carbon nanotubes and fullerenes. It can be synthesized by different techniques such as micromechanical cleavage, exfoliation and chemical vapour deposition. It has also been commonly synthesized from the reduction in graphene oxide. The reduction of graphene oxide again is also carried out either by using hazardous chemicals or by green approach. This chapter also discusses the properties of graphene/reduced graphene oxide, which makes them a potential material in diverse applications. The potential fields of applications of graphene/reduced graphene oxide include medicine, electronics, energy devices, sensors, environmental and many more.

P. Chamoli · K. K. Raina
Department of Physics, DIT University, Dehradun, Uttarakhand, India
e-mail: pchamoli83@gmail.com

K. K. Raina
e-mail: kkraina@gmail.com

P. Chamoli · S. Banerjee · K. K. Kar (✉)
Advanced Nanoengineering Materials Laboratory, Materials Programme, Indian Institute of Technology Kanpur, Kanpur 208016, India
e-mail: kamalkk@iitk.ac.in

S. Banerjee
e-mail: somabanerjee27@gmail.com

K. K. Kar
Advanced Nanoengineering Materials Laboratory, Material Science Programme and Department of Mechanical Engineering, Indian Institute of Technology Kanpur, Kanpur 208016, India

© Springer Nature Switzerland AG 2020

155

K. K. Kar (ed.), *Handbook of Nanocomposite Supercapacitor Materials I*,
Springer Series in Materials Science 300,
https://doi.org/10.1007/978-3-030-43009-2_5

5.1   Introduction

Carbon is the foundation of organic chemistry and the key material for life. Due to flexibility of bond formation, carbon-based systems are composed of a number of structures of wide physical properties. Graphene/reduced graphene oxide, the exciting material of modern age, has found interest by the scientific community of interdisciplinary fields due to the remarkable physical properties and chemical tunability opening the provisions of vast applications in the field of physics, chemistry and materials science. It plays a decisive role to determine the electronic properties of carbon allotrope and can be considered as a honeycomb structure made of hexagons. Graphene, the mother of all carbon allotrope, has been unknowingly produced by someone, who writes with a pencil. However, it has been recognized much later. The main reason behind the fact is that the existence of graphene in the free state is questionable and no experimental tool has been available to detect one-atom-thick graphene layer from the pencil debris.

A group of researchers at Manchester University, Andre Geim and Konstantin Novoselov, have successfully isolated a single layer of graphene by mechanically cleaving graphite crystal [1]. Graphene is the 'thinnest' 2D material and exhibits extraordinary electrical, optical, mechanical and thermal properties [1, 2]. It has inspired the development of many envisaged modern devices due to the excellent electrical conductivity, transparency along with flexibility such as roll-up and flexible electronics, energy storage materials, energy conversion materials, polymer composites and transparent conducting film/electrodes [3, 4]. Graphene, the wonder material, exhibits a theoretical specific surface area of 2630 m^2 g^{-1} with excellent intrinsic mobility of 200,000 cm^2 v^{-1} s^{-1}. It exhibits high Young's modulus and thermal conductivity values of <1 TPa and <5000 W m^{-1} K^{-1}, respectively, with good optical transmittance (<97.7%) [5–7]. The excellent electronic properties of graphene make them a suitable material as transparent conductive electrodes [8].

5.2   Structure of Graphene

Graphene is the structural building block of all graphitic forms, including 3D graphite, 2D graphene, 1D carbon nanotubes and 0D fullerenes. It is one atomic thick layer of sp^2 hybridized carbon atoms arranged in a honeycomb lattice with C–C bond lengths of 1.42 Å and partially filled pi-orbitals above and below the plane of sheet [9]. Graphite is the form of numerous graphene sheets stacked together with an interplanar spacing of 3.35 Å. You will be happy to know that approximately three million layers of stacked graphene sheets are present in a 1-mm-thick graphite crystal [2]. The categorization of graphene has been made on the basis of electronic properties and generally prefixed by 'monolayer', 'bilayer' or 'few-layer' (generally excepted to be <10 layers) graphene.

Monolayer graphene generally exists without stacking of sheets, and 4–6 few-layer graphene can have a number of stacking arrangements, including ABAB (named as Bernal stacking), ABCABC (called rhombohedral stacking) and sometimes AAA stacking. In general, graphene is a planar structure of carbon atom arranged in a hexagonal lattice and has been seen as a triangular lattice with a basis of two atoms per unit cell (Fig. 5.1). The lattice vectors can be written as [10]:

$$a_1 = \frac{a}{2}\left(3, \sqrt{3}\right) \quad a_2 = \frac{a}{2}\left(3, -\sqrt{3}\right) \tag{5.1}$$

where a ~ 1.42 Å is the C–C bond distance.

The reciprocal-lattice vectors are represented by

$$b_1 = \frac{2\pi}{3a}\left(1, \sqrt{3}\right) \quad b_2 = \frac{2\pi}{3a}\left(1, -\sqrt{3}\right) \tag{5.2}$$

The two points K and K' are at the corners of the graphene Brillouin zone, named as Dirac points and positions in momentum space are represented by

$$K = \left(\frac{2\pi}{3a}, \frac{2\pi}{3\sqrt{3}a}\right) \quad K' = \left(\frac{2\pi}{3a}, -\frac{2\pi}{3\sqrt{3}a}\right) \tag{5.3}$$

The three nearest-neighbour vectors in real space are written as

$$\delta_1 = \frac{a}{2}\left(1, \sqrt{3}\right) \quad \delta_2 = \frac{a}{2}\left(1, -\sqrt{3}\right) \quad \delta_3 = a(1, 0) \tag{5.4}$$

with the six second nearest neighbours are located at

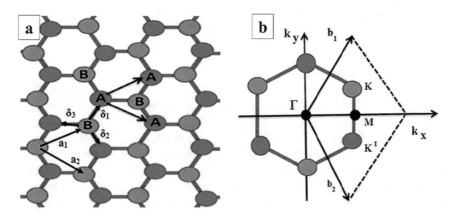

Fig. 5.1 **a** Triangular sublattice of graphene A having three nearest neighbours in sublattice and B vice versa, **b** K and K' as the Dirac points. Redrawn and reprinted with permission from [10]

$$\delta'_1 = \pm a_1 \quad \delta'_2 = \pm a_2 \quad \delta'_3 = \pm(a_2 - a_1) \tag{5.5}$$

5.3 Synthesis of Graphene/Reduced Graphene Oxide

Since graphene was isolated using the scotch tape method [1], many processes have been explored/demonstrated to produce few-to-single-layer graphene. The main concern is to synthesize high carrier mobility and low defect density in graphene. Hence, different synthesis approaches of continuous growth of few-layer graphene have been explored such as exfoliation, chemical vapour deposition and wet chemical approach. Few of them have been discussed here. Moreover, chemical vapour deposition gives a great impact over thermal exfoliation due to the defect-free synthesis of graphene. Wet chemical approach is another widely method of graphene production due to the provisions of large-scale synthesis at a low cost.

5.3.1 Exfoliation Approach

5.3.1.1 Micromechanical Cleavage

A scotch tape method developed by Geim and Novoselov is known as 'micromechanical cleavage', involves adhesive tape assisted peeling off a piece of graphite [11]. In this method, the monolayer graphene has been produced with high structural quality. In fact, this approach provides the best graphene structure in terms of high purity, low defects and optoelectronic properties. Unfortunately, this method is not feasible for large-scale production and hence limited to fundamental research applications only. Micromechanical cleavage is schematically represented in Fig. 5.2.

5.3.1.2 Solution and Thermal Exfoliation

Separation of few layers of graphene from 3D layered bulk graphite gains tremendous interest due to the increased electrical conductivity from 3D bulk scale to the 2D nanoscale. Hence, different exfoliation approaches have been demonstrated to produce graphene nanosheets. Initially, the mechanical exfoliation process uses HOPG (highly oriented pyrolitic graphite) as a precursor, subjected to an oxygen plasma etching to create 5-μm-deep mesas, and scotch tape was used to repeatedly peel graphite flakes from the mesas [12]. In addition, exfoliation of wet chemically derived graphite layers is another route for separation of individual sheets of graphene, mainly attractive for a large area sheet fabrication with a possible large volume of production [13–15]. The most common disadvantage of wet techniques is the formation of by-products, which contaminate the graphite layers. Hence, the

Scotch tape

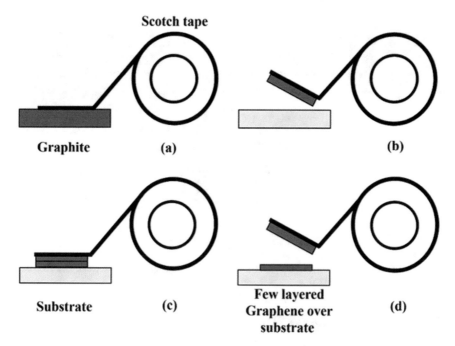

Fig. 5.2 Schematic representation of micromechanical cleavage technique, **a** and **b** adhesive tape is used to cleave the top few layers of bulk graphite, **c** graphitic tape pressed on desired substrate, **d** few layers remain on the substrate. Redrawn and reprinted with permission from [11]

development of a one-step procedure could yield in high per cent, cost-effective and efficient large-scale monolayer graphene. Several organic solvents have been studied so far to exfoliate graphene oxide (GO) assisted by thermal reduction as schematically represented in Fig. 5.3.

The most common solvent used for the reduction in GO is propylene carbonate (PC) due to excellent dispersion ability. This technique could lead to large-scale production of graphene in an easy one-step procedure [16]. In particular, the improvement of reduction in the presence of PC has been demonstrated employing the increase of ultrasonication time. Moreover, a liquid-phase exfoliation has been demonstrated to produce high quality and defect-free large-area graphene sheets upon treatment with N-methyl-2-pyrrolidone (NMP) operating at ~800 °C [17]. Therefore, thermal treatments have been heavily used as the next step following wet-processing techniques for thin film preparation. In addition, microwave exfoliation has achieved great interest for thermal treatment. Microwave irradiation is an effective and attractive method for graphene synthesis with controllable size and shape as compared to conventional heating due to the homogeneous distribution of heat energy.

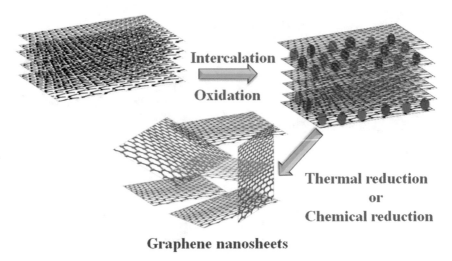

Intercalation

Oxidation

Thermal reduction

or

Chemical reduction

Graphene nanosheets

Fig. 5.3 Intercalation or exfoliation process for synthesis of graphene nanosheets. Drawn based on the concept taken from [15]

5.3.2 Chemical Vapour Deposition

Many methods have been reported for the production of graphene to date. Chemical vapour deposition (CVD) and its variants with transition metal substrates retain a great attention for large-area continuous film production and commercialization of graphene [18, 19]. CVD technique deposits solid thin film through the vapour or gases by chemical reactions on desire substrates. Figure 5.4 shows the schematic of a tube-furnace CVD instrument for the growth of graphene on the desire substrate. It is composed of a gas delivery system consisting of necessary valves, mass flow controllers, pressure controller that regulate the flow rates of the gases and a gas mixing unit, which is responsible for mixing of various gases uniformly before they are subjected to the reactor, a reactor and gas removal system. During the synthesis process, reactive gas species are fed into the reactor by the gas delivery system in which the chemical reaction takes place. The heater helps to achieve high temperature, and solid materials are uniformly deposited on the desire substrates.

The by-products of reaction and non-reacted gases are removed by the gas delivery system. Growth of graphene by CVD on polycrystalline metals such as Ir, Ni, Pt, Pd, Rh, Cu and Co has been thoroughly studied and investigated [20]. CVD processes have reported that graphene is obtained by nucleation of active carbon formed by the decomposition of precursor materials on the metal surface. The metal surface plays the role of an active catalyst material. Moreover, catalyst-free growth of graphene on Si, W, Mo, SiO_2 and Al_2O_3 has also been demonstrated by ECR-CVD and PECVD methods [21–24]. Furthermore, continuous defect-free graphene films at low temperature have been reported by MPCVD [25, 26]. Graphene film growth over polycrystalline Ni by CVD has primarily been performed at atmospheric pressure,

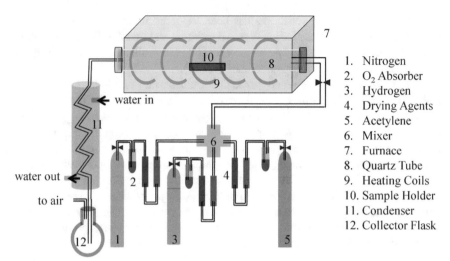

1. Nitrogen
2. O_2 Absorber
3. Hydrogen
4. Drying Agents
5. Acetylene
6. Mixer
7. Furnace
8. Quartz Tube
9. Heating Coils
10. Sample Holder
11. Condenser
12. Collector Flask

Fig. 5.4 Schematic illustration of chemical vapour deposition set-up

yielding graphene continuous films [26–28]. In contrast, annealing of the polycrystalline nickel [29] and an increase in deposition temperature [30] strongly promote the synthesis of large grain sizes of Ni (111) over Ni (100). Moreover, many reports have been published so far on the effect on synthesis parameters such as growth temperature, gas mixing ratio and growth time in the chemical vapour deposition of acetylene/hydrogen. These are the deciding parameters for producing few-layer graphene continuous film with minimal defects [26, 31].

In addition, the growth of few-layer graphene continuous film on nickel surface has also been fabricated based on diffusion and segregation phenomena of carbon from underlying amorphous carbon or nanodiamond films [32, 33]. Hence, bilayer graphene films have been fabricated directly on self-assembled monolayers (SAMs) or insulating substrates at low pressure [34]. In addition, few-layer continuous graphene film on SiO_2/Si from SAMs has been fabricated at atmospheric pressure and transfer-free graphene growth with uniform thickness has been achieved [35]. Meanwhile, a uniform and low-defect bilayer graphene has been synthesized on evaporated polycrystalline nickel films after exposing methane at 1000 °C with optimized process parameters such as growth time, annealing profile and flow rates of different gases. Raman mapping shows the ratio of 2D to G peak intensities (I_{2D}/I_G) is in the range of 0.9–1.6 over 96% of the 200 × 200 μm areas and the average ratio of D to G peak intensity (I_D/I_G) is about 0.1, confirms the bilayer formation [36].

Furthermore, the growth of graphene on several hexagonal surfaces has also been studied. Hexagonal substrates are frequently referred for growth of graphene due to the less lattice mismatch. Co (0001) and Ni (111) show less than 1% lattice mismatch [37, 38]. In contrast, the lattice mismatch between graphene and Pt (111) [39], Pd (111) [40], Ru (111) [41] and Ir (111) [42] is greater than 1%, and therefore, the growth is incommensurate. Recently, graphene growth on relatively polycrystalline

Ni [43, 44] and Cu [45] substrates has been triggered interest in fabrication and optimizing CVD conditions for large-area deposition and transfer. Moreover, copper has also been studied as a catalyst to grow several carbon allotropes such as graphite [46], diamond [47] and carbon nanotubes [48, 49].

The growth of graphene on copper is due to the surface and independent of diffusion from bulk. Based on earlier literature, a possible model for the nucleation and three-stages of growth mechanism of graphene on copper can be suggested. As the growth time is higher, the graphene domains increase. A facile technique for the growth of graphene directly on substrates through a simple thermal annealing process has been demonstrated [50]. Transparent graphene film has been prepared successfully and transferred by using a roll-to-roll transfer technique with the help of a thermal release tape as a support polymer. A large-area touch screen has been prepared with the graphene so transferred over the substrate. The thermal adhesive tape plays a key role in this transfer process of these very large-area graphene films from copper to plastic substrate materials [51].

5.3.3 Graphene/Reduced Graphene Oxide from Graphene Oxide

Large-scale production of graphene is a great challenge. One of the best cost-effective methods for mass production of graphene is the reduction of graphene oxide (GO) in the presence of a reducing agent. GO has been first prepared by Benjamin C. Brodie by repeatedly treating the graphite with potassium chlorate in the presence of nitric acid [52]. Later, the most commonly used and less hazardous method has been proposed by Hummers and Offeman, which involves the oxidation of graphite by treating graphite with a mixture of sulphuric acid, sodium nitrate and potassium permanganate. This is the most popular chemical route of graphene/reduced GO synthesis, nowadays. During oxidation of graphite, individual sheets of GO have been designed with oxygen-containing functional groups such as carbonyl, hydroxyl and epoxy on both sides of the plane as well as the edges [53]. Hence, GO exhibits negatively charged features and electrostatically stabilized in water, alcohols and certain organic solvents. The maximum lateral size of GO sheets is strongly dependent on the size of initial graphite crystals used for graphite oxidation. Due to nonstoichiometric composition, the determination of GO structure is a challenge and strongly depends on the synthetic methods. Generally, oxygen is present on the basal plane of GO in the form of hydroxyl, epoxy, carboxyl, carbonyl groups, etc. and is also located at the planar edges as shown in Fig. 5.5. Hence, GO exhibits electrical insulation and can produce homogeneous colloidal suspensions.

Therefore, to restore the electronic property of graphene, reduction (by chemical or thermal route) is required to remove the oxygen functional groups as shown in Fig. 5.6.

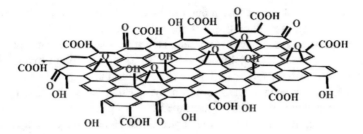

Fig. 5.5 Surface functional groups of graphene oxide. Redrawn and reprinted with permission from [54]

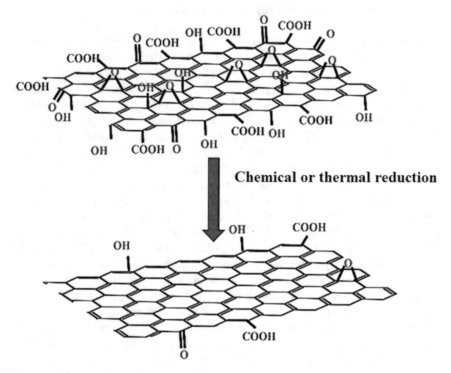

Fig. 5.6 Schematic illustration of reduction in graphene oxide into graphene sheet. Redrawn and reprinted with permission from [55]

5.3.3.1 Hazardous Approach of Graphene Oxide Reduction

During chemical reduction by reducing agent, the colour of GO dispersion (initially light yellow) in water turns into blackish colour and strongly suggests the removal of oxygen species. Generally, the chemical reduction in GO sheets has been performed with different reducing agents. For examples, these are hydrazine (N_2H_4) [56], sodium borohydride ($NaBH_4$) [57], hydriodic acid (HI) [58], hydroquinone

[59], etc. However, the chemical reduction in GO has many drawbacks such as chemical toxicity (these reducing agents not only toxic but also explosive in nature) and therefore hazardous to our environment. Hence, less toxic reducing agents such as metal nanoparticles [60], sulphuric acid (H_2SO_4) [61], sodium hypophosphite (Na_3PO_2) [62], bismuth ferrite ($BiFeO_3$) [63], aluminium iodide (AlI_3) [64] and phenolic acids have been studied as alternatives. However, such reducing agents imprint some amount of damaging chemical traces on the graphene surface and are not suitable enough for bio-related applciations. Hence, the green approach of reduction comes into play.

5.3.3.2 Green Approach of Graphene Oxide Reduction

Developments and optimizations have been explored to reduce GO by eco-friendly techniques [65–67]. For example, Cote et al. have used a chemical-free room-temperature flash reduction process for the reduction in GO by photothermal heating using photographic camera flash [68]. Flash irradiation triggers the deoxygenation reaction of GO and rapidly creates a fused polymer composite. They have prepared composite patterns such as interdigitated electrode arrays by using photomask on flexible substrates [68]. Ghadim et al. have used nanosecond pulsed laser (having wavelength of 532 nm and average power of 0.3 W) for the effective and environment-friendly reduction in GO sheets in ammonia solution (pH $\sim$ 9) at room temperature [69]. Based on X-ray photoelectron spectroscopy (XPS) analysis, the O/C ratio of the GO sheets decreases from 49 to 21% after 10 min of laser irradiation. These results help further promotion and application of pulsed lasers for the reduction in GO in environment-friendly way [69]. In addition, Zhou et al. have reported a simple, clean and controlled hydrothermal dehydration route to convert GO to a stable graphene solution [70]. Compared to chemical reduction processes using hydrazine, the present 'water-only' route has the combined advantages of removing oxygen-containing functional groups from GO and fixing the aromatic structures for tunable optical limiting performance [70]. Moreover, Ai et al. have adopted a simplistic and efficient way for the preparation of soluble reduced graphene oxide (rGO) sheets [71]. In this process, reduction in GO has been done by reducing agent, dimethyl-formamide (DMF). The as-prepared rGO sheets exhibit a high reduction level, good conductivity, and are well dispersed in many solvents. Synthesized rGO sheets show potential applications in the construction of high-performance graphene-based NO gas sensing devices [71]. Xu et al. have adopted a simple approach for the preparation of graphene–metal particle nanocomposites in a water–ethylene glycol system [72]. GO is used as a precursor and metal nanoparticles (Au, Pt and Pd) as building blocks. The nanoparticles adsorb on GO sheets and play a vital role in catalytic reduction in GO. Cyclic voltammogram analyses have indicated its potential application in direct methanol fuel cells [72]. Consequently, Akhavan et al. have synthesized GO platelets by using a chemical exfoliation method to deposit anatase TiO_2 thin films [73]. Post-annealing of the GO/TiO_2 thin films at 400 °C in air partially produces Ti–C bond between the platelets. The annealed GO/TiO_2 thin films are immersed in ethanol, and photocatalytic reduction has commenced. After 4 h of photocatalytic reduction, the concentration of the C=O bond has been reduced to 85%, indicating the effective reduction in the GO platelets to graphene nanosheets. The GO/TiO_2 thin film has been utilized as nanocomposite photocatalysts for the degradation of E. coli in an aqueous solution under solar light irradiation [73]. Consequently, biological approaches have been used to synthesized rGO from GO. For example, vitamin C [74], sugar [75], glucose [76], bovine serum albumin [77], melatonin [78], green tea [79], ginseng [80], curcumin [81], bacteria [82] and bacteriorhodopsin [83] have been explored in the literature. Hence, green methodologies are promising for the reduction in GO with a minimized environmental impact.

5.4 Properties of Graphene

5.4.1 Electronic Properties

Graphene has excellent structural flexibility that is well reflected in electrical properties of it [84]. The sp^2 hybridization between s and p orbitals generates a trigonal planar structure leading to the formation of a σ bond between the carbon atoms separated by 1.42 Å. The σ bands in graphene provide the structural robustness in graphene. These bands retain a filled shell and form deep valence bands. The p orbitals, which are unaffected, remain perpendicular to the planar structure. This helps in the formation of covalent bonds with the neighbouring carbon atoms generating π bands. The p orbitals have one extra electron showing the half-filled configuration. The half-filled configuration plays a crucial role in determining the physics behind the system. The tight-binding character in graphene leads to large Coulomb energies, which in turn show strong collective effects such as magnetism and insulating behavior as a result of correlation gaps [85]. Linus Pauling proposes that the electronic properties of graphene are on the basis of resonant valence bond (RVB) structure. However, this theoretical understanding of electronic properties of graphene if compared with the band structure of graphene, it is found to play a role of a semimetal with Dirac electrons. Here, it is worthy to mention, most of the current experimental studies on the electronic properties of graphene are based on its band structure. However, the electron–electron interaction in graphene is a matter of intense research.

The energy dispersion of graphene was formulated, and tight-binding Hamiltonian was described to explain electrons in graphene [86]. By using tight-binding Hamiltonian, the energy dispersion relation has been represented as [87]:

$$E^{\pm}\left(k_x, k_y\right) = \pm\gamma_0\sqrt{1 + 4\cos\frac{\sqrt{3}k_x a}{2}\cos\frac{k_y a}{2} + 4\cos^2\frac{k_y a}{2}} \qquad (5.6)$$

where γ_0 is the first-neighbour π-orbitals energy (2.9 − 3.1 eV), a ~ 1.42 Å is the lattice constant and $k = (k_x, k_y)$ vectors denote the first Brillouin zone electronic momenta.

This energy dispersion will have six points, where the conduction and valence band touch each other called Dirac points. The ballistic transport at this point with a mean free path of 300–500 nm takes place as shown in Fig. 5.7. This makes graphene a semi-metal. The linear dispersion near the Dirac points will then be:

$$E_{\pm}(\mathbf{k}) = \pm \hbar v_F |k| \tag{5.7}$$

Here, $\hbar$ is the reduced Planck constant and v_F is the Fermi velocity (~10^6 m/s) [10, 87].

Mobility of graphene has been reported between 2000 and 20,000 cm^2 V^{-1} s^{-1} on different substrates at low temperatures and carrier concentrations in the order of 10^{12} cm^{-2} [89]. Moreover, the suspended single-layer graphene shows mobility over 200,000 cm^2 V^{-1} s^{-1} at electron densities of ~2×10^{11} cm^{-2}. This result is found when single-layer graphene suspension (~150 nm) has been deposited over a Si/SiO$_2$ gate electrode by electron beam lithography. Significant enhancement in the mobility has been achieved and widths of the characteristic Dirac peaks have been reduced by a factor of 10 compared to non-suspended devices [90]. Meanwhile, the Drude model assumes that the conductivity of graphene is in the range from 0.0032 to 0.00032 S in unsuspended graphene and the order of 0.0032 S for suspended graphene. The theoretical value of minimal conductivity is in the order of e^2/hπ. However, in practice, the minimal conductivity has been achieved π times larger than the value of e^2/h [1, 91].

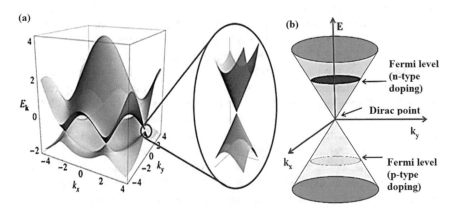

Fig. 5.7 **a** Graphene band structure as two cones touching at Dirac point. Reprinted with permission from [10] and **b** the position of the Fermi level determining nature of doping and transport carrier (redrawn and reprinted with permission from [88])

5.4.2   Optical Properties

Monolayer graphene has shown 97.7% of light transmittance [92]. The absorption of the graphene depends on the fine-structure constant (α) and increases linearly with the number of layers of graphene stacked together (experimentally demonstrated up to five layers). Fresnel equation has been applied with a universal optical conductance ($G_0 = e^2/4\hbar \approx 6.08 \times 10^{-5}\ \Omega^{-1}$) to give the absorbance of freestanding monolayer graphene [87]:

$$T = (1 + 0.5\pi\alpha)^{-2} \approx 1 - \pi\alpha \approx 97.7\% \tag{5.8}$$

where $\alpha\left(= e^2/(4\pi\varepsilon_0\hbar c) = G_0/(\pi\varepsilon_0 c) \approx 1/137\right)$ is   fine-structure   constant. Graphene reflects <0.1% of the incident light in the visible region, rising to ~2% for ten layers. Hence, we can take the optical absorption of graphene to be proportional to the number of layers, each absorbing $A \approx 1 - T \approx \pi\alpha \approx 2.3\%$ over the visible spectrum. The absorption spectrum of monolayer graphene gives a peak at 270 nm in the ultraviolet region due to the exciton-shifted Van Hove singularity in the graphene density of states. In few-layered graphene, interband transitions at lower energies can be seen. Figure 5.8 gives an overview of different transparent conductors including single-walled carbon nanotubes (SWNTs), ITO, ZnO/Ag/ZnO and TiO$_2$/Ag/TiO$_2$ and graphene with their transmittance. It is evident that the transmittance for SWNTs ~ 70–80%, ITO ~ 70–90%, ZnO/Ag/ZnO ~ 90%, TiO$_2$/Ag/TiO$_2$ ~ 60–70% and for graphene ~ 95% have been achieved. Among these transparent conductors, graphene has higher transmittance over a wider range of wavelengths [87].

The optical transition in graphene has been measured by infrared spectroscopy, and the study reveals that the optical transition is gate dependent [93]. Near the Dirac point of a graphene sheet, the density of states remain low. Hence, a change of Fermi level owing to gate leads to variation in charge density to a great extent followed by

Fig. 5.8 Transmittance of graphene compared to that of other materials. Redrawn and reprinted with permission from [87]

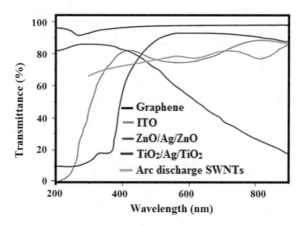

modified transition characteristics. The process of relaxation and recombination of electron and hole pairs in graphene have been governed by the carrier concentration and are noticed on a timescale of tens of picoseconds [94, 95]. Graphene can be extensively used in high-speed optoelectronics due to the remarkable carrier transport velocity under electric filed [96].

5.4.3 Thermal Properties

Heat removal in the electronic industry has become a crucial issue. Therefore, a rapid growth of interest has been seen in the thermal properties of material by the scientific communities around the globe. A variety of carbon allotropes having different thermal conductivity have been studied and occupy a unique place in particular applications. Thermal conductivity of different carbon materials has been found from 0.01 W m^{-1} K^{-1} in amorphous carbon to above 2000 W m^{-1} K^{-1} in diamond at room temperature or one-atom-thick 2D graphene as shown in Fig. 5.9 [97]. An interesting fact can be noticed by evaluating the thermal conductivities of few layers of graphene as a function of layer thickness or number of atomic planes. Here, the transport phenomena are restricted by two properties, one intrinsic and other extrinsic. The extrinsic properties of few-layer graphene are dependent on the defect scattering. The increase in the atomic plane of graphene leads to a major change in

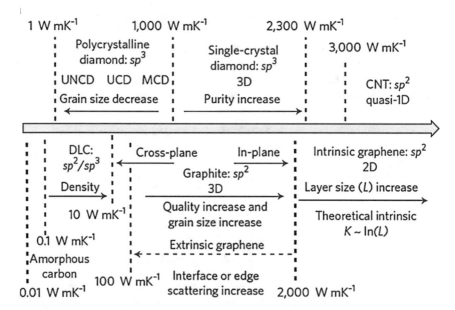

Fig. 5.9 Thermal conductivity of carbon allotropes and their derivatives. Reprinted with permission from [97]

phonon dispersion, and more and more phase space is obtainable for scattering to proceed. Hence, the scattering of phonon from top to bottom boundaries has been restricted if the constant number of atomic planes is maintained over the entire length of graphene layer [98].

The thermal conductivity of graphene has been measured first using Raman optothermal techniques in non-contact mode [5, 97]. The measurements are carried out with suspended graphene layers obtained by exfoliation of bulk graphite. This method of determination is essential to obtain the thermal conductivity of atomically thin graphene. Results report that thermal conductivity may exceed 3000 W m^{-1} K^{-1} at room temperature for larges flakes of graphene [98]. The thermal conductivity values of graphene are reported to be better than bulk graphite in this study. The room-temperature thermal conductivity of CNTs has been found 3000–3500 W m^{-1} K^{-1}. The suspended single-layer graphene has shown thermal conductivity in the range of $4.84 \pm 0.44 \times 10^3$ to $5.30 \pm 0.48 \times 10^3$ W m^{-1} K^{-1} [99]. The thermal conductivity of non-doped graphene is not very much dependant on electronic contribution due to low carrier concentration. The thermal conductivity, in this case, is governed by phonon transport specifically termed as diffusive and ballistic conduction at high and low temperatures, respectively [100]. Molecular dynamic simulation study reveals the relation as $\kappa \propto 1/T$ for pure graphene >100 K. At room temperature, 6000 W m^{-1} K^{-1} has been predicted, which is quite high compared to graphitic carbon under the same condition [101]. In another work, the thermal conductivity of single-layer graphene flakes obtained by exfoliation is calculated to be around 5000 W m^{-1} K^{-1} [102]. Graphene is suspended in a channel, and laser beam is focused from the center to support. The loss of heat through the air is nominal compared to the conduction of heat in graphene [103]. In this study, the frequency of Raman G peak is calculated in terms of laser excitation power. The thermal conductivity of the graphene has been evaluated from the slope of the trend line. These values are remarkably good and suggest that graphene is a very good material from a thermal point of view. Thermal conductivity of graphene can be tuned to a wide range by the introduction of disorder or edge roughness into the honeycomb lattice. Hence, the excellent heat-conduction properties of graphene make it a promising candidate for all proposed electronic and photonic applications.

5.4.4 Mechanical Properties

Inherent strength makes graphene so special. It is not only extraordinary strong ($E = 1.0$ TPa) but also light in weight (0.77 mg m^{-2}) [1]. The elastic properties and intrinsic breaking strength of monolayer graphene have been measured by nanoindentation in an atomic force microscope. Using an atomic force microscope, effective spring constants have been measured for stacks of graphene sheets (less than 5 with sheets of thicknesses between 2 and 8 nm) suspended over photolithographically defined trenches in SiO$_2$ (schematically shown in Fig. 5.10). Graphene shows nonlinear elastic stress-strain response, yields second-order elastic stiffness of 340 N m^{-1}

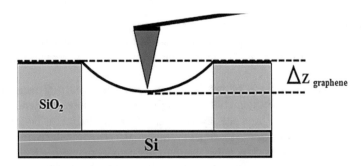

Fig. 5.10 Schematic illustration of AFM tip pushing down a suspended graphene sheet. Redrawn and reprinted with permission from [104]

and third-order elastic stiffness of 690 N m^{-1}. In addition, the breaking strength is found 42 N m^{-1} and Young's modulus of $E = 1.0$ TPa. These experiments establish graphene as the strongest material ever measured [104].

The mechanical properties of single-layer graphene have been studied by numerical simulations, i.e. molecular dynamics [105–107]. AFM has also been used for a circular membrane of few-layer graphene by force–volume measurements [108]. Determination of mechanical properties of graphene by AFM studies is commonly carried out by indentation method [109]. A mean elastic modulus of 0.25 TPa has been reported in this measurement for chemically modified graphene. Study also remains, where the mechanical properties of graphene oxide platelets have been determined by fabricating paper-like material [110]. The results report a fracture strength and elastic modulus of <120 MPa and <32 GPa, respectively. The improvement in mechanical performance of these paper-like materials has been evidenced due to the chemical crosslinking between individual platelets with divalent ions and polyallylamine [111, 112].

The excellent mechanical properties of graphene broaden up its application both as individual and reinforcing materials for the preparation of composites. The exceptionally good mechanical properties of graphene are due to the stable sp^2 bonds that face the in-plane deformations. Nicholl et al. demonstrate a non-contact experimental approach using interferometric profilometry [113, 114]. In this approach, both CVD and exfoliated graphene have been subjected to pressurize by electrostatic way through the application of a voltage between the membrane and gating chip. The study reveals that graphene exhibits in-plane stiffness of 20–100 N m^{-1} at room temperature and has been softened considerably due to the out-of-plane crumpling effect. The defect-free single layer of graphene is considered to be the strongest material since others have studied the intrinsic strength of graphene membrane of 42 N m^{-1}, which equates to 130 GPa [7]. The effect of defects in graphene on intrinsic strength and stiffness has also been explored [114, 115]. In this study, the defects are introduced by modified plasma methods with AFM nanoindentation technique to subsequently measure the stiffness and strength of defective graphene. The defects are of sp^3 type and vacancy dominated determined from I_D/I_{D0} ratio of Raman peaks.

Fig. 5.11 Strain–strain curves for cracked graphene at different crack sizes; #3: 518 nm and #5: 1256 nm. Reprinted with permission from [117]

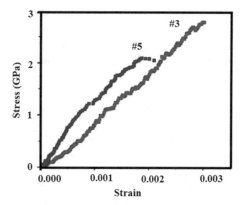

The quantification of defects is also done based on Raman studies by considering the ratio of I_D/I_G and I_{2D}/I_G ratios.

Interlayer shear mechanical properties of a multilayer graphene in wrinkled form have been predicted by simulation [116]. Results show that the mechanical properties of the wrinkled graphene have been improved compared to flat counterpart due to favourable geometrical locking. Graphene layers of different aspect ratios have been chosen. The shape of the wrinkled sheet remains unaltered when the aspect ratio is 0.77, and however, a change in the interlayer distance has been evidenced. The geometrical locking in the structure prevents the flipping over the deformation, and hence, the separation between two layers of the graphene takes place, which leads to a change in its interlayer distance. The wrinkled graphene is expected to reach tensile strength of 12 GPa due to interlayer shear modulus and strength. Determination of fracture toughness is another important mechanical properties of graphene that is relevant to engineering application. An in situ micromechanical testing device has been developed by Zhang et al. [117]. The fracture toughness of graphene synthesized in CVD mode has been determined by using a nanoindenter clubbed with scanning electron microscopy. A central crack has been induced using the focused ion beam, and brittle fracture has been noticed once the load is applied. The fracture stress is found to be increased with crack length. The critical strain release rate attained in this study is 15.9 J m^{-2}, and a fracture toughness of about 4 MPa has been reported. Figure 5.11 displays the engineering stress–strain curves for cracked graphene for crack sizes of 518 and 1256 nm.

5.5 Doping of Hetero-atoms into Graphene Lattice

Chemical doping is an efficient process, which provides the opportunity to tailor the electronic, chemical, optical and magnetic properties of materials [118, 119]. Graphene doped with different atoms (i.e. boron (B), nitrogen (N), sulphur (S), silicon (Si), etc.) results change in electronic properties and chemical reactivity of honeycomb lattice [120–123]. In particular, N-doping in graphene lattice produces additional *n*-type carriers for applications in high-frequency semiconductor devices and also enhances catalysis for energy conversion and storage [124–126]. In addition, N-doping shows biocompatibility and therefore is favourable for biosensing applications [127–129]. Recently, the highly crystalline monolayer N-doped graphene nanosheets have been synthesized and composed of two quasi-adjacent substitutional nitrogen atoms within the same graphene sub-lattice (N_2^{AA}) with improved Raman scattering [130]. In addition, magnetic properties for foreign atoms such as B-atoms, S-atoms and F-atoms have been studied. Especially, B-doped graphene with strong electron-withdrawing capability has been studied, which gives a catalyst-free thermal annealing approach in the presence of boron oxide for oxygen reduction in fuel cell application [131]. B-atoms of ~3.2 at.% have been successfully doped into graphene lattice [131]. Furthermore, S-doped graphene has been studied by magnesiothermic reduction in easily available, low-cost, non-toxic CO_2 (obtained from Na_2CO_3 and Na_2SO_4) as carbon and sulphur sources for the synthesis of efficient metal-free electrocatalysts for oxygen reduction reaction (ORR). The S-doped graphene exhibits high activity for ORR with a four-electron pathway having superior durability and tolerance to methanol crossover to 40% [132].

5.6 Applications

The area of application of graphene is broad and diverse in true sense. Graphene grabs different field of application in recent years due to the promising properties as discussed earlier. In the field of medicine, graphene has been extensively researched by scientists and shows its potential in drug delivery, tissue engineering, bio-imaging, biomicrobiotics, etc. Graphene extends its applicability in modern energy devices both in the field of generation and storage [133]. Graphene is suitable to be used as transistors, transparent conducting electrodes, spintronics, conductive ink, quantum dots, electronic devices and many more due to amazing electronic properties [134]. Graphene has also been used as lubricants, structural materials, waterproof coating, catalysts, metamaterials, plasmonics and many more indicating its capability to be suitable in the wide area of applications. The unique band structure of graphene makes the charge carrier bipolar in nature that can be effectively tuned by the application of electric field [12]. Hence, graphene has been successfully investigated by various research groups as field-effect transistors [91, 135]. The field-effect mobility of graphene is about an order of magnitude higher than silicon. Monolayer graphene is an excellent material to detect different gases and biomolecules and has been studied extensively as sensors [136–139]. The charge transfer between graphene and adsorbed molecules is accountable for chemical response. When a molecule gets adsorbed on the surface of graphene, the location experiences a charge transfer. This leads to a change in Fermi level and electronic resistance of graphene. Graphene and graphene-based fillers have been widely used for the preparation of polymer nanocomposites for possible use in wide application areas [140–142]. Polymer matrices of interest for the preparation of polymer nanocomposites with graphene are polystyrene, polyvinyl alcohol, polyester, epoxy, polycarbonate, polypropylene and many more [143–147].

5.7 Concluding Remarks

Graphene is a unique material in true sense. It is a material of choice for the modern day, a truly two-dimensional carbon material. Graphene has rare electronic excitation that moves in a curved space and interesting combinations of semiconductor and metal with the additional benefit of properties of a soft material. Graphene has different properties than conventional metal and semiconductors due to unusual electronic properties. The electron therein is undisturbed by the presence of disorder with a very long mean free path. Graphene has a robust but flexible structural arrangement. The properties of a graphene sheet can be modified by changing the number of layers, controlling the geometry, chemical doping and application of electric and magnetic field, etc. This chapter discusses the structure, synthesis and properties of graphene/reduced graphene oxide in detail. Graphene owing to its unique properties is found applications in different fields from chemical sensors to transistors. The application window of this material can be widened by chemical or structural modification of graphene to alter its functionality. Moreover, graphene can be easily synthesized from graphite, which is quite abundant on earth. This extends the availability of raw material for graphene synthesis for extensive research in this field.

Acknowledgements The authors acknowledge the financial support provided by Department of Science and Technology, India (DST/TMD/MES/2K16/37(G)), for carrying out this research work. Authors are thankful to Ms Tanvi Pal for drafting a figure.

References

1. A.K. Geim, K.S. Novoselov, Nat. Mater. **6**, 183 (2007)
2. S.P. Lonkar, Y.S. Deshmukh, A.A. Abdala, Nano Res. **8**, 1039 (2015)
3. S. Banerjee, P. Benjwal, M. Singh, K.K. Kar, Appl. Surf. Sci. **439**, 560 (2018)
4. R. Kumar, E.T. da Silva, R.K. Singh, R. Savu, A.V. Alaferdov, L.C. Fonseca, L.C. Carossi, A. Singh, S. Khandka, K.K. Kar, J. Colloid Interface Sci. **515**, 160 (2018)
5. A.A. Balandin, S. Ghosh, W. Bao, I. Calizo, D. Teweldebrhan, F. Miao, C.N. Lau, Nano Lett. **8**, 902 (2008)
6. W. Cai, Y. Zhu, X. Li, R.D. Piner, R.S. Ruoff, Appl. Phys. Lett. **95**, 123115 (2009)
7. C. Lee, X. Wei, J.W. Kysar, J. Hone, Science **321**, 385 (2008)
8. X. Li, Y. Zhu, W. Cai, M. Borysiak, B. Han, D. Chen, R.D. Piner, L. Colombo, R.S. Ruoff, Nano Lett. **9**, 4359 (2009)

9. H. O. Pierson, *Handbook of carbon, graphite, diamonds and fullerenes: Processing, properties and applications* (William Andrew, 2012)
10. A.C. Neto, F. Guinea, N.M. Peres, K.S. Novoselov, A.K. Geim, Rev. Mod. Phys. **81**, 109 (2009)
11. K. S. Novoselov, A. H. Castro Neto, Phys. Scr. **2012**, 014006 (2012)
12. K.S. Novoselov, A.K. Geim, S.V. Morozov, D. Jiang, Y. Zhang, S.V. Dubonos, I.V. Grigorieva, A.A. Firsov, Science **306**, 666 (2004)
13. A. Ciesielski, P. Samorì, Chem. Soc. Rev. **43**, 381 (2014)
14. M. Acik, Y.J. Chaba, J. Mater. Sci. Res. **2**, 101 (2013)
15. https://www.princeton.edu/~cml/html/research/TEGO.html. Accessed 10 January 2019
16. Y. Zhu, M.D. Stoller, W. Cai, A. Velamakanni, R.D. Piner, D. Chen, R.S. Ruoff, ACS Nano **4**, 1227 (2010)
17. S.Y. Oh, S.H. Kim, Y.S. Chi, T.J. Kang, Appl. Surf. Sci. **258**, 8837 (2012)
18. W. Choi, I. Lahiri, R. Seelaboyina, Y.S. Kang, Crit. Rev. Solid State Mater. Sci. **35**, 52 (2010)
19. http://majalah1000guru.net/2013/05/carbon-nanotubes-material-cerdas/. Accessed 10 January 2019
20. M. Batzill, Surf. Sci. Rep. **67**, 83 (2012)
21. H. Medina, Y.C. Lin, C. Jin, C.C. Lu, C.H. Yeh, K.P. Huang, K. Suenaga, J. Robertson, P.W. Chiu, Adv. Funct. Mater. **22**, 2123 (2012)
22. H.J. Song, M. Son, C. Park, H. Lim, M.P. Levendorf, A.W. Tsen, J. Park, H.C. Choi, Nanoscale **4**, 3050 (2012)
23. T. Takami, S. Ogawa, H. Sumi, T. Kaga, A. Saikubo, E. Ikenaga, M. Sato, M. Nihei, Y. Takakuwa, e-J. Surf. Sci. Nanotechnol. **7**, 882 (2009)
24. Y. Wang, X. Xu, J. Lu, M. Lin, Q. Bao, B. Ozyilmaz, K.P. Loh, ACS Nano **4**, 6146 (2010)
25. Y. Kim, W. Song, S. Lee, C. Jeon, W. Jung, M. Kim, C.-Y. Park, Appl. Phys. Lett. **98**, 263106 (2011)
26. R.S. Edwards, K.S. Coleman, Acc. Chem. Res. **46**, 23 (2013)
27. L. Gomez De Arco, Y. Zhang, C.W. Schlenker, K. Ryu, M.E. Thompson, C. Zhou, ACS Nano **4**, 2865 (2010)
28. A. Reina, X. Jia, J. Ho, D. Nezich, H. Son, V. Bulovic, M.S. Dresselhaus, J. Kong, Nano Lett. **9**, 30 (2008)
29. Y. Zhang, L. Gomez, F.N. Ishikawa, A. Madaria, K. Ryu, C. Wang, A. Badmaev, C. Zhou, J. Phys. Chem. Lett. **1**, 3101 (2010)
30. S. Thiele, A. Reina, P. Healey, J. Kedzierski, P. Wyatt, P.-L. Hsu, C. Keast, J. Schaefer, J. Kong, Nanotechnol. **21**, 015601 (2009)
31. S.J. Chae, F. Güneş, K.K. Kim, E.S. Kim, G.H. Han, S.M. Kim, H.J. Shin, S.M. Yoon, J.Y. Choi, M.H. Park, Adv. Mater. **21**, 2328 (2009)
32. J.M. Garcia, R. He, M.P. Jiang, P. Kim, L.N. Pfeiffer, A. Pinczuk, Carbon **49**, 1006 (2011)
33. M. Zheng, K. Takei, B. Hsia, H. Fang, X. Zhang, N. Ferralis, H. Ko, Y.-L. Chueh, Y. Zhang, R. Maboudian, Appl. Phys. Lett. **96**, 063110 (2010)
34. Z. Yan, Z. Peng, Z. Sun, J. Yao, Y. Zhu, Z. Liu, P.M. Ajayan, J.M. Tour, ACS Nano **5**, 8187 (2011)
35. H.J. Shin, W.M. Choi, S.M. Yoon, G.H. Han, Y.S. Woo, E.S. Kim, S.J. Chae, X.S. Li, A. Benayad, D.D. Loc, Adv. Mater. **23**, 4392 (2011)
36. A. Umair, H. Raza, Nanoscale Res. Lett. **7**, 437 (2012)
37. Y. Bando, Y. Takahashi, E. Ueta, N. Todoroki, T. Wadayama, J. Electrochem. Soc. **162**, F463 (2015)
38. J. Hamilton, J. Blakely, Surf. Sci. **91**, 199 (1980)
39. P. Sutter, J.T. Sadowski, E. Sutter, Phys. Rev. B **80**, 245411 (2009)
40. S.-Y. Kwon, C.V. Ciobanu, V. Petrova, V.B. Shenoy, J. Bareno, V. Gambin, I. Petrov, S. Kodambaka, Nano Lett. **9**, 3985 (2009)
41. P.W. Sutter, J.-I. Flege, E.A. Sutter, Nat. Mater. **7**, 406 (2008)
42. J. Coraux, M. Engler, C. Busse, D. Wall, N. Buckanie, F.-J. M. Zu Heringdorf, R. Van Gastel, B. Poelsema, T. Michely, New J. Phys. **11**, 023006 (2009)

43. Z.-Y. Juang, C.-Y. Wu, A.-Y. Lu, C.-Y. Su, K.-C. Leou, F.-R. Chen, C.-H. Tsai, Carbon **48**, 3169 (2010)
44. H.J. Park, J. Meyer, S. Roth, V. Skákalová, Carbon **48**, 1088 (2010)
45. X. Li, L. Colombo, R.S. Ruoff, Adv. Mater. **28**, 6247 (2016)
46. O. Carneiro, M. Kim, J. Yim, N. Rodriguez, R. Baker, J. Phys. Chem. B **107**, 4237 (2003)
47. L. Constant, C. Speisser, F. Le Normand, Surf. Sci. **387**, 28 (1997)
48. L. Ding, A. Tselev, J. Wang, D. Yuan, H. Chu, T.P. McNicholas, Y. Li, J. Liu, Nano Lett. **9**, 800 (2009)
49. W. Zhou, Z. Han, J. Wang, Y. Zhang, Z. Jin, X. Sun, Y. Zhang, C. Yan, Y. Li, Nano Lett. **6**, 2987 (2006)
50. C. Jia, J. Jiang, L. Gan, X. Guo, Sci. Rep. **2**, 707 (2012)
51. S. Bae, H. Kim, Y. Lee, X. Xu, J.-S. Park, Y. Zheng, J. Balakrishnan, T. Lei, H.R. Kim, Y.I. Song, Nat. Nanotechnol. **5**, 574 (2010)
52. M. Aliofkhazraei, N. Ali, W. I. Milne, C. S. Ozkan, S. Mitura, J. L. Gervasoni, *Graphene science handbook: Fabrication methods* (CRC Press, 2016)
53. D. Lee, L. De Los Santos V, J. Seo, L. L. Felix, A. Bustamante D, J. Cole, C. Barnes, J. Phys. Chem. B **114**, 5723 (2010)
54. O.C. Compton, S.T. Nguyen, Small **6**, 711 (2010)
55. H. Li, C. Bubeck, Macromol. Research **21**, 290 (2013)
56. S. Park, R.S. Ruoff, Nat. Nanotechnol. **4**, 217 (2009)
57. H.J. Shin, K.K. Kim, A. Benayad, S.M. Yoon, H.K. Park, I.S. Jung, M.H. Jin, H.K. Jeong, J.M. Kim, J.Y. Choi, Adv. Funct. Mater. **19**, 1987 (2009)
58. I.K. Moon, J. Lee, R.S. Ruoff, H. Lee, Nat. Commun. **1**, 73 (2010)
59. C. Li, L. Li, L. Sun, Z. Pei, J. Xie, S. Zhang, Carbon **89**, 74 (2015)
60. P. Labhane, L. Patle, V. Huse, G. Sonawane, S. Sonawane, Chem. Phys. Lett. **661**, 13 (2016)
61. Y. Hong, Z. Wang, X. Jin, Sci. Rep. **3**, 3439 (2013)
62. C. Liu, G. Han, Y. Chang, Y. Xiao, Y. Li, M. Li, H. Zhou, Chin. J. Chem. **34**, 89 (2016)
63. Z.-T. Hu, J. Liu, X. Yan, W.-D. Oh, T.-T. Lim, Chem. Eng. J. **262**, 1022 (2015)
64. C. Liu, F. Hao, X. Zhao, Q. Zhao, S. Luo, H. Lin, Sci. Rep. **4**, 3965 (2014)
65. M. Aunkor, I. Mahbubul, R. Saidur, H. Metselaar, Rsc Adv. **6**, 27807 (2016)
66. P. Chamoli, M.K. Das, K.K. Kar, Curr. Graphene Sci. **1**, 000 (2017)
67. P. Chamoli, M.K. Das, K.K. Kar, J. Phys. Chem. Solids **113**, 17 (2018)
68. L.J. Cote, R. Cruz-Silva, J. Huang, JACS **131**, 11027 (2009)
69. E.E. Ghadim, N. Rashidi, S. Kimiagar, O. Akhavan, F. Manouchehri, E. Ghaderi, Appl. Surf. Sci. **301**, 183 (2014)
70. Y. Zhou, Q. Bao, L.A.L. Tang, Y. Zhong, K.P. Loh, Chem. Mater. **21**, 2950 (2009)
71. K. Ai, Y. Liu, L. Lu, X. Cheng, L. Huo, J. Mater. Chem. **21**, 3365 (2011)
72. C. Xu, X. Wang, J. Zhu, J. Phys. Chem. C **112**, 19841 (2008)
73. O. Akhavan, E. Ghaderi, J. Phys. Chem. C **113**, 20214 (2009)
74. M. Fernández-Merino, L. Guardia, J. Paredes, S. Villar-Rodil, P. Solís-Fernández, A. Martínez-Alonso, J. Tascón, J. Phys. Chem. C **114**, 6426 (2010)
75. C. Zhu, S. Guo, Y. Fang, S. Dong, ACS Nano **4**, 2429 (2010)
76. O. Akhavan, E. Ghaderi, S. Aghayee, Y. Fereydooni, A. Talebi, J. Mater. Chem. **22**, 13773 (2012)
77. J. Liu, S. Fu, B. Yuan, Y. Li, Z. Deng, JACS **132**, 7279 (2010)
78. A. Esfandiar, O. Akhavan, A. Irajizad, J. Mater. Chem. **21**, 10907 (2011)
79. O. Akhavan, M. Kalaee, Z. Alavi, S. Ghiasi, A. Esfandiar, Carbon **50**, 3015 (2012)
80. O. Akhavan, E. Ghaderi, E. Abouei, S. Hatamie, E. Ghasemi, Carbon **66**, 395 (2014)
81. S. Hatamie, O. Akhavan, S.K. Sadrnezhaad, M.M. Ahadian, M.M. Shirolkar, H.Q. Wang, Mater. Sci. Eng., C **55**, 482 (2015)
82. O. Akhavan, E. Ghaderi, Carbon **50**, 1853 (2012)
83. O. Akhavan, Carbon **81**, 158 (2015)
84. P. Chamoli, M.K. Das, K.K. Kar, J. Appl. Phys. **122**, 185105 (2017)
85. P. Phillips, Ann. Phys. **321**, 1634 (2006)

86. P.R. Wallace, Phys. Rev. **71**, 622 (1947)
87. F. Bonaccorso, Z. Sun, T. Hasan, A. Ferrari, Nat. Photonics **4**, 611 (2010)
88. P. Avouris, Nano Lett. **10**, 4285 (2010)
89. Y.-W. Tan, Y. Zhang, K. Bolotin, Y. Zhao, S. Adam, E. Hwang, S.D. Sarma, H. Stormer, P. Kim, Phys. Rev. Lett. **99**, 246803 (2007)
90. K.I. Bolotin, K. Sikes, Z. Jiang, M. Klima, G. Fudenberg, J. Hone, P. Kim, H. Stormer, Solid State Commun. **146**, 351 (2008)
91. Y. Zhang, Y.-W. Tan, H.L. Stormer, P. Kim, Nature **438**, 201 (2005)
92. R.R. Nair, P. Blake, A.N. Grigorenko, K.S. Novoselov, T.J. Booth, T. Stauber, N.M. Peres, A.K. Geim, Science **320**, 1308 (2008)
93. F. Wang, Y. Zhang, C. Tian, C. Girit, A. Zettl, M. Crommie, Y.R. Shen, Science **320**, 206 (2008)
94. P.A. George, J. Strait, J. Dawlaty, S. Shivaraman, M. Chandrashekhar, F. Rana, M.G. Spencer, Nano Lett. **8**, 4248 (2008)
95. F. Rana, P.A. George, J.H. Strait, J. Dawlaty, S. Shivaraman, M. Chandrashekhar, M.G. Spencer, Phys. Rev. B **79**, 115447 (2009)
96. F. Xia, T. Mueller, Y.-M. Lin, A. Valdes-Garcia, P. Avouris, Nat. Nanotechnol. **4**, 839 (2009)
97. A.A. Balandin, Nat. Mater. **10**, 569 (2011)
98. D.G. Papageorgiou, I.A. Kinloch, R.J. Young, Carbon **95**, 460 (2015)
99. E. Pop, V. Varshney, A.K. Roy, MRS Bull. **37**, 1273 (2012)
100. C. Yu, L. Shi, Z. Yao, D. Li, A. Majumdar, Nano Lett. **5**, 1842 (2005)
101. S. Berber, Y.-K. Kwon, D. Tománek, Phys. Rev. Lett. **84**, 4613 (2000)
102. S. Ghosh, Appl. Phys. Lett. **92**, 151911 (2008)
103. I.-K. Hsu, M.T. Pettes, A. Bushmaker, M. Aykol, L. Shi, S.B. Cronin, Nano Lett. **9**, 590 (2009)
104. I. Frank, D. M. Tanenbaum, A. M. van der Zande, P. L. McEuen, J. Vacuum Sci. & Technol. B: Microelectronics and Nanometer Structures Processing, Measurement, and Phenomena **25**, 2558 (2007)
105. K.N. Kudin, G.E. Scuseria, B.I. Yakobson, Phys. Rev. B **64**, 235406 (2001)
106. C. Reddy, S. Rajendran, K. Liew, Nanotechnol. **17**, 864 (2006)
107. G. Van Lier, C. Van Alsenoy, V. Van Doren, P. Geerlings, Chem. Phys. Lett. **326**, 181 (2000)
108. M. Poot, H.S. van der Zant, Appl. Phys. Lett. **92**, 063111 (2008)
109. C. Gómez-Navarro, M. Burghard, K. Kern, Nano Lett. **8**, 2045 (2008)
110. D.A. Dikin, S. Stankovich, E.J. Zimney, R.D. Piner, G.H. Dommett, G. Evmenenko, S.T. Nguyen, R.S. Ruoff, Nature **448**, 457 (2007)
111. S. Park, D.A. Dikin, S.T. Nguyen, R.S. Ruoff, J. Phys. Chem. C **113**, 15801 (2009)
112. S. Park, K.-S. Lee, G. Bozoklu, W. Cai, S.T. Nguyen, R.S. Ruoff, ACS Nano **2**, 572 (2008)
113. R.J. Nicholl, H.J. Conley, N.V. Lavrik, I. Vlassiouk, Y.S. Puzyrev, V.P. Sreenivas, S.T. Pantelides, K.I. Bolotin, Nat. Commun. **6**, 8789 (2015)
114. D.G. Papageorgiou, I.A. Kinloch, R.J. Young, Prog. Mater Sci. **90**, 75 (2017)
115. A. Zandiatashbar, G.-H. Lee, S.J. An, S. Lee, N. Mathew, M. Terrones, T. Hayashi, C.R. Picu, J. Hone, N. Koratkar, Nat. Commun. **5**, 3186 (2014)
116. H. Qin, Y. Sun, J.Z. Liu, Y. Liu, Carbon **108**, 204 (2016)
117. P. Zhang, L. Ma, F. Fan, Z. Zeng, C. Peng, P.E. Loya, Z. Liu, Y. Gong, J. Zhang, X. Zhang, Nat. Commun. **5**, 3782 (2014)
118. P. Chamoli, M.K. Das, K.K. Kar, Mater. Res. Exp. **4**, 015012 (2017)
119. P. Chamoli, S.K. Singh, M. Akhtar, M.K. Das, K.K. Kar, Physica E: Low-dimensional Syst. Nanostruct. **103**, 25 (2018)
120. J. Dai, J. Yuan, P. Giannozzi, Appl. Phys. Lett. **95**, 232105 (2009)
121. L. Panchakarla, K. Subrahmanyam, S. Saha, A. Govindaraj, H. Krishnamurthy, U. Waghmare, C. Rao, Adv. Mater. **21**, 4726 (2009)
122. X. Wang, X. Li, L. Zhang, Y. Yoon, P.K. Weber, H. Wang, J. Guo, H. Dai, Science **324**, 768 (2009)
123. Y. Zou, F. Li, Z. Zhu, M. Zhao, X. Xu, X. Su, Eur. Phys. J. B **81**, 475 (2011)

124. T. Cui, R. Lv, Z.-H. Huang, H. Zhu, J. Zhang, Z. Li, Y. Jia, F. Kang, K. Wang, D. Wu, Carbon **49**, 5022 (2011)
125. R. Lv, T. Cui, M.S. Jun, Q. Zhang, A. Cao, D.S. Su, Z. Zhang, S.H. Yoon, J. Miyawaki, I. Mochida, Adv. Funct. Mater. **21**, 999 (2011)
126. L. Qu, Y. Liu, J.-B. Baek, L. Dai, ACS Nano **4**, 1321 (2010)
127. J. Carrero-Sanchez, A. Elias, R. Mancilla, G. Arrellin, H. Terrones, J. Laclette, M. Terrones, Nano Lett. **6**, 1609 (2006)
128. A.L. Elías, J.C. Carrero-Sánchez, H. Terrones, M. Endo, J.P. Laclette, M. Terrones, Small **3**, 1723 (2007)
129. Y. Wang, Y. Shao, D.W. Matson, J. Li, Y. Lin, ACS Nano **4**, 1790 (2010)
130. R. Lv, Q. Li, A.R. Botello-Méndez, T. Hayashi, B. Wang, A. Berkdemir, Q. Hao, A.L. Elías, R. Cruz-Silva, H.R. Gutiérrez, Sci. Rep. **2**, 586 (2012)
131. Z.-H. Sheng, H.-L. Gao, W.-J. Bao, F.-B. Wang, X.-H. Xia, J. Mater. Chem. **22**, 390 (2012)
132. J. Wang, R. Ma, Z. Zhou, G. Liu, Q. Liu, Sci. Rep. **5**, 9304 (2015)
133. Y. Zhu, S. Murali, W. Cai, X. Li, J.W. Suk, J.R. Potts, R.S. Ruoff, Adv. Mater. **22**, 3906 (2010)
134. B. H. Nguyen, V. H. Nguyen, Adv. Natural Sci.: Nanosci. Nanotechnol. **7**, 023002 (2016)
135. K.S. Novoselov, A.K. Geim, S. Morozov, D. Jiang, M. Katsnelson, I. Grigorieva, S. Dubonos, A.A. Firsov, Nature **438**, 197 (2005)
136. N. Mohanty, V. Berry, Nano Lett. **8**, 4469 (2008)
137. Y. Ohno, K. Maehashi, Y. Yamashiro, K. Matsumoto, Nano Lett. **9**, 3318 (2009)
138. S.K. Singh, P. Azad, M. Akhtar, K.K. Kar, Mater. Res. Exp. **4**, 086301 (2017)
139. L. Tang, Y. Wang, Y. Li, H. Feng, J. Lu, J. Li, Adv. Funct. Mater. **19**, 2782 (2009)
140. B.Z. Jang, A. Zhamu, J. Mater. Sci. **43**, 5092 (2008)
141. K. K. Kar, *Composite materials: processing, applications, characterizations* (Springer, 2016)
142. K. K. Kar, S. Rana, J. Pandey, *Handbook of polymer nanocomposites processing, performance and application* (Springer, 2015)
143. K. K. Kar, A. Hodzic, *Carbon nanotube based nanocomposites: recent developments* (Research Publishing Services, 2011)
144. M. Fang, K. Wang, H. Lu, Y. Yang, S. Nutt, J. Mater. Chem. **19**, 7098 (2009)
145. S. Ganguli, A.K. Roy, D.P. Anderson, Carbon **46**, 806 (2008)
146. K. Kalaitzidou, H. Fukushima, L.T. Drzal, Compos. Sci. Technol. **67**, 2045 (2007)
147. T. Ramanathan, A. Abdala, S. Stankovich, D. Dikin, M. Herrera-Alonso, R. Piner, D. Adamson, H. Schniepp, X. Chen, R. Ruoff, Nat. Nanotechnol. **3**, 327 (2008)

Chapter 6
Characteristics of Carbon Nanotubes

Soma Banerjee and Kamal K. Kar

Abstract Carbon nanotubes (CNTs) are the materials of modern age having diverse applications in every sector. This article discusses the structure, synthesis, properties, purification, and application aspects of CNTs. CNTs can be classified based on tube structure and shapes. This chapter elaborates single-walled nanotubes (SWNTs), multi-walled nanotubes (MWNTs), and double-walled nanotubes (DWNTs) in brief. In addition, researches are going on worldwide to generate CNTs of variable shapes such as coiled, waved, bent, beaded, junction CNTs. CNTs can be synthesized by chemical and physical routes. The most common method of CNT production is chemical vapour deposition (CVD) technique that can produce pure CNTs in large quantities. This article also opens up the present aspects of growth mechanism of CNTs. The unique combination of electrical, thermal, and mechanical properties of CNTs makes them a potential performer in the number of fields such as supercapacitors, fuel cells, energy devices, high strength composites, biomedical, chemical. However, the application of CNTs is still somehow limited and yet to reach the desired plateau due to the purification, large-scale production, and toxicity issues.

6.1 Introduction

Since prehistoric times, carbon has been used in different forms in various aspects of art and technology [1–3]. Starting from the use of charcoal and soot in cave paintings to modern use of carbon nanomaterials, carbon has a remarkable impression in every

S. Banerjee · K. K. Kar (✉)
Advanced Nanoengineering Materials Laboratory, Materials Science Programme, Indian Institute of Technology Kanpur, Kanpur 208016, India
e-mail: kamalkk@iitk.ac.in

S. Banerjee
e-mail: somabanerjee27@gmail.com

K. K. Kar
Advanced Nanoengineering Materials Laboratory, Department of Mechanical Engineering, Indian Institute of Technology Kanpur, Kanpur 208016, India

© Springer Nature Switzerland AG 2020   179
K. K. Kar (ed.), *Handbook of Nanocomposite Supercapacitor Materials I*,
Springer Series in Materials Science 300,
https://doi.org/10.1007/978-3-030-43009-2_6

sphere of life. The use of other forms of carbon such as graphite, carbon black, and charcoal is extensive in the field of drawing, writing, printing, etc. Charcoal has played a tremendous role in the first technology of human being, i.e. smelting and working of metals. Later on, in the advanced stage of eighteenth century, charcoal has been replaced by coke, which lead to an industrial revolution. In 1896, the first synthetic graphite has been synthesized by American Edward Acheson. Afterward in the twentieth century, and activated carbon has grabbed the attention due to the ability to purify waste water. The invention of carbon fiber in 1950s opens up the development of an interesting lightweight and ultra-strong carbon material to the engineers. Diamonds are industrially explored at around 1950s by the development of successful synthesis methodology by General Electrics. In 1980s, the research in the field of carbon science became mature and grabbed Nobel Prizes. In 1985, Harry Kroto and group have invented the first all carbon-based molecules named as buckminsterfullerene [4]. This development in carbon science opened up further synthesis of fullerene-related nanotubes.

In the year 1991, Ijima discovered the magical carbon nanomaterial, carbon nanotubes (CNTs) containing more than one graphitic layer having an inner diameter of 4 nm. In 1993, Bethune et al. and Ijima et al. have independently reported the development of single-walled nanotube (SWNT) [5, 6]. The discovery of SWNTs is remarkable since their structures are seemed to be close to theoretical prediction. The extraordinary properties of SWNTs have grabbed more attention the researcher compared to multi-walled nanotube (MWNTs). CNTs are tube-shaped carbon materials having a diameter in nanometres. The graphitic layers of CNTs are rolled-up chicken wire having carbon at the apex of a hexagon also called graphene. CNTs made of same graphite sheets can have different properties due to the difference in length, helicity, thickness, and number of layers, etc. Depending on the structural arrangements, they can be metallic or semiconductor in nature. MWNTs are CNTs composed of many graphitic shells having adjacent shell separation of 0.34 nm and diameter of 1 nm. Lengths of the nanotubes are of several microns and graphene sheets can be rolled in a number of ways generating different types of nanotube based on their structural arrangements. The high length-to-diameter ratio of these materials provides an interesting combination of properties catching the attention of researchers of interdisciplinary fields. The spatial arrangement of carbon atoms generates various carbon networks [7, 8]. Figure 6.1 represents different forms of carbon nanomaterials.

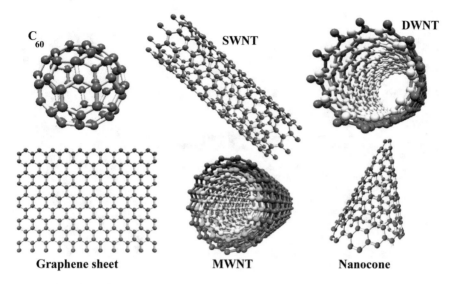

Fig. 6.1 Different forms of carbon nanomaterials

6.2 Structure of Carbon Nanotubes

6.2.1 Based on Tube Structure

6.2.1.1 Single-Walled Nanotubes

Single-walled nanotube (SWNT) can be imagined as a perfect graphene sheet consisting of sp^2 hybridized carbon in a hexagon, which is rolled in the form of a cylinder with the confirmation that the hexagonal rings located in contact are combined coherently. The tips of the tube are closed by a cap made of hemifullerene of a specific diameter. Single-walled nanotubes (SWNTs) are the nanomaterial of unique properties and behavior and a beautiful material for fundamental physicists and experimental chemists. Geometrically, the SWNT tube diameter can be anything; however, calculations reveal that the SWNT collapses into flatten ribbon-like structure and energetically the ribbon-like structure becomes more favourable once the diameter of the tube reaches 2.5 nm [9]. SWNTs of diameter as low as 0.4 nm have been reported [10]. For the tube length of SWNT, no such restriction is known; however, the length of the tubes will be governed by the method of synthesis and parameters of synthesis such as thermal gradient, residence time. The length of the SWNTs can be a few micrometres to millimetre range. These specific features of SWNT makes them a material of choice due to huge aspect ratio. A simple way of rolling of graphene sheet to form CNT is shown in Fig. 6.2.

From the structure of SWNTs, two different aspects can be discussed. First, all the carbons are a part of hexagonal rings and are in equivalent positions except at

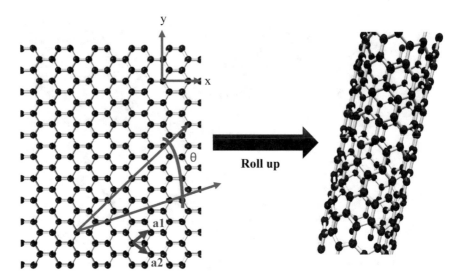

Fig. 6.2 Rolling up of graphene sheet to form nanotube. Drawn based on the concept taken from [11]

the tip of the nanotubes, where $(6 \times 5 = 30)$ atoms of carbons are present forming pentagonal rings. Hence, for ideal cases, the reactivity will be exceptional at the tip part of the tube rather at the hexagonal part of it. Again, the carbon atoms in the aromatic ring are not exactly planar indicating the fact that the carbon atoms are not having pure sp^2 hybridization. There remains some sp^3 character as well, and this will be dependent on the radius of curvature of the tube. As for example, fullerenes C60 molecule having radius of curvature 0.35 nm possesses 10% sp^3 character [12]. These structural features make SWNT different from graphene, and SWNT is expected to have a bit more reactivity compared to planar graphene. These factors are again responsible for the overlapping of bands, and hence, a unique electrical behavior has been evidenced for SWNTs.

As discussed earlier, the graphene sheets can be rolled in a number of ways to form SWNTs giving rise to the formation of Zigzag, armchair, and helical SWNTs as represented in Fig. 6.3. Some of the nanotubes will possess planes of symmetry both perpendicular and parallel to the tube axis, while others will not. The latter is commonly named as chiral nanotubes since they are not able to superimpose on their mirror image. The way to roll a graphene sheet is mathematically expressed by the vector of helicity, C_h (OA), and angle of helicity, θ; i.e.,

$$OA = C_h = na_1 + ma_2 \tag{6.1}$$

where

$$a_1 = ax\sqrt{3}/2 + ay/2 \tag{6.2}$$

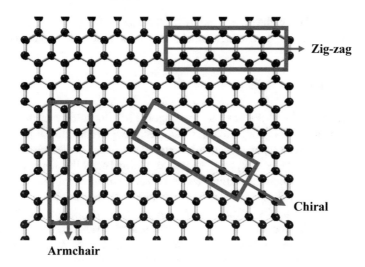

Fig. 6.3 SWNT structures of different types: zigzag, armchair, and helical nanotube. Drawn based on the concept taken from [14]

and

$$a_2 = ax\sqrt{3}/2 - ay/2 \tag{6.3}$$

where $a = 2.46$ Å
and

$$\cos\theta = (2n + m)/2\sqrt{(n^2 + m^2 + nm)} \tag{6.4}$$

The vector of helicity is considered as perpendicular to the nanotube axis, whereas the angle of helicity is taken at the zigzag axis. Hence, the zigzag type nanotubes are emerged due to the vector of helicity. The diameter (D) of a SWNT is related to C_h by the following equation.

$$D = |C_h|/\pi = a_{C=C}\sqrt{3(n^2 + m^2 + nm)}/\pi$$
$$1.41\,\text{Å} \leq a_{C=C} \leq 1.44\,\text{Å}$$
$$\text{(graphite)} \quad (C_{60}) \tag{6.5}$$

The parameters of helicity, C_h, θ, and tube diameter all are expressed as a function of integer values n and m. Hence, SWNT nomenclature is given as per the integer values as (n, m). The integer value, in other words, the naming of SWNT, can be easily given by calculating the number of hexagons separating the edges of C_h vector following a_1 and a_2 [13]. In general, it can be concluded that when the vector of helicity remains perpendicular to any of the three C=C bond directions, it generates zigzag type SWNTs. On the other hand, when the vector of helicity remains parallel to C=C bond directions, it generates armchair types nanotubes denoted as (n, n).

6.2.1.2 Multi-walled Nanotubes

Multi-walled nanotube (MWNT) can be imagined as concentric carbon nanotube composed of many SWNTs of increasing diameter arranged coaxially as per the Russian doll model. These types of nanotubes are synthesized by electric arc method without the use of a catalyst or bycatalyst-assisted thermal cracking of gaseous hydrocarbons or by CO disproportionation reactions. The number of tubes can be in any numbers starting from two. The inter-tube distance is the same as that of the inter-graphene distance of 0.34 nm a little more compared to the graphite of 0.335 nm. An increased radius of curvature during the formation of MWNT restricts the regular arrangement of carbons as in graphite. MWNTs can also be of faceted morphologies developed by thermal treatment at a very high temperature of around 2500 °C. Another common structural feature of MWNT is herringbone texture. Here, graphene forms an angle with the axis of the carbon nanotubes. The angle formed will be dependent on processing conditions such as morphology of the catalyst, composition of the reaction environment, etc. The angle may be varied from zero to 90° leading to the change of nanotube morphology to nanofibers [15]. Catalyst-induced thermal cracking of hydrocarbons and CO disproportionation is the common method used for the synthesis of herringbone MWNTs. The aspect ratio of MWNTs is small compared to SWNTs and often allows to predict tube ends when imaged under transmission electron microscopy. Again, imaging shows that the nanotube tips are associated with catalyst crystals when grown by the method of electric arc discharge.

A special attention is to be given to double-walled carbon nanotubes, abbreviated as DWNT having similar kinds of morphology and properties as that of SWNT with the additional benefit of improved chemical resistance properties. The synthesis of DWNT is proposed in 2003 by CCVD technique using selective reducing oxide solution in the presence of methane and hydrogen [16]. DWNTs can be modified more easily by the functionalization approach compared to SWNTs. In case of SWNTs, the functionalization leads to breaking of carbon–carbon double bonds creating holes in the structure of the nanotubes, which thereby leads to alteration in mechanical and electrical properties. For DWNTs, the outer walls are only modified by grafting of chemical entities at the nanotube surface to generate new properties in CNTs.

6.2.1.3 Nanotorus

The word nanotorus is theoretically explained as the CNTs made in the form of a doughnut, i.e., bent like a torus. These types of CNTs have extraordinary magnetic properties having 1000 times of magnetic moments. The radius of torus determines the magnetic moments and other properties of the doughnut such as thermal stability. These materials can be promising in nanophotonics [17].

6.2.1.4   Nanobuds

Carbon nanobuds can be defined as the new form of carbon made of two allotropes of carbon, namely CNT and fullerene. They are formed by covalent bonding between the CNTs and fullerenes. The fullerenes are bonded to the outer sidewalls of a CNT to form a bud over it. These materials are found to be excellent field emitters. The other utility of nanobuds is they can act like an anchor to matrix materials when used as a functional filler in the composites. The buds may prevent the slippage of CNTs inside the matrix of a composite, and hence, a remarkable improvement in mechanical properties can be achieved [18].

6.2.1.5   Nanohorns

Nanohorns are another interesting form of CNT discovered first by Haris and Ijima [19]. The single-walled nanohorn is defined as a horn-type SWNT with a conical tip [20]. The main utility of single-walled nanohorns is that they can be synthesized without catalysts and hence can be of very high purity. These materials are promising for energy storage applications and as an electrode material due to the high surface area and electronic properties. These materials have been extensively studied in other applications such as adsorption, gas storage, fuel cells, magnetic resonance analysis, catalyst supports, electrochemistry [21].

6.2.2   Based on Shapes

CNTs are studied extensively in recent days due to the provision to be tailored in a number of ways based on diameter, number of walls, tube length, crystallinity, purity, alignment, etc. [22, 23]. A perfect nanotube is composed of crystalline structure of hexagonal network. In the presence of defects, nanotubes get curved. The nanotubes of different morphologies have their potential in number of applications due to the unique variation in properties with shape. Till date, different shapes of CNTs are synthesized such as coiled, waved, branched, bent, beaded, straight.

6.2.2.1   Straight Carbon Nanotubes

Several methods generate vertically aligned CNTs by catalytic CVD [24]. The vertically aligned CNTs are the nanotubes that are oriented perpendicular to the substrate. Two major researches are to be mentioned for the development of CNT arrays—first by Fan et al. [25] and other by Hata et al. [26]. Fan et al. have developed blocks of MWNTs that have grown perpendicular to the substrate [25]. They have used porous silicon substrate with catalyst pattern via e-beam evaporation. Hata et al. have synthesized SWNT arrays of few millimetre height using water-assisted CVD technique

[26]. Study reveals that SWNT of DWNTs will collapse and generate stacks of parallel graphene layers when diameters become greater than 5 nm. Aligned CNT arrays are suitable in advanced devices such as light emitters, field-effect transistors, logic circuits, sensors [27–31].

6.2.2.2 Waved Carbon Nanotubes

A single nanotube generally bends during the process of growth if no external force is applied. The bends in nanotubes arise due to the presence of pentagon–heptagon defect pairs by means of local mechanical deformation forces. A nanotube can be elastically deformed by the application of small bending stress, and when the local curvature exceeds a critical value, it is buckled [31–33]. Again, when a nanotube grows, the interaction with other nanotubes may generate bending. In addition, limited growing space and its own weight can also be factors affecting the bending of a nanotube. This kind of regular bending may result in wavy-kind morphology in CNTs. Again, it is believed that wavy structures in nanotubes are generated when two types of catalysts are present over the substrate. One may be of higher catalytic activity than other leading to a high growth rate of CNTs. Due to the van der Walls forces, the nanotubes when touch each other stick together. The growth of CNT arrays is limited for catalyst present over the surface of the substrate with a relatively slow rate of growth. The nanotube growing at a higher rate is forced to bend. Hence, the extent of wavy nature will be dependent on the variation in the growth rate of these two groups. These kinds of structures of nanotubes are important for assembling of CNT sheets or yarns [34, 35].

6.2.2.3 Branched Carbon Nanotubes

Branched CNTs of the type Y-junction are formed by inserting non-hexagonal rings in a network of a hexagon in particular to the zones, where three arms of Y are clubbed together [36, 37]. Other structural arrangements of nanotubes also follow the same rule of protecting sp^2 hybridized carbon [38, 39]. The only difference lies in kind, placement, and number of non-hexagonal units. This opens up the provision of variation in angles from Y to T shapes [40, 41]. These structures have the potential for electrical applications since metallic and semiconducting tubes can be converted to CNT junctions for building different parts of nanometric-integrated circuits [42]. CNTs produced by the arc discharge process show T, Y, and L junctions [43]. Synthesis of most of the junctional CNTs is based on catalytic CVD [44–47]. Branched SWNTs have potential in advanced electronic devices such as nano-transistors, nano-interconnect, nano-diode [31].

6.2.2.4 Regularly Bent Carbon Nanotubes

Regular bends in CNTs are useful for their applications in the field of nanocircuit interconnects, high-resolution AFM tips, and many more. Nanotubes can be forced to be aligned during their growth by application of external force generated from the electric field [48]. The external forces can also be created by other means such as interaction with substrate, gas flow, etc. [49, 50]. A zigzag morphology of nanotubes may contain 2–4 sharp and alternating bends at 90° angle. A DC plasma-induced CVD may generate CNTs of such structural morphology [51]. The change of direction of electric field in the growth region induces bending in nanotubes.

6.2.2.5 Coiled Carbon Nanotubes

Coiled CNTs are formed when a pair of heptagon and pentagon arrange in a periodic fashion inside a hexagonal network of carbon [52]. Helical structures of carbon have been evidenced in the early fifties [53]. Coiled CNTs are synthesized in the year of 1994 [54]. In recent practice, coiled CNTs are synthesized by catalytic CVD technique on iron coated with indium tinoxide substrate [55]. This process produces more than 95% of helical CNTs of different diameters and pitches. Study reveals that indium and tin have a major role in the formation of coiled CNTs [56]. A study on the thermodynamic model explains the reason behind the formation of coiled structures in CNTs [57]. They suggest that interaction between the particular metal catalysts and growing nanostructures plays remarkable roles in deciding the possibility of the helical growth of CNTs. Coiled CNTs are useful for applications such as sensors, electrical inductors, nanoscale mechanical springs, electromagnetic wave absorber.

6.2.2.6 Beaded Carbon Nanotubes

Beaded CNTs are formed in different patterns, and structurally, they can be either amorphous or polycrystalline graphite. The beads can be present either with or after the formation of CNTs [58, 59]. Beaded CNTs can be occasionally formed from the interior of arc deposition. The beads are amorphous phases of carbon glass. They occur during the growth of the nanotubes since the carbon-coated CNTs are viscous liquid form and cooling leads the increment in viscosity to a great degree that beading stagnated. Bead formation in CNTs may take place during low-temperature CVD process [59]. CNTs with beads can be useful as fillers to be incorporated in composites. Incorporation of them in composites is effective to improve mechanical and electrical properties of the composites since the presence of beads prevents slippage of nanotubes [31].

6.3   Synthesis of Carbon Nanotubes

The oldest method of synthesis of CNTs is electric arc discharge, which has been used in the sixties for the production of carbon fibers as well. Later on, in 1990, this very old method has been used for the production of fullerenes and subsequently for the development of MWNT and SWNT. Other popular methods of synthesis of CNTs are laser ablation, chemical vapour deposition (CVD), etc. The laser ablation process is technically similar to that of the arc discharge method; only the difference remains in purity and quality of CNT produced. The complete synthesis routes of CNTs can be classified under three main classes such as physical, chemical, and other methods as represented in Fig. 6.4. Other methods of CNT production such as electrolysis, solar techniques, hydrothermal process, etc., will also be discussed in this chapter.

6.3.1   Physical Methods

6.3.1.1   Arc Discharge

In the arc discharge method, two highly pure graphite electrodes are used. The anode material used is either a pure graphite or metal. In case the metal anode is used, metal is mixed with graphite and kept in a hole made at the centre of the anode. The electrodes are shortly contacted, and arc is hit. The synthesis has been carried out in an inert atmosphere or an environment of reactant gas under low-pressure range (30–130 or 500 torr). The distance between the two electrodes is reduced until current flows (50–150 A). A consistent gap is maintained between the cathode and anode by

Fig. 6.4 Different routes of synthesis of CNTs

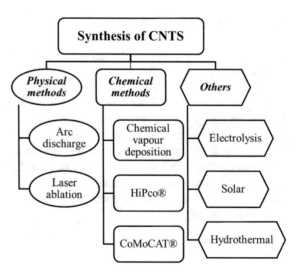

Fig. 6.5 Outline of arc
discharge method

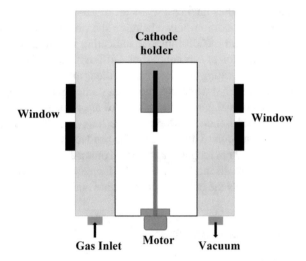

Fig. 6.5 Outline of arc discharge method

changing the anode position. A plasma is hence formed, and the stabilization of this plasma has been done by controlling distance between the cathode and anode. The reaction time varies from a few seconds to a few minutes. The arc discharge method is schematically represented in Fig. 6.5.

Different products are formed at different parts of the reactor. For examples, huge quantities and types of soot are formed on the wall of the reactor, web-like carbon morphology between the cathode and wall of the chamber, around the cathode deposit sponge collaret, at cathode end deposition of grey hard materials, etc. The product may contain amorphous and polyhedral carbons, encapsulated metallic particles, etc. [60]. The metals used for the synthesis of CNTs are Fe, Mo, Co, Y, Ni, etc., one at a time or in mixtures. In the absence of catalysts, soot and deposits are the major products in this process. Fullerene can be found in soot, whereas the MWNT and graphite carbon nanoparticles have been evidenced in the carbon deposits. In this process, nanotubes formed are of diameter in the range of 1–3 nm having an outer diameter of 2–25 nm. The tube lengths are usually of about one micron with closed tips of nanotubes. For the synthesis of SWNTs, they are devoid of any catalysts and either remain as isolated or in bundles. SWNTs are made of a diameter of about 1.1–1.4 nm with a length of a few microns.

The product quality of arc discharge method depends on several factors such as concentration and dispersion of carbon vapour in inert gas, catalyst composition, reactor temperature, presence of hydrogen and addition of promoter [61–63]. These are the major factors controlling the types of nanotubes (SWNT or MWNT), nucleation and growth mechanism of nanotubes, inner and outer diameter, length of the nanotubes. The study reveals that the use of hydrogen inside the reactor reduces the formation of carbon nanoparticles [61]. In addition, other major factors affecting the quality of nanotubes are pressure of reactant gas and type of discharge methods. Shi et al. showed that for helium gas environment at a pressure of 700 torr produces

SWNT bundles having a diameter of 20–30 nm and length of 15 μm [64]. Under the same condition only, changing the pressure to 300 torr produces a mixture of fullerene and metallofullerene indicating the influence of reactant gas pressure on the production of nanotubes [64]. Different ways of arc discharge such as intermittent or plasma rotating arc discharge methods are used for the development of nanotubes. Intermittent arc discharge produces carbon nano-onions with nanotubes [65]. The main technical part in the process of arc discharge is that it uses pulse duration for several milliseconds, which is quite long compared to the microsecond pulse duration for the pulsed arc process. The products of this process may be straight MWNT of 100–500 nm long and aggregated onion-like carbon nanoparticles. For the plasma rotating arc discharge method, the increase in rotation speed improves the yield of nanotubes as the collector comes more and more close to plasma [66].

6.3.1.2 Laser Ablation

SWNT has been synthesized by Guo et al. by utilizing the pulse laser ablation method [67]. In 1996, a successful approach has been taken by Smalley for the mass production of SWNT by the laser ablation method [68]. Nanotubes obtained by the laser ablation process are of the high purity of about 90%, and their structures are found to be better graphitized as compared to those obtained by arc discharge method. However, the major disadvantage of this method remains a small carbon deposit. Laser ablation method favours the growth of SWNTs, and MWNTs can only be formed under special reaction conditions. In a typical synthesis method, a laser beam of 532 nm has been focused on a composite target made of metal and graphite. The target has been placed in a high-temperature furnace at 1200 °C. The laser beam scans across the surface of the target material to control the smooth and uniform face of vaporization. The soot produced has been swept by argon flow from the high-temperature zone of the furnace to the copper collector. The target material has been formed by uniform mixing of metal and graphite in the form of a rod in three steps: first, a paste is formed by mixing metal, graphite, and carbon cement at room temperature and subsequently introduced in the mould; second, the mould has been kept in a hydraulic press and baked at 130 °C for 5 h at a constant pressure. In the final step, the curing of rod has been performed by heating at 810 °C for 8 h under argon atmosphere following by heating at 1200 °C for another 8 h. The metal used in the process is Co, Ni, Ni/Pt, Co/Cu, etc. The synthesis of nanotubes by laser ablation method has been schematically represented in Fig. 6.6.

The nanotubes formed by laser ablation method are accompanied by amorphous carbon, fullerene, catalyst particles, and graphite. Other impurities found are silicon and hydrocarbons, which may be from unknown sources [67]. The product quality of this process depends on a number of parameters such as ratio and types of metal catalyst particles [69, 70], ambient gas type and pressure [71, 72], laser-related parameters [73], and furnace temperature [74]. The properties of the nanotubes much dependent on laser parameters such as peak power, continuous wave (cw) versus pulse, oscillation wavelength, rate of repetition. Other contributing factors are composition of

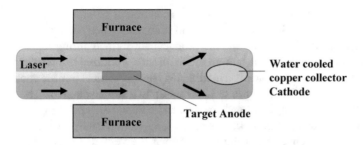

Fig. 6.6 Schematic representation of laser ablation method for synthesis of CNT

target material, pressure inside the chamber and its chemical composition, substrate types, distance between the target and substrate, etc. To date, the relation between excitation wavelength and growth mechanism of nanotubes is not clear. Possibilities are UV laser creates new nanoparticle materials suggesting a new growth mechanism. The UV laser is much better than infrared laser, which is more effective in photothermal ablation. Lebel et al. reported the synthesis of SWNTs from the UV laser ablation method. The target material used is graphite doped with Co/Ni metal catalyst and examined for the reinforcing ability in polyurethane [75].

Both arc discharge and laser ablation processes produce good quality of nanotubes; however, due to the several drawbacks as mentioned below, the industrial production of nanotubes by these processes is limited. These two processes are somehow energy-extensive methods hence uneconomical for large-scale production of nanotubes. The nanotubes prepared by those methods are impure and require extensive purification steps. These additional purification steps are difficult as well as expensive in view of large-scale production. Hence, other synthesis methods are developed for large production of high-quality pure carbon nanotubes.

6.3.2 Chemical Methods

6.3.2.1 Chemical Vapour Deposition

Chemical vapour deposition (CVD) method of synthesis of CNTs is based on the catalyst-induced decomposition of carbon monoxide or hydrocarbons with the help of supported transition metals as catalysts. The chemical reaction proceeds in a flow furnace kept at atmospheric pressure. There are mainly two types of furnace—one is vertical, and the other is horizontal. The horizontal furnace is the most popular one. In horizontal furnace configuration, the catalyst particles are kept in a quartz or ceramic boat placed inside a quartz tube. The reaction mixture is composed of a hydrocarbon source and inert gas, which are then passed over the catalyst to commence the reaction. The temperature of the furnace is kept from 500 to 1100 °C. After the desired reaction time, the furnace is cooled down to ambient temperature.

The vertical configuration is employed for the synthesis of large quantities of carbon fibers and nanotubes. Here, the catalyst and source of carbon are together introduced at the top side of the furnace and the products are formed during flight, which are subsequently collected at the bottom of the furnace. Fluidized bed reactor is another popular form, which is typically a modified version of vertical furnace configuration. Figure 6.7 shows all three types of CVD set-up in a simple diagram. The supported catalysts are placed in the furnace centre, and carbon feedstock in gaseous form is introduced via upward flow. In this process, supported catalyst experiences much longer time residence inside the furnace than the vertical furnace types [76]. In a generalized way, nanotubes grow in a CVD process by dissociation of hydrocarbons in the presence of transition metal catalysts. The carbon gas precipitated from the catalyst particles leads to the progressive formation of tubular carbon solids in sp^2 structure [77].

Chemical vapour deposition technique is of different types, such as catalytic CVD (CCVD) can be thermal or plasma-enhanced (PE) techniques, which are a common standardized method of CNTs production in recent days [78]. Other common types of CVD techniques are water-aided CVD [79, 80], oxygen-assisted CVD [81], radiofrequency CVD [82], hot-filament CVD [83], and microwave plasma CVD [84, 85].

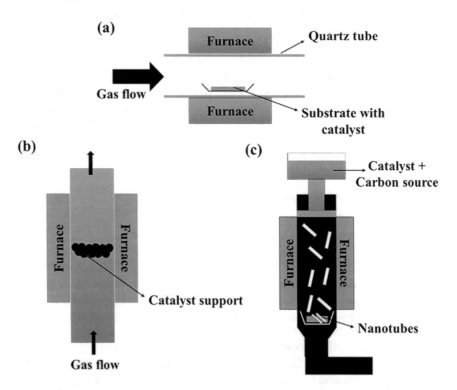

Fig. 6.7 Schematic representation of simplified CVD set-up, **a** horizontal furnace, **b** fluidized bed reactor, and **c** vertical furnace

Catalysts play a major role leading to the decomposition of the carbon sources either by plasma treatment (PECVD) or by heat effect (thermal CVD). PECVD can again be operated at different modes such as radio, direct current, diffusion, microwave named as RF-PECVD, DC-PECVD, D-PECVD, MW-PECVD, respectively. The most common catalysts are transition metals, i.e. Fe, Co, and Ni [86]. In some cases, these traditional catalysts are mixed with other metals such as Au [87]. The examples of carbon sources are various hydrocarbons and their mixtures, ethanol, etc. [88–90]. For gaseous source material, the growth of CNTs is dependent on the reactivity and gas-phase concentration of the intermediates generated by decompositions of hydrocarbons. The common substrates used for the synthesis of CNTs by CVD method are Ni, SiO_2, Cu, Si, stainless steel, graphite, tungsten foil, etc. [91, 92]. A special substrate of mesoporous silica can also be used, where it may also play the role of a template material controlling the growth of nanotubes [93].

The quality of nanotubes formed in the CVD depends on a number of parameters such as operating temperature and pressure, hydrocarbon types, volume and its concentration, metal catalyst types, nature and its pre-treatment process, nature of support materials, and the overall reaction time [94]. CCVD is considered as the most successful method of CNT production, and the quality of the CNT is much better compared to the laser ablation method. The development of pure CNT with the provisions of large-scale production is the main advantage of this technique.

6.3.2.2 High-Pressure Carbon Monoxide Process (HiPco®)

This method is developed for the synthesis of CNTs in 1999 [95]. This method stands different than others in the sense that here catalyst has been introduced in gaseous form. Both catalyst and hydrocarbon in gaseous form feed into the reactor or furnace. This method is particularly suitable for large-scale synthesis of CNTs since the synthesized nanotubes will be free of catalytic support and the reaction may proceed continuously. In general, CO is used as a source of hydrocarbon gas, which reacts with $Fe(CO)_5$ to generate SWNT. This process is popularly known as HiPco® process. SWNTs are also produced with a slightly modified HiPco® process where a mixture of benzene, ferrocene, and $Fe(C_5H_5)_2$ undergoes reaction in a flow of hydrogen to form SWNT [96]. In these two processes, catalyst nanoparticles are generated by thermal decomposition of organometallic compounds, e.g. ferrocene and $Fe(CO)_5$.

6.3.2.3 CoMoCAT® Process

In another method known as CoMoCAT process, cobalt and molybdenum are used as a catalyst with CO gas for the synthesis of CNTs [97]. In this method, SWNTs are grown by the disproportionation reaction of CO forming C and CO_2 in the presence of this special catalyst at 700–950 °C under a total pressure of 1–10 atm in the flow of pure CO. This process is successful to grow a considerable amount of SWNT in a few hours with a selectivity ratio higher than 80%. About 0.25 g of SWNTs are formed per gram of catalyst. The specially designed catalyst combination is effective to induce synergism in catalytic performance. The nanotubes formed by HiPco process produce a number of bands indicating more variation in the diameter of nanotubes compared to the CoMoCAT process. Again, the diameter distribution by the HiPco process remains significantly greater compared to that in the CoMoCAT process. Hence, this process has a great potential to be scaled up for the large-scale industrial production of SWNTs.

6.3.3 Others

6.3.3.1 Electrolysis

One of the unorthodox methods of CNT production is by electrolysis developed by Hsu et al. in the year of 1995 [98]. In this method, an alkali or alkaline earth metal is electrodeposited on a graphite cathode followed by nanotube production by the interaction of the electrodeposited metal on the cathode. The temperature for the synthesis of the nanotubes varies from the electrolyte to electrolyte; for example, for NaCl and LiCl as electrolytes, the temperature is kept just above the melting point of the salt. Once the electrolysis is completed, the carbonaceous materials are collected by the dissolution of ionic salts in water followed by filtration. During the electrolysis, cathode erodes. In this method, the main impurities are carbon nanoparticles of different structures, amorphous carbon, carbon-encapsulated metals, carbon filaments, etc. Electrolysis produces mainly MWNTs; however, SWNTs are also grown [99]. In this study, nanotubes are prepared by electrolysis using NaCl at 810 °C under argon atmosphere by electrolytic conversion of graphite to nanotubes. SWNTs obtained in this method are comparable to that synthesized from other standard methods, and the diameter remains in the range of 1.3–1.6 nm.

6.3.3.2 Solar

Another method of nanotube production is using solar energy. This process is utilized around 1996 mainly for the production of fullerenes. Laplaze et al. have demonstrated the method of production of carbon nanotubes using solar techniques [100]. In this method, carbon and catalyst, in mixed form are vaporized by incident solar energy. A high-flow vacuum pump is used to maintain the inside pressure. Argon or helium is used as the buffer gas to prevent the condensation of the mixed vapours by swiping its interior surface. The solar energy is focused on the target material with the help of a parabolic mirror placed above the chamber. The vaporization temperature is in the range of 2627–2727 °C. The vaporized mixture of carbon and catalyst along with the buffer gas is then aimed to pass between the target and the graphite tube at the backside of the reactor. SWNTs are reported to be produced by solar method in grams using 50 kW solar reactor [101]. The experimental set-up is placed in a solar furnace of 1 MW capacity. The effects of various parameters such as length of the target, location of sample collector, buffer gas have been investigated with respect to the quality of the nanotubes produced.

6.3.3.3   Hydrothermal

Hydrothermal or sonochemical process is another alternative synthesis methodology used for the development of various carbon-based nanoarchitectures such as nanorods, nanobelts, nanowires, nano-onions, MWNTs. The main advantages of this process over others are, the starting materials are stable at ambient temperatures and are abundant, and it is a low-temperature economical process; the temperature range is about 150–180 °C, and there is no need of hydrocarbons or carrier gas, etc. MWNTs are successfully prepared by the hydrothermal method from a mixture of polyethylene and water in the presence of Ni as catalyst. The temperature and pressure are kept in the range of 700–800 °C and 60–100 MPa [102]. MWNTs of closed and open tube ends of more than 100 carbons are produced by this method. One of the specific features of MWNTs produced from the hydrothermal method remains that the nanotubes are of small wall thickness and a large core diameter of about 20–800 nm. In another study, graphitic nanotubes are prepared by the use of ethylene glycol with Ni catalyst under the temperature and pressure condition of 730–800 °C and 60–100 MPa [103]. TEM analysis of the nanotubes reveals the fact that the nanotubes are long and have wide internal channels, and the catalyst particles are located at the tip positions. The wall thickness of nanotubes prepared by this method remains in the range of 7–25 nm with an outer diameter of 50–150 nm. This method is also suitable to produce nanotubes with a thin wall having the internal diameter in the range of 10–1000 nm.

The growth mechanisms of CNTs are still not known exactly. New synthesis methods are focused to be developed with the aim of high yield, purity, and fewer defects at low cost. Among the methods commonly employed for the synthesis of nanotubes, CVD is found to be best for large-scale production of MWNTs. Variants of CVD processes such as CoMoCAT® and HiPco® may be scaled up for large-scale industrial production of CNTs. Production of SWNTs via CVD method still needs attention since the yield is only in grams and they are mixed with helical CNTs.

6.4   Growth Mechanism of Carbon Nanotubes

The growth mechanisms of CNTs are still not concrete. Based on reaction parameters and post-analysis of products, several groups have proposed different mechanisms. MWNT grows without the use of any catalyst particles in arc discharge and laser ablation methods, whereas CVD method requires the use of metal particles as catalysts. However, to grow SWNTs, the need for catalysts is essential. Hence, the growth mechanism of CNTs varies from methodology to methodology depending on the number of parameters. The growth of CNTs synthesized by CVD techniques depends on catalyst particle size and concentration, growth time and temperature, pressure, gas flow rate, etc. The common mechanisms proposed for the growth of CNT are *vapour-solid-solid (VSS)* and *vapour-liquid-solid (VLS)* mechanisms.

6.4.1 Vapour-Solid-Solid (VSS) Mechanism

In this mechanism, the growth of CNTs can be explained by four steps [104]. At the beginning, nanoparticles of metal catalysts are formed on the substrate, which are deformed either by laser ablation or by heating to a thin metallic film. In the next step, decomposition of hydrocarbon gas proceeds over the catalyst surface leading to the release of hydrogen and carbon. In the third step, carbon diffuses through the metal particles and precipitated. Finally, in the last step, overcoating and deactivation of metal catalysts lead to the termination of nanotube growth. The growth of CNT is commonly described by two mechanisms involving the metal particles called as tip-growth and base-growth mechanisms. In the tip-growth mechanism, metal particles detach and move towards the head of a nanotube. Continuous growth process reveals that the nanotubes grow at the support side having the metal particles lifted at the tip side of the tubes. The movement of catalyst particles towards upside is due to the diffusion and osmotic pressure leading to the continuous deposition of carbon below the catalyst particles. In this case, the interaction between the catalyst and substrate remains weak, and hence, hydrocarbon decomposition takes place on the top of the metal particle surface followed by diffusion of carbons through the metal. This leads to precipitation of CNTs across the bottom of the metal particles and pushes off the entire metal particles from the substrate. This process continues as long as the metal surface is available for the decomposition of hydrocarbon and the concentration gradient is present with progressive growth of length of nanotubes. Once the metal particle is completely covered with carbon, the catalytic activity is over and termination takes place. The complete growth process by tip-growth mechanism is presented in Fig. 6.8.

Another common mechanism of nanotube formation is by the base-growth mechanism. In this case, the interaction between the substrate and metal particles is strong. At the very beginning, decomposition and diffusion of carbon follow the same path as that of the tip-growth mechanism. However, the strong interaction between the catalyst and substrate prevents the CNT precipitation to push successfully the catalyst particles towards upwards, and hence, the growth of nanotubes begins from the apex of the metal since there exists least interaction with the substrate and are farthest to the substrate side. Carbon crystallizes in the form of a hemispherical dome and extends further as a seamless cylinder of graphite. Subsequent decomposition of carbon takes place at the periphery of the metal, and carbon diffuses upward. In this mechanism since growth of the nanotubes proceeds from root of the catalyst particles hence, named as base-growth mechanism.

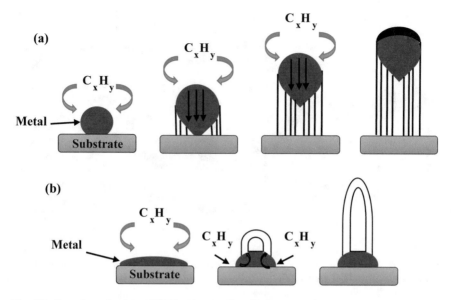

Fig. 6.8 Growth mechanism of CNT, **a** tip growth and **b** base growth

6.4.2 Vapour-Liquid-Solid (VLS) Mechanism

This mechanism consists of three stages, e.g. nucleation, precipitation and deposition. In the arc discharge method, when metal particles get evaporated with carbon, form carbon-metal alloys on the cathode side. At the start, the alloy remains in liquid state due to the high temperature at the cathode surface and reduction of melting point takes place during alloy formation. Since, the soot produced in this case is in a carbon-rich atmosphere, the alloy in liquid phase contains higher amount of carbon over its solubility limit as that in solid state. Thereafter, as the temperature of the cathode falls the alloy begins to segregate. This mechanism is acceptable to a great level; however, the state of alloy in liquid form at the process temperature of about 600–900 °C is still debatable. Again, one important point to be considered in this respect is that the decomposition of hydrocarbon over the catalyst surface is a highly exothermic reaction, which may add some value to the statement of liquid state of metals. This mechanism has been supported by scientists for the growth of SWNTs [105]. Figure 6.9 displays the growth mechanism of SWNTs at different temperatures. However, for MWNTs and large metal particles, this point is still to be explained. As the bigger particles are expected to be preferably in solid phase, a different mechanism is to be followed for MWNT and SWNT. As a concluding remark, we come to our initial statement; the growth mechanism is still a matter of debate, and the different outlooks are present based on the reaction and process parameters.

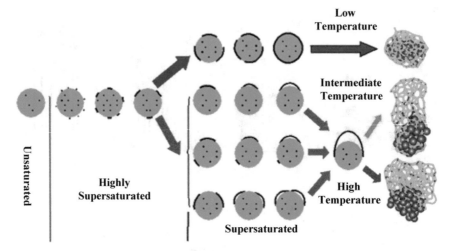

Fig. 6.9 VLS mechanism showing growth of SWNTs at different temperatures. Redrawn and reprinted with permission from [105]

6.5 Purification of Carbon Nanotubes

Depending on the method of preparation of CNTs, different approaches are taken to purify the synthesized CNTs. In addition to the large-scale synthesis hurdles, the purifications of CNTs are also a challenging aspect. In general, the purification procedures of CNTs can be broadly classified into three classes, namely gas phase, liquid phase, and intercalation methods. In the gas-phase purification process of CNTs, it has been evidenced that the process of oxidation remains easier for defective nanotubes compared to relatively perfect ones. NASA Glenn Research centre, in 2002, has introduced a new gas-phase purification process using high-temperature oxidation and repeated the treatment with nitric and hydrochloric acids for SWNTs [106]. This process is found to be suitable due to the improvement in nanotube stability. The liquid-phase purification methods mainly include filtration, dissolution, microfiltrations, centrifugal, and chromatographic method of separations. For liquid-phase purification, the CNTs undergo hydrogen peroxide-based liquid-phase oxidation to remove the amorphous carbons. Intercalation of CNTs is another method to obtain clean CNTs using $CuCl_2$ to oxidize the nanoparticles. In this technique, in the first stage, cathode is subjected to immersion in molten form of copper and potassium chloride mixture for some days under temperature. The excess chlorides are then removed by the process of ion exchange. The intercalated copper and potassium chloride metals are then heated in a reactive environment of hydrogen or helium. Afterward, the oxidation of the intercalated compound has been performed in flowing air with simultaneous heat treatment. This easy method, however, suffers from drawbacks of loss of CNTs at the oxidation stage with chances of contamination from intercalates.

The main steps followed for the purifications of CNTs are removal of big graphite particles and agglomerations by filtration, dissolution of the CNTs in proper solvents to remove the catalyst particles, removal of amorphous carbon parts from CNTs, microfiltration and chromatography for size separation [107]. SWNTs can be purified by ultrasound assistant microfiltrations and size exclusion chromatography. The catalyst particles and amorphous carbons are successfully removed producing clean nanotubes [82, 84]. Again, the above-mentioned common impurities from SWNTs are also removed by treating with nitric and hydrofluoric acids as well [83, 108]. Based on the diameters of SWNTs, density gradient ultracentrifugation has been used [109]. Size exclusion chromatography has also been utilized for the purifications of SWNTs in a recent study [110]. A combined method of ion-exchange chromatography and DNA dispersion has been used for the purification of SWNT based on the concept of chirality. Common methods of fluorination and bromination have also been utilized to purify MWNTs and SWNTs by suspending CNTs in suitable organic solvents [111, 112].

The main impurities present in CNTs are graphite, fullerenes, amorphous carbon and metal catalyst particles. The common industrial techniques used for the purification of nanotubes are oxidations, acid reflux, etc. Figure 6.10 shows the different routes of purification of CNTs. However, major difficulties in these techniques are its adverse effect on the structure of the nanotubes, insolubility of nanotubes also limiting the provisions of liquid chromatography. Removal of impurities from CNTs affects its surface area, porosity, pore-volume, introduction of additional functional groups over the CNT surface, removal of blockage at the entrance of the pore by selective decomposition of functional groups, etc. All of these have a significant effect on the ultimate properties of the nanotubes.

Oxidation of nanotubes is done with the aim to remove carbon-containing impurities [113]. However, the major drawback remains that the process of oxidation oxidizes the nanotubes as well with impurities, likely to make some damage in its structure. However, the fact is, the damage on the nanotubes is quite less as compared to impurities since the impurities have a more open structures with more defects to attack. Again, another benefit of oxidizing the impurities is that catalyst particles help the process since they are attached to these impurities. The purification of nanotubes by oxidation depends on factors like oxidizing agent, temperature, oxidation time, and metal content. Metallic catalyst particles are removed from CNTs by acid

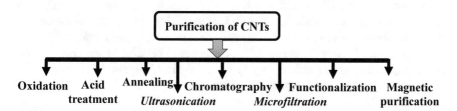

Fig. 6.10 Purification of CNTs

treatment. At first, the metal surface has been exposed by oxidative treatment or sonication, which is subsequently introduced to acidic treatment and solvation. CNTs at this point remain suspended. The best part of acidic purification by nitric acid is that the acid will only affect the metal particles and the nanotubes remain intact [114]. The mild acidic treatment has also similar utility as nitric acid reflux; however, the metal particles are needed to be exposed completely to the acids to solvate [115]. Annealing of nanotubes is carried out at high temperatures (600–1600 °C) to consume the defect and rearrange the nanotubes [116]. High temperature also helps in pyrolyse short fullerenes and graphitic carbonaceous materials. During heating under vacuum, the metal particles are melted and can also be removed easily from the nanotube as observed by Kajiura et al. [117]. Ultrasonication is one more common technique used for purification of nanotubes where particle separation is due to ultrasonic vibration. The agglomerated particles are vibrated and dispersed. The separation of the particles is very much dependant on the type of surfactant, solvent and reagents. The stability of dispersed nanotubes is affected by the nature of solvent; for e.g., the nanotubes are found to be more stable in poor solvents, when they are still attached to the metal particles of the catalyst. The magnetic impurities present in CNTs can be easily removed by using a permanent magnet. In this method, the catalytic magnetic particles are effectively removed from graphitic shells [118]. The CNTs in suspended form are mixed with ZrO_2, $CaCO_3$, etc., in an ultrasonic bath. The magnetic particles are trapped by a permanent magnet later. This process of magnetic separation is useful since no large equipment is needed and laboratory size quantity of CNTs free of magnetic impurities can be prepared very easily by this process. The particle-based impurities such as catalyst particles, carbon nanoparticles, fullerenes can be effectively removed from CNTs by microfiltration [114]. Fullerenes are removed by microfiltration approach by treating with CS_2 solution. The filter traps CS_2 insoluble. The fullerenes solvated in CS_2 is only able to pass through the filter [115]. Functionalization is another approach to prepare clean CNTs. Functionalization makes CNTs soluble by attachment of suitable chemical entities to the nanotubes. This enables to separate the CNTs easily from insoluble impurities such as metallic catalyst particles [119]. Chemistry of functionalization of CNTs has been summarized in Fig. 6.11. CNTs can also be cleaned by other functionalization where the structure of nanotubes is kept intact and they remain soluble for preferable chromatographic separation [116]. After the purification step is over, the functional groups are usually removed by heat treatment or annealing. All the purification processes employed for CNTs lead to a little bit of structural changes in CNTs. This changes the properties of CNTs as well. Hence, during purification of nanotubes, the motto should be to remove the carbonaceous impurities and metals with no or minimal change in its tube structure. Extra care is to be taken when choosing the time, temperature, and chemicals. Again, economical and large-scale techniques are of utmost importance for successful large-scale synthesis of pure CNTs.

Fig. 6.11 Chemistry of
functionalization of CNTs.
Drawn based on the concept
taken from [120]

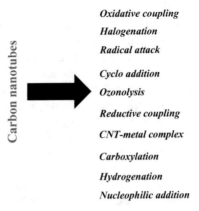

Oxidative coupling

Halogenation

Radical attack

Cyclo addition

Ozonolysis

Reductive coupling

CNT-metal complex

Carboxylation

Hydrogenation

Nucleophilic addition

6.6   Properties of Carbon Nanotubes

In recent years, CNTs have progressively attracted the attention of scientific community due to the excellent combination of electrical, mechanical, thermal, etc., properties. The alteration in tube structure enables tailoring of the nanotube properties making it a material of choice in different application domains.

6.6.1   Electrical

The electrical properties of CNTs are dependent on its structure to a great extent. Nanotubes with armchair configuration behave like metal, and when a voltage is applied at the end of these nanotubes, current flows. The other two configurations, namely zigzag and chiral, behave as semiconductors. In these two forms, the nanotubes conduct electrical current when extra energy is supplied to generate free electrons from carbon. Semiconducting CNTs are of importance in electrical devices, transistors and integrated circuit, etc. Again, the electrical resistance of CNTs can be changed remarkably when attached to other molecules. These specific electrical properties of CNTs are useful for their applications as sensors to detect carbon monoxide. Hence, the structural arrangements of carbons in CNTs open provisions to be explored as interconnects and functional materials to enhance conductivity in polymeric composites. Researchers are also aiming to fabricate electrically conducting nanowires entirely made of CNTs. However, challenges remain to overcome the resistivity at the connection of two nanotubes, where the conductivity is reduced by orders of magnitude compared to the individual one.

The electrical conductivity of nanotubes having current carrying capacity is a thousand times better than copper as reported [121]. An isolated nanotube exhibits electrical conductivity of 2×10^7 S m^{-1} and ampacity of 10^{13} A m^{-2} [122, 123]. Based on the method of producing, CNT materials can be of eight different types. They are broadly classified into three main categories—the first-class deals with vertically aligned CNTs made of CVD technique also named as CNT forest. They may be densified in the form of pillars, spun into fibers, drawn in form of a mat. The second category fabricates CNTs from the solution phase. They can be made in the form of a mat by filtration of the CNT dispersed liquid, named as buckypaper or can be fabricated in the form of single fiber as well. Porous CNT foams can also be produced from aqueous gel. The last category is obtained by direct spinning of CNT aerogels producing direct spun fibers and mats depending on the fabrication procedure [124]. Below in Fig. 6.12, variation of electrical conductivity of different categories of CNTs has been shown with respect to density.

Other than the method of fabrication, the transfer of phonon or electrical conductance in CNTs also depends on the length of the nanotubes [125]. As per the theory of percolation, electron transfer is achieved with a lesser number of nanotubes when

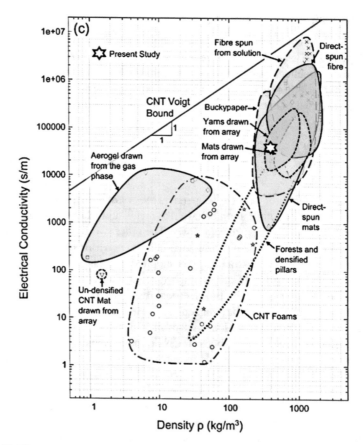

Fig. 6.12 Electrical conductivity as a function of density of different CNT materials. Reprinted with permission from [124]

long CNTs are used. The percolation threshold and length of the nanotubes are governed by the equation $N_c = 5.71/L_s^2$, where N_c and L_s represent the percolation threshold and length of nanotube [126–128]. Studies report that much better electrical conductivity has been achieved for transparent films made of long CNTs [129, 130]. Long CNTs are also beneficial to improve the properties of CNT/polymer composites due to fewer nanotube junctions [131, 132]. CNTs are of high aspect ratio, and hence, a low loading of nanotubes will be sufficient to achieve desired conductivity. The high aspect ratio of nanotubes makes the electrical conductivity of these materials comparable to that of conventional additives such as stainless steel fiber, carbon black, and carbon fibers, etc.

6.6.2   Mechanical

CNTs are stiffest and stronger fibers ever produced. The Young's modulus of CNTs can be as high as five times more than steel with a tensile strength of 50 times better than steel. These unique properties of CNTs combined with low density make them a potential candidate for structural applications. Common practice is being incorporation of these materials to the polymer matrix to fabricate high strength composites for structural and electrical applications. In CNTs, each carbon in a graphite sheet is connected with a strong chemical bond to three neighbouring carbon atoms. Hence, CNTs are of strong basal plane elastic modulus and is a fiber of high strength. The elastic modulus of SWNTs remains higher than steel making them a highly resistive material. Pressing at the tube ends although it leads to bending of the tubes; however, the nanotubes soon return back to its original state as soon as the force is removed. This particular character of CNTs opens up the possibilities to be used as tips of high-resolution scanning probe microscopy.

The mechanical properties of nanotubes can be explained from the general properties of a graphene sheet. The interaction is weak van der Waals force acting among the sheets of the graphene. The interlayer distance of a typical sheet of graphene is 0.335 nm [133]. The interlayer elastic modulus of MWNTs is comparable to that of graphite of 1.04 TPa having a shear strength of 0.48 MPa, which is of the same order as that for graphite [134]. Lei et al. developed a mathematical model to calculate the mechanical properties of SWNTs, where carbon–carbon bonds are considered as Euler beams [135]. Figure 6.13 displays the effect of radius in nm on Young's modulus of SWNTs. Model predicts that the Young modulus of SWNT gradually lowers with tube radius. The modulus of armchair configuration attains a bit higher compared to zigzag one. As the radius increases further, the two horizontal lines try to become constant and approach the value of 1.0424 TPa, i.e. Young's modulus of the sheet of graphene.

Overney et al. obtained Young's modulus of SWNTs of about 1.5 TPa approximately equal to graphite [136]. Soon after, other works on the calculation of Young's modulus of nanotubes appear close to 1 TPa irrespective of tube type and diameter [137]. These results are fitted in MD simulations by Yakobson et al. taking thickness

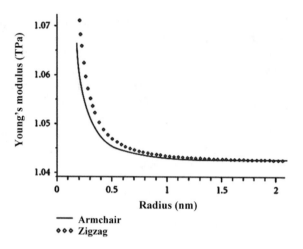

Fig. 6.13 Variation in Young's modulus with nanotube radius in nm as predicted by mathematical model. Redrawn and reprinted with permission from [135]

and modulus as fitting parameters and obtained a thickness of 0.066 nm with Young's modulus of 5.5 TPa [33]. Yokobson et al. proposed the dislocation theory explaining the mechanical relaxation of nanotubes under tensile load using molecular dynamics [138]. For tensile strain greater than 5%, dislocation consisting of pentagon and heptagon (5-7) dipole via bond rotation is observed to be the most favourable relaxation in CNTs under tensile strain. The dislocation of type 5-7-7-5 separates in two pairs of 5-7 resulting in ductile transformation under stress. Again, the dislocation of the type 5-7-7-5 may grow a crack leading to brittle fracture as well. The ductile transformation leads to the formation of tubes of smaller diameter and new chiral symmetry. The determination of mechanical properties of individual SWNTs has not been explored yet, and mechanics related to CNT is to be further investigated in detail [139, 140] In addition, the investigation on functionalization of nanotubes and associated interface mechanics at the nanoscale are still very challenging [141, 142].

6.6.3 Thermal

CNTs are good thermal conductors along the tube however act as an insulator lateral to the tube axis. The conductivity measurement of SWNT at room temperature shows a value of 3500 W m^{-1} K^{-1} compared to thermally conductive cooper of 385 W m^{-1} K^{-1} indicating its huge potential [143]. Study of thermal conductivity of CNTs confirms that the value goes to about 3000 W/mK for MWNTs [144]. The direct measurement of thermal conductivity of single nanotubes is difficult due to the technological constraints of testing in nanoscale [145]. This is the reason why thermal conductivity measurements of CNTs are mostly based on simulations and indirect experimental measurement procedures [146, 147]. The thermal transport in CNT is assumed to take place by phonon conduction process. The phonon transport in CNTs is governed by different parameters, e.g. boundary for surface scattering, mean free path length for phonons, number of phonon active modes, etc. [148, 149]. The thermal conductivity of nanotubes depends on a number of parameters such as atomic arrangements to form the tubes, diameter, and length, structural defects and morphology, and the presence of impurities [150, 151]. Again, the presence of crystallographic defects strongly affects the thermal properties of the nanotubes. Defects increase the scattering of phonons increasing the rate of phonon relaxation; hence, a reduction in thermal conductivity is expected. CNTs can be superconductive below 20 K due to strong C–C bonds of graphene. This strong C–C bond provides exceptionally high strength and stiffness in nanotubes. The aggregation tendency of SWNT compels to measure the conductivity of single nanotubes in the form of mats [152]. Hone et al. determined the thermal conductivity of SWNT at room temperature for highly pure mats composed of nanotubes bundles. The thermal conductivities, in this case, vary in the range from 1750 to 5800 W/mK [153]. However, it is observed that the contact resistance between the nanotubes lowers the thermal conductivity of SWNT by two orders of magnitude as compared to that measured for densely packed mat of SWNTs [154]. The presence of junctions in nanotubes affects the thermal conductivity to a great extent due to the increase in thermal resistance locally [155]. This effect, in particular, reduces the thermal conductivity of junction CNTs since lattice defects are present in the form of non-hexagonal rings of carbons at the junction of the nanotubes. Results reveal that the thermal conductivity of X-junction nanotubes decreases by 20–80% compared to that of straight ones [156]. The effect of tube length on thermal conductivity of CNTs is shown in Fig. 6.14. The plots are obtained by taking the model length of (3,3) CNTs at 200 K. The high thermal conductivity of CNTs makes them a material of choice for sensing, electronic and actuating devices. The incorporation of nanotubes is expected to improve the thermo-mechanical properties of the composites when introduced as a functional filler in the polymer matrix.

Fig. 6.14 Thermal conductivity as a function of tube length. Reprinted with permission from [156]; data used in this figure was taken from [157]

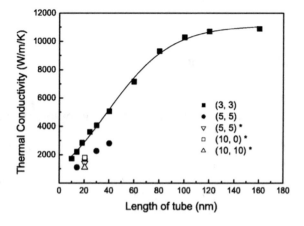

6.7   Characteristics of Carbon Nanotubes

Below we are mentioning some structural characteristic of MWNTs of interest

- Branching, if any
- Number of walls or tubes
- Concentric cylinders
- Open or capped nanotube ends
- Width or diameter of inner walls
- Ring size and connectivity
- Morphology of tubes, i.e. straight, bent, or buckled
- Molecular weight
- Catalyst particle size
- Structural defects present.

In addition to the above-mentioned structural parameters of CNTs, some important characterization of CNTs which are of general interest is as follows. These characterizations of CNTs are in practice as can be evidenced from the informal survey of vendor Webpage.

- Colour
- Purity
- Catalyst types and content
- Density
- Ash content
- % of metal oxide
- Amorphous carbon content
- Metallic impurities
- Diameter and length
- Surface area
- Raman D/G ratio indicating defects.

6.8   Application of Carbon Nanotubes

CNTs are one of the most rapid-growing nanometric materials of the modern age. The structural arrangement and hence the unique properties of CNTs make them suitable for a number of applications. Many studies are focused on the diverse application of CNTs starting from material science, biomedical, electronic, energy storage, and even CNTs as fillers as well [158]. CNTs are also useful for its high conductivity, adsorption properties, which makes them useful for applications like supercapacitors, fuel cells, energy devices, high strength composites, etc. [159–166]. CNTs because of its adsorption properties are explored in waste water treatment [167]. Below we have included a brief overview of various fields of applications of CNTs with examples

(Fig. 6.15). Instead of all these attractive combinations of properties, the use of CNTs is somewhat restricted due to the high cost and non-renewable nature. Large-scale production of pure CNTs is a point of consideration to the researcher of this field. Efforts are taken to overcome associated challenges of CNTs and to synthesize these sensational nanomaterials at low costs.

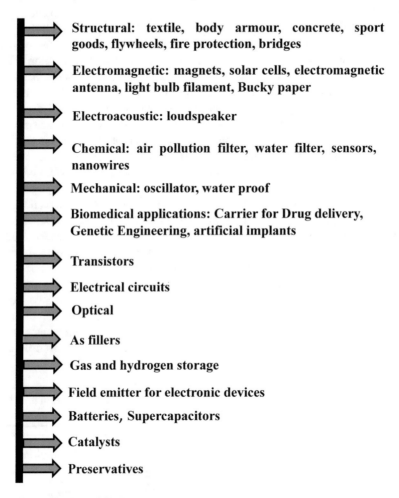

Structural: textile, body armour, concrete, sport goods, flywheels, fire protection, bridges

Electromagnetic: magnets, solar cells, electromagnetic antenna, light bulb filament, Bucky paper

Electroacoustic: loudspeaker

Chemical: air pollution filter, water filter, sensors, nanowires

Mechanical: oscillator, water proof

Biomedical applications: Carrier for Drug delivery, Genetic Engineering, artificial implants

Transistors

Electrical circuits

Optical

As fillers

Gas and hydrogen storage

Field emitter for electronic devices

Batteries, Supercapacitors

Catalysts

Preservatives

Fig. 6.15 Applications of CNTs

6.9 Challenges and Future Scope

6.9.1 Dispersion of Carbon Nanotubes

One of the major challenges associated with CNTs is they tend to agglomerate due to the weak intermolecular forces. This leads to considerable difficulties to disperse in a polymeric matrix and even in solvents. The disperse of CNTs can be improved for their possible dispersion inside a polymer matrix during the synthesis of polymer nanocomposites by functionalization of CNTs. Functionalization reduces the bundle formation and agglomeration of tubes of CNTs and ensures dispersion. Functionalization has also been an effective approach to purify CNTs as discussed earlier to enhance the degree of reactivity and homogeneous dispersion. The ollowing are the approaches taken to disperse the CNTs, either by covalent or non-covalent functionalization. The chemistry of functionalization has been included in Fig. 6.16.

Non-covalent functionalization includes hydrophobic and Pi–Pi interaction based on van der Waal's forces. This functionalization approach proves to be beneficial due to the nominal damage in CNT structure. Some common examples of non-covalent functionalization include the use of surfactants, protein interactions, and CNT wrapping [168]. In the case of covalent functionalization, the functional groups are attached to the tips and sidewalls of nanotubes. This functionalization leads to the change in nanotube structure permanently and irreversibly. Chemical entities such as carboxyl, fluorine, dichloro groups are added to the chain ends or surface of the nanotubes [120, 169, 170]. The main advantage of this functionalization is that here the nanotubes can be chemically bonded with the polymer matrix leading to the homogeneous dispersion of CNTs for the best possible utilization of unique features of CNTs. However, these types of functionalization may induce some defects in CNTs as well. Again, the dispersion of CNTs can also be improved by using

Fig. 6.16 Functionalization of CNTs. Drawn based on the concept taken from [168]

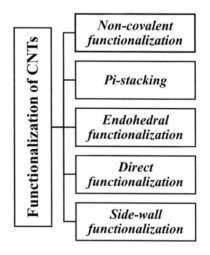

surfactants [171]. Some common surfactants are sodium dodecyl sulphate, dodecyl-benzene sodium sulphonate, and polyethylene glycol. The Pi-stacking interaction of aromatic ring is responsible to promote adsorption of the surfactants. Again, in addition to benzene rings, the naphthenic groups also ensure a good affinity between surfactant and CNT.

6.9.2   Toxicity of Carbon Nanotubes

The immense potential of CNTs in different application areas opens up their importance in a number of industries. However, certain issues such as toxicity restrict their use. CNTs adversely affect human health specifically attacking the pulmonary systems a primary route of exposure [172]. Toxicity of CNTs and their adverse effect on human health is a serious issue, which is under the lens [173–176]. Hence, major initiatives are to be taken to resolve this issue. Overexposure of nanotubes leads to oxidative stress and inflammation as well [177]. Study reveals that the length and rigidity of CNTs have a renowned effect on pro-inflammation [178]. Further research is in need to explore the influence of nanotubes on physiochemical properties [179].

6.10   Concluding Remarks

This chapter elaborates the structure, properties, synthesis, purification, application aspects of the magical carbon nanomaterial, CNT. Research in worldwide is continued on various aspects of CNTs, such as synthesis of nanotubes of a particular diameter, precise control over its chirality, controlling the number of sidewalls of CNTs specifically for MWNTs, etc. Although the progress in this field is enormous, still these factors need to be addressed to optimize the synthesis of CNTs as per requirements. Again, in view of CNT growth, the synthesis of isolated nanotubes of similar diameter and chirality over the entire length of the nanotube is a big challenge. The detailed investigation into the growth parameters of CNTs will essentially provide valuable information in this respect. Finally, the large-scale synthesis of low-cost pure nanotubes of a particular structural parameter is definitely a future challenge to the research community.

Acknowledgements The authors acknowledge the financial support provided by the Department of Science and Technology, India, (DST/TMD/MES/2K16/37(G)) for carrying out this research work. Authors are thankful to Ms Tanvi Pal for drafting a figure.

References

1. G. Collin, in CFI. *Ceramic forum international* (Göller, 2000), pp. 28
2. J. Emsley, Astron. **31**, 87 (2003)
3. K.K. Kar, *Composite Materials* (Springer, Berlin, Heidelberg, 2017)
4. H.W. Kroto, J.R. Heath, S.C. O'Brien, R.F. Curl, R.E. Smalley, Nature **318**, 162 (1985)
5. D. Bethune, C.H. Kiang, M. De Vries, G. Gorman, R. Savoy, J. Vazquez, R. Beyers. Nature **363**, 605 (1993)
6. S. Iijima, T. Ichihashi, Nature **363**, 603 (1993)
7. A. Dasgupta, L.P. Rajukumar, C. Rotella, Y. Lei, M. Terrones, Nano Today **12**, 116 (2017)
8. K.K. Kar, J.K. Pandey, S. Rana, *Handbook of Polymer Nanocomposites. Processing, Performance and Application* (Springer Berlin Heidelberg, Berlin, Heidelberg, 2015)
9. J. Tersoff, R. Ruoff, Phys. Rev. Lett. **73**, 676 (1994)
10. N. Wang, Z.-K. Tang, G.-D. Li, J.S. Chen, Nature **408**, 50 (2000)
11. A.L. Kalamkarov, A. Georgiades, S. Rokkam, V.P. Veedu, M.N. Ghasemi-Nejhad, Int. J. Solids Struct. **43**, 6832 (2006)
12. R. Haddon Science **261**, 1545 (1993)
13. N. Hamada, S-i. Sawada, A. Oshiyama Phys. Rev. Lett. **68**, 1579 (1992)
14. A. Annu, B. Bhattacharya, P.K. Singh, P.K. Shukla, H.-W. Rhee, J. Alloys Compd. **691**, 970 (2017)
15. N.M. Rodriguez, A. Chambers, R.T.K. Baker, Langmuir **11**, 3862 (1995)
16. E. Ganesh, Inter. J. Innov. Technol. Explor. Eng. **2**, 311 (2013)
17. Q. Shi, Z. Yu, Y. Liu, H. Gong, H. Yin, W. Zhang, J. Liu, Y. Peng, Optics Commun. **285**, 4542 (2012)
18. X. Wu, X.C. Zeng, Nano Lett. **9**, 250 (2008)
19. S. Iijima, Nature **354**, 56 (1991)
20. P.J. Harris, S.C. Tsang, J.B. Claridge, M.L.H. Green, Chem Soc. Faraday Trans. **90**, 2799 (1994)
21. S. Zhu, G. Xu, Nanoscale **2**, 2538 (2010)
22. P. Agnihotri, S. Basu, K.K. Kar, Carbon **49**, 3098 (2011)
23. R. Sharma, K.K. Kar, Electrochim. Acta **156**, 199 (2015)
24. A. Huczko, Appl. Phys. A **74**, 617 (2002)
25. S. Fan, M.G. Chapline, N.R. Franklin, T.W. Tombler, A.M. Cassell, H. Dai, Science **283**, 512 (1999)
26. K. Hata, D.N. Futaba, K. Mizuno, T. Namai, M. Yumura, S. Iijima, Science **306**, 1362 (2004)
27. A. Bachtold, P. Hadley, T. Nakanishi, C. Dekker, Science **294**, 1317 (2001)
28. J. Kong, N.R. Franklin, C. Zhou, M.G. Chapline, S. Peng, K. Cho, H. Dai, Science **287**, 622 (2000)
29. J. Misewich, R. Martel, P. Avouris, J.C. Tsang, S. Heinze, J. Tersoff, Science **300**, 783 (2003)
30. S.J. Tans, A.R. Verschueren, C. Dekker, Nature **393**, 49 (1998)
31. M. Zhang, J. Li, Mater. Today **12**, (12) (2009)
32. S. Iijima, C. Brabec, A. Maiti, J. Bernholc, J. Chem. Phys. **104**, 2089 (1996)
33. B.I. Yakobson, C. Brabec, J. Bernholc, Phys. Rev. Lett. **76**, 2511 (1996)
34. M. Zhang, K.R. Atkinson, R.H. Baughman, Science **306**, 1358 (2004)
35. M. Zhang, S. Fang, A.A. Zakhidov, S.B. Lee, A.E. Aliev, C.D. Williams, K.R. Atkinson, R.H. Baughman, Science **309**, 1215 (2005)
36. L. Chernozatonskii, Phys. Lett. A **172**, 173 (1992)
37. G.E. Scuseria, Chem. Phys. Lett. **195**, 534 (1992)
38. V. Meunier, M.B. Nardelli, J. Bernholc, T. Zacharia, J.-C. Charlier, Appl. Phys. Lett. **81**, 5234 (2002)
39. G. Treboux, P. Lapstun, K. Silverbrook, Chem. Phys. Lett. **306**, 402 (1999)
40. A.N. Andriotis, M. Menon, D. Srivastava, L. Chernozatonskii, Appl. Phys. Lett. **79**, 266 (2001)

41. M. Menon, D. Srivastava, Phys. Rev. Lett. **79**, 4453 (1997)
42. V.N. Popov, Mater. Sci. Eng.: R: Reports **43**, 61 (2004)
43. D. Zhou, S. Seraphin, Chem. Phys. Lett. **238**, 286 (1995)
44. C. Luo, L. Liu, K. Jiang, L. Zhang, Q. Li, S. Fan, Carbon **46**, 440 (2008)
45. B. Satishkumar, P.J. Thomas, A. Govindaraj, C.N.R. Rao, Appl. Phys. Lett. **77**, 2530 (2000)
46. L.F. Su, J.N. Wang, F. Yu, Z.M. Sheng, Chem. Vap. Deposition **11**, 351 (2005)
47. D. Wei, Y. Liu, L. Cao, L. Fu, X. Li, Y. Wang, G. Yu, D. Zhu, Nano Lett. **6**, 186 (2006)
48. A. Ural, Y. Li, H. Dai, Appl. Phys. Lett. **81**, 3464 (2002)
49. S. Han, X. Liu, C. Zhou, J. Am. Chem. Soc. **127**, 5294 (2005)
50. S. Huang, X. Cai, J. Liu, J. Am. Chem. Soc. **125**, 5636 (2003)
51. J.F. AuBuchon, L.-H. Chen, A.I. Gapin, D.-W. Kim, C. Daraio, S. Jin, Nano Lett. **4**, 1781 (2004)
52. R. Gao, Z.L. Wang, S. Fan, J. Phys. Chem. B **104**, 1227 (2000)
53. W. Davis, R. Slawson, G. Rigby, Nature **171**, 756 (1953)
54. S. Amelinckx, X. Zhang, D. Bernaerts, X.F. Zhang, V. Ivanov, J.B. Nagy, Science **265**, 635 (1994)
55. M. Zhang, Y. Nakayama, L. Pan, Japn. J. Appl. Phys. **39**, L1242 (2000)
56. W. Wang, K. Yang, J. Gaillard, P.R. Bandaru, A.M. Rao, Adv. Mater. **20**, 179 (2008)
57. P. Bandaru, C. Daraio, K. Yang, A.M. Rao, J. Appl. Phys. **101**, 094307 (2007)
58. W.A. De Heer, P. Poncharal, C. Berger, J. Gezo, Z. Song, J. Bettini, D. Ugarte, Science **307**, 907 (2005)
59. Y. Nakayama, M. Zhang, Japn. J. Appl. Phys. **40**, L492 (2001)
60. C. Journet, W. Maser, P. Bernier, et al Nature **388**, 756 (1997)
61. L.P. Biró, C.A. Bernardo, G. Tibbetts, P. Lambin, *Carbon Filaments and Nanotubes: Common Origins, Differing Applications?* (Springer Science & Business Media, 2012)
62. H. Qiu, Z. Shi, L. Guan, L. You, M. Gao, S. Zhang, J. Qiu, Z. Gu, Carbon **44**, 516 (2006)
63. Y.-H. Wang, S.-C. Chiu, K.-M. Lin, Y.-Y. Li, Carbon **42**, 2535 (2004)
64. Z. Shi, Y. Lian, F.H. Liao, X. Zhou, Z. Gu, Y. Zhang, S. Iijima, H. Li, K.T. Yue, S-L. Zhang, J. Phys. Chem. Solids **61**, 1031 (2000)
65. K. Imasaka, Y. Kanatake, Y. Ohshiro, J. Suehiro, M. Hara, Thin Solid Films **506**, 250 (2006)
66. S.J. Lee, H.K. Baik, J-E. Yoo, J.H. Han, Diamond Relat. Mater. **11**, 914 (2002)
67. T. Guo, P. Nikolaev, A.G. Rinzler, D. Tomdnek, D.T. Colbert, R.E. Smalley, J. Phys. Chem. **99**, 10694 (1995)
68. J.H. Hafner, M.J. Bronikowski, B.R. Azamian, P. Nikolaev, A.G. Rinzler, D.T. Colbert, K.A. Smith, R.E. Smalley, Chem. Phys. Lett. **296**, 195 (1998)
69. H. Kataura, Y. Kumazawa, Y. Maniwa, Y. Ohtsuka, R. Sen, S. Suzuki, Y. Achiba, Carbon **38**, 1691 (2000)
70. M. Zhang, M. Yudasaka, S. Iijima, Chem. Phys. Lett. **336**, 196 (2001)
71. D. Nishide, H. Kataura, S. Suzuki, K. Tsukagoshi, Y. Aoyagi, Y. Achiba, Chem. Phys. Lett. **372**, 45 (2003)
72. M. Yudasaka, T. Komatsu, T. Ichihashi, Y. Achiba, S. Iijima, J. Phys. Chem. B **102**, 4892 (1998)
73. H. Zhang, Y. Ding, C. Wu, Y. Chen, Y. Zhu, Y. He, S. Zhong, Physica B: Conden. Matt. **325**, 224 (2003)
74. N. Braidy, M. El Khakani, G. Botton, Carbon **40**, 2835 (2002)
75. L.L. Lebel, B. Aissa, M.A. El Khakani, D. Therriault, Compos. Sci. Technol. **70**, 518 (2010)
76. K.B. Teo, C. Singh, M. Chhowalla, Encyclopedia of nanoscience and nanotechnology **10**, 1 (2003)
77. A. Szabó, C. Perri, A. Csató, G. Giordano, D. Vuono, J.B. Nagy, Materials **3**, 3092 (2010)
78. S.A. Steiner III, T.F. Baumann, B.C. Bayer, R. Blume, M.A. Worsley, W.J. MoberlyChan, E.L. Shaw, R. Schlög, A.J. Hart, S. Hofmann, B.L. Wardle, J. Am. Chem. Soc. **131**, 12144 (2009)
79. R. Smajda, J. Andresen, M. Duchamp, R. Meunier, S. Casimirius, K. Hernádi, L. Forró, A. Magrez, Physica status solidi (b) **246**, 2457 (2009)

80. H. Tempel, R. Joshi, J.J. Schneider, Mater. Chem. Phys. **121**, 178 (2010)
81. H.-R. Byon, H.-S. Lim, H.-J. Song, H.-C. Choi, Bull. Korean Chem. Soc. **28**, 2056 (2007)
82. Y. Xu, E. Dervishi, A.R. Biris, A.S. Biris, Mater. Lett. **65**, 1878 (2011)
83. D. Varshney, B.R. Weiner, G. Morell, Carbon **48**, 3353 (2010)
84. B. Brown, C.B. Parker, B.R. Stoner, J.T. Glass, Carbon **49**, 266 (2011)
85. H.D. Kim, J.-H. Lee, W.S. Choi, J. Korean Phys. Soc. **58**, 112 (2011)
86. O. Lee, J. Jung, S. Doo, S.-S. Kim, T.-H. Noh, K.-I. Kim, Y.-S. Lim, Metals Mater. Int. **16**, 663 (2010)
87. R. Sharma, S.-W. Chee, A. Herzing, R. Miranda, P. Rez, Nano Lett. **11**, 2464 (2011)
88. M. Palizdar, R. Ahgababazadeh, A. Mirhabibi, B. Alireza, P. Rik, P. Shima, J. Nanosci. Nanotechnol. **11**, 5345 (2011)
89. T. Tomie, S. Inoue, M. Kohno, Y. Matsumura, Diamond Relat. Mater. **19**, 1401 (2010)
90. Z. Yong, L. Fang, Z. Zhi-Hua, Micron **42**, 547 (2011)
91. A. Afolabi, A. Abdulkareem, S. Mhlanga, S.E. Iyuke, J. Exp. Nanosci. **6**, 248 (2011)
92. S. Dumpala, J.B. Jasinski, G.U. Sumanasekera, M.K. Sunkara, Carbon **49**, 2725 (2011)
93. J. Zhu, M. Yudasaka, S. Iijima, Chem. Phys. Lett. **380**, 496 (2003)
94. M. Paradise, T. Goswami, Mater. Des. **28**, 1477 (2007)
95. P. Nikolaev, M.J. Bronikowski, R.K. Bradley, F. Rohmund, D.T. Colbert, K.A. Smith, R.E. Smalley, Chem. Phys. Lett. **313**, 91 (1999)
96. Z. Tang, L. Zhang, N. Wang, X.X. Zhang, G.H. Wen, G.D. Li, J.N. Wang, C.T. Chan, P. Sheng, Science **292**, 2462 (2001)
97. D. Resasco, W. Alvarez, F. Pompeo, L. Balzano, J.E. Herrera, B. Kitiyanan, A. Borgna, J. Nanopart. Res. **4**, 131 (2002)
98. W. Hsu, J. Hare, M. Terrones, H.W. Kroto, D.R.M. Walton, P.J.F. Harris, Nature **377**, 687 (1995)
99. J. Bai, A.-L. Hamon, A. Marraud, B. Jouffrey, V. Zymla, Chem. Phys. Lett. **365**, 184 (2002)
100. D. Laplaze, P. Bernier, W. Maser, G. Flamant, T. Guillard, A. Loiseau, Carbon **36**, 685 (1998)
101. D. Luxembourg, G. Flamant, D. Laplaze, Carbon **43**, 2302 (2005)
102. Y. Gogotsi, J.A. Libera, M. Yoshimura, J. Mater. Res. **15**, 2591 (2000)
103. Y. Gogotsi, N. Naguib, J. Libera, Chem. Phys. Lett. **365**, 354 (2002)
104. V. Jourdain, H. Kanzow, M. Castignolles, A. Loiseau, P. Bernier, Chem. Phys. Lett. **364**, 27 (2002)
105. F. Ding, K. Bolton, A. Rosen, J. Phys. Chem. B **108**, 17369 (2004)
106. B.K. Kaushik, M.K. Majumder, Carbon nanotube: Properties and applications, in *Carbon Nanotube Based VLSI Interconnects* (Springer, 2015)
107. S. Patole, P. Alegaonkar, H-C. Lee, J-B. Yoo, Carbon **46**, 1987 (2008)
108. J. Prasek, J. Drbohlavova, J. Chomoucka, J. Hubalek, O. Jasek, V. Adam, R. Kizek, J. Mater. Chem. **21**, 15872 (2011)
109. N. Muradov, Int. J. Hydrogen Energy **26**, 1165 (2001)
110. J. Pinilla, R. Moliner, I. Suelves, M.J. Lázaro, Y. Echegoyen, J.M. Palacios, Int. J. Hydrogen Energy **32**, 4821 (2007)
111. N. Fotopoulos, J. Xanthakis, Diamond Relat. Mater. **19**, 557 (2010)
112. S. Naha, I.K. Puri, J. Phys. D Appl. Phys. **41**, 065304 (2008)
113. G. Hajime, F. Terumi, F. Yoshiya, O. Toshiyuki, *Method of purifying single wall carbon nanotubes from metal catalyst impurities* (Honda Giken Kogyo Kabushiki Kaisha, Japan, 2002)
114. E. Borowiak-Palen, T. Pichler, X. Liu, M. Knupfer, A. Graff, O. Jost, W. Pompe, R.J. Kalenczuk, J. Fink, Chem. Phys. Lett. **363**, 567 (2002)
115. S. Bandow, A. Rao, K. Williams, A. Thess, R.E. Smalley, P.C. Eklund, J. Phys. Chem. B **101**, 8839 (1997)
116. V. Georgakilas, D. Voulgaris, E. Vazquez, M. Prato, D.M. Guldi, A. Kukovecz, H. Kuzmany, J. Am. Chem. Soc. **124**, 14318 (2002)
117. H. Kajiura, S. Tsutsui, H. Huang, Y. Murakami, Chem. Phys. Lett. **364**, 586 (2002)
118. L. Thiên-Nga, K. Hernadi, E. Ljubović, S. Garaj, L. Forró, Nano Lett. **2**, 1349 (2002)

119. S. Niyogi, H. Hu, M. Hamon, P. Bhowmik, B. Zhao, S.M. Rozenzhak, J. Chen, M.E. Itkis, M.S. Meier, R.C. Haddon, J. Am. Chem. Soc. **123**, 733 (2001)
120. H-C. Wu, X. Chang, L. Liu, F. Zhaoa, Y. Zhao, J. Mater. Chem. **20**, 1036 (2010)
121. P.G. Collins, P. Avouris, Sci. Am. **283**, 62 (2000)
122. T.W. Ebbesen, H.J. Lezec, H. Hiura, J.W. Bennett, H.F. Ghaemi, T. Thio, Nature **382**, 54 (1996)
123. B.Q. Wei, R. Vajtai, P.M. Ajayan, Appl. Phys. Lett. **79**, 1172 (2001)
124. J.C. Stallard, W. Tan, F.R. Smail, T.S. Gspann, A.M. Boies, N.A. Fleck, Extreme Mechanics Lett. **21**, 65 (2018)
125. S. Sakurai, F. Kamada, D.N. Futaba, M. Yumura, K. Hata, Nanoscale Res. Lett. **8**, 546 (2013)
126. L. Hu, D.S. Hecht, G. Gruner, Nano Lett. **4**, 2513 (2004)
127. E. Bekyarova, M.E. Itkis, N. Cabrera, B. Zhao, A. Yu, J. Gao, R.C. Haddon, J. Am. Chem. Soc. **127**, 5990 (2005)
128. H.E. Unalan, G. Fanchini, A. Kanwal, A.D. Pasquier, M. Chhowalla, Nano Lett. **6**, 677 (2006)
129. Z.R. Li, H.R. Kandel, E. Dervishi, V. Saini, Y. Xu, A.R. Biris, D. Lupu, G.J. Salamo, Langmuir **24**, 2655 (2008)
130. G. Gruner, J. Mater. Chem. **16**, 3533 (2006)
131. X. Wang, Q. Jiang, W. Xu, W. Cai, Y. Inoue, Y. Zhu, Carbon **53**, 145 (2013)
132. S. Frankland, A. Caglar, D. Brenner, M. Griebel, J. Phys. Chem. B **106**, 3046 (2002)
133. M.F. Yu, J. Eng. Mater. Technol. **126**, 271 (2004)
134. B.T. Kelly, *Physics of Graphite* (Applied Science, London, 1981)
135. X. Lei, T. Natsuki, J. Shi, Q.-Q. Ni, J. Nanomater. **2011**, 1 (2011)
136. G. Overney, W. Zhong, D. Tom´anek, Zeitschrift f¨ur Physik D **27**, 93 (1993)
137. J.P. Lu, J. Phys. Chem. Solids **58**, 1649 (1997)
138. B.I. Yakobson, J. Electrochem. Soc. **97–42**, 549 (1997)
139. A.M. Fennimore, T.D. Yuzvinsky, W.Q. Han, M.S. Fuhrer, J. Cumings, A. Zettl, Nature **424**, 408 (2004)
140. P.A. Williams, S.J. Papadakis, A.M. Patel, M.R. Falvo, S. Washburn, R. Superfine, Phys. Rev. Lett. **89**, 255502 (2002)
141. C. Velasco-Santos, A.L. Martinez-Hernandez, F.T. Fisher, R. Ruoff, V.M. Castaño, Chem. Mater. **15**, 4470 (2003)
142. R.J. Chen, H.C. Choi, S. Bangsaruntip, E. Yenilmez, X. Tang, Q. Wang, Y.-L. Chang, H. Dai, J. Am. Chem. Soc. **126**, 1563 (2004)
143. E. Pop, D. Mann, Q. Wang, K. Goodson, H. Dai, Nano Lett. **6**, 96 (2006)
144. P. Kim, L. Shi, A. Majumdar, P.L. McEuen, Phys. Rev. Lett. **87**, 215502/1–4 (2001)
145. H. Xie, A. Cai, X. Wang, Phys. Lett. A **369**, 120 (2007)
146. M.A. Stroscio, M. Dutta, D. Kahn, K.W. Kim, Superlattices Microstruct. **29**, 405 (2001)
147. M. Grujicic, G. Cao, B. Gersten, Mater. Sci. Eng., B **107**, 204 (2004)
148. J. Maultzsch, S. Reich, C. Thomsen, E. Dobardžić, I. Milošević, M. Damnjanović, Solid State Commun. **121**, 471 (2002)
149. H. Ishii, N. Kobayashi, K. Hirose, Physica E **40**, 249 (2007)
150. A. Kasuya, Y. Saito, Y. Sasaki, M. Fukushima, T. Maedaa, C. Horie, Y. Nishina, Mater. Sci. Eng. A **217/218**, 46 (1996)
151. V.N. Popov, Carbon **42**, 991 (2004)
152. J.L. Sauvajol, E. Anglaret, S. Rols, L. Alvarez, Carbon **40**, 1697 (2002)
153. J. Hone, M. Whitney, C. Piskoti, A. Zettl, Phys. Rev. B **59**, R2514 (1999)
154. J. Hone, M. Whitney, A. Zettl, Synth. Met. **103**, 2498 (1999)
155. A. Cummings, M. Osman, D. Srivastava, M. Menon, Phys. Rev. B **70**, 115405/1–6 (2004)
156. F.Y. Meng, S. Ogata, D.S. Xu, Y. Shibutani, S.Q. Shi, Phys. Rev. B **75**, 205403/1–6 (2007)
157. M.A. Osman, D. Srivastava, Nanotechnol. **12**, 21 (2001)
158. A. Helland, P. Wick, A. Koehler, K. Schmid, C. Som, Environ. Health Perspect. **115**, 1125 (2007)
159. R.H. Baughman, A.A. Zakhidov, W.A. De Heer, Science **297**, 787 (2002)

160. A. Cao, H. Zhu, X. Zhang, X. Li, D. Ruan, C. Xu, B. Wei, J. Liang, D. Wu, Chem. Phys. Lett. **342**, 510 (2001)
161. C. Jayesh, K.K. Kar, J. Mater. Chem. A **4**, 9910 (2016)
162. S.K. Singh, P. Azad, M.J. Akhtar, K.K. Kar, A.C.S. Appl, Nano Mater. **9**, 94746 (2018)
163. C. Jayesh, K.K. Kar, RSC Adv. **5**, 34335 (2015)
164. R. Sharma, K.K. Kar, Mater. Lett. **137**, 150 (2014)
165. J. Cherusseri, R. Sharma, K.K. Kar, Carbon **105**, 113 (2016)
166. R. Sharma, A.K. Yadav, V. Panwar, K.K. Kar, J. Reinforced Plast. Compos. **34**, 941 (2015)
167. S. Kar, R. Bindal, S. Prabhakar, P.K. Tewari, K. Dasgupta, D. Sathiyamoorthy, Int. J. Nuclear Desalination **3**, 143 (2008)
168. A. Hirsch, Angew. Chem. Int. Ed. **41**, 1853 (2002)
169. J.-H. Kim, B.-G. Min, Carbon lett. **11**, 298 (2010)
170. K.J. Saeed, Chem. Soc. Pak. **32**, 559 (2010)
171. H. Wang, Current Opinion in Colloid Interface Sci **14**, 364 (2009)
172. B. Satishkumar, A. Govindaraj, M. Nath, C.N.R. Rao, J. Mater. Chem. **10**, 2115 (2000)
173. Y.-M. Choi, D.-S. Lee, R. Czerw, P.-W. Chiu, N. Grobert, M. Terrones, M. Reyes-Reyes, H. Terrones, J.-C. Charlier, P.M. Ajayan, S. Roth, D.L. Carroll, Y.-W. Park, Nano Lett. **3**, 839 (2003)
174. G. Oberdörster, J. Internal, Medicine **267**, 89 (2010)
175. A.A. Shvedova, V.E. Kagan, B. Fadeel, Annu. Rev. Pharmacool. Toxicol. **50**, 63 (2010)
176. P.P. Simeonova, Nanomedicine **4**, 373 (2009)
177. A. Nemmar, H. Vanbilloen, M. Hoylaerts, P.H. Hoet, A. Verbruggen, B. Nemery, American J. Respiratory. Crit. Care Med. **164**, 1665 (2001)
178. C.A. Poland, R. Duffin, I. Kinloch, A. Maynard, W.A.H. Wallace, A. Seaton, V. Stone, S. Brown, W. MacNee, K. Donaldson, Nature Nanotechnol. **3**, 423 (2008)
179. C-W. Lam, J.T. James, R. McCluskey, S. Arepalli, R.L. Hunter, Crit. Rev. Toxicol. **36**, 189 (2006)

Chapter 7
Characteristics of Carbon Nanofibers

Raghunandan Sharma and Kamal K. Kar

Abstract Carbon nanofibers (CNFs) are fibrous nanostructures of sp^2-hybridized carbon having partial structural similarity with carbon nanotubes (CNTs). This chapter reviews the structure, synthesis techniques, properties and applications of CNFs. CNFs can be produced either by vapor phase growth using chemical vapor deposition or by carbonization of pre-synthesized polymer nanofibers. Both techniques offer an easy synthesis of CNFs as compared to that of CNTs. The synthesis techniques affect their structure as well as properties significantly. CNFs, owing to their graphitic structure, offer unique properties such as high mechanical strength, corrosion resistance, thermal and electronic conductivities. In view of their unique properties, CNFs have drawn significant attention toward their use in advanced applications such as electrochemical power generation and storage, high-strength composites, sensors, adsorbents. Promising advancements project CNFs as the futuristic materials for lithium-ion batteries, supercapacitors, fuel cells and solar cells.

7.1 Introduction

Nanostructured allotropes of carbon have played a vital role in the conceptual as well as the technological evolution of nanoscience and nanotechnology. Their exceptional physical properties have opened up diverse scientific interests toward their applications as advanced materials. Carbon nanostructures consisting of sp^2-hybridized carbon networks are one of the most studied nanomaterials during the last two decades or more. As the nanomaterials have at least one dimension in the nanoscale (few nanometers, typically 1–100 nm), they can be classified as zero-dimensional (0-D),

R. Sharma · K. K. Kar (✉)
Advanced Nanoengineering Materials Laboratory, Materials Science Programme, Indian Institute of Technology Kanpur, Kanpur, UP 208016, India
e-mail: kamalkk@iitk.ac.in

K. K. Kar
Advanced Nanoengineering Materials Laboratory, Department of Mechanical Engineering, Indian Institute of Technology Kanpur, Kanpur, UP 208016, India

© Springer Nature Switzerland AG 2020   215
K. K. Kar (ed.), *Handbook of Nanocomposite Supercapacitor Materials I*,
Springer Series in Materials Science 300,
https://doi.org/10.1007/978-3-030-43009-2_7

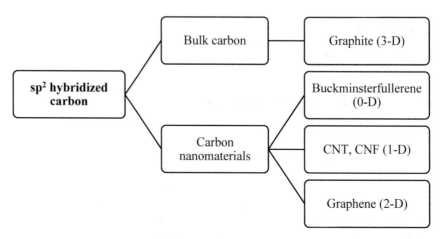

Fig. 7.1 Allotropes of sp^2-hybridized carbon

one-dimensional (1-D) and two-dimensional (2-D) materials (Fig. 7.1). The sp^2-hybridized carbon nanomaterials exist in all the above dimensions, namely 0-D (buckminsterfullerene [1]), 1-D (carbon nanotube (CNT), carbon nanofiber (CNF) [2]) and 2-D (graphene [3]), along with the three-dimensional (3-D) bulk allotrope graphite. Among these, 1-D nanomaterials (CNTs, CNFs) have attained special attention due to their unique properties and diverse applications [4, 5]. The scope of the present chapter is to review the recent advancements in synthesis, properties and applications of CNFs.

7.1.1 Structure

CNF is an allotrope of carbon having a fibrous structure with an aspect ratio of ~100 or more. Contrary to the well-known CNTs, where graphitic carbon layers are rolled up to form a cylindrical hollow structure, CNFs are solid structures. CNFs may have either amorphous or crystalline structure. Crystalline CNFs consist of graphitic layers, where carbon atoms are arranged in a honeycomb lattice as shown in Fig. 7.2a. The graphitic layers are stacked with an interlayer separation of 0.34 nm [6].

Based on the arrangement of the graphitic layers, CNFs can be categorized as platelet, herringbone/fishbone and tubular types, where the graphitic layers are arranged parallel, perpendicular and at an angle to the fiber axis, respectively [8]. The herringbone/fishbone structure can be explained by a cup-stacked model (Fig. 7.2b) [7]. Cones of graphitic carbon are stacked on each other to form the fiber structure that is different than the seamless cylindrical structure of MWCNTs (Fig. 7.2c). As proposed by Yoon et al., these three structures of CNF originate from more basic structures such as carbon nanorods (CNRs) and carbon nanoplates (CNPs), which are more basic structures consisting of a few graphitic layers [9, 10]. While CNRs,

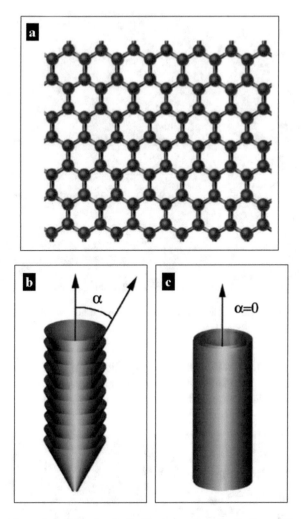

Fig. 7.2 **a** Structure of graphitic layer of carbon, **b** arrangement of graphitic layers in her-ringbone/fishbone CNF and **c** in multiwalled CNT (MWCNT). Reprinted with permission from [7]

having a diameter or ~2.5 nm and a length of 15–100 nm, consist of multilayered graphene (8–10 layers), CNPs are formed by the association of a few CNRs. The final structure and the diameter of CNF are governed by the varying arrangements of the CNRs and CNPs. Figure 7.3 shows the high-resolution transmission electron microscope (HRTEM) images of the platelet and herringbone/fishbone-structured CNFs.

More recent studies have revealed that vapor-grown CNFs (VGCNFs) have a two-layered structure. The inner layer consists of an ordered cone-helix structure, whereas the outer layer is composed of disordered MWCNT like graphitic layers [12,

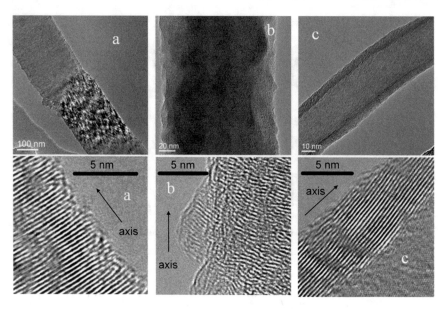

Fig. 7.3 HRTEM images of **a** platelet, **b** herringbone and **c** hollow tubular structure of CNF. Reprinted with permission from [11]

13]. As shown in Fig. 7.4a, the defective graphic sheets of the outer layer remain parallel to the fiber axis and at an angle to the inner layers. The cone-helix structure consists of a graphitic sheet rolled in a form of cone in such a way that it overlaps

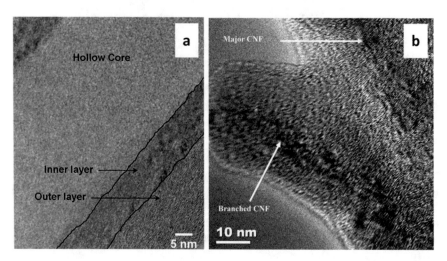

Fig. 7.4 **a** HRTEM image of two-layered structure of VGCNFs showing inner layer, outer layer and hollow core. Reprinted with permission from [15], **b** HRTEM image of CNFs showing platelet arrangement in a branched structure. Reprinted with permission from [14]

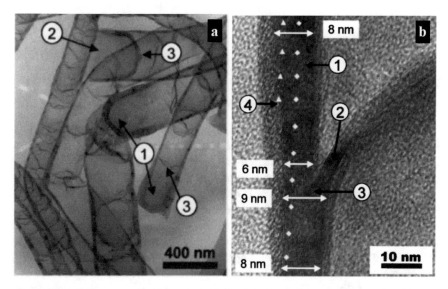

Fig. 7.5 HRTEM image of **a** bamboo-shaped CNFs synthesized by CVD showing segmented structure and **b** high-resolution image of bamboo-shaped carbon nanostructures showing the joint between the outer wall and the bamboo compartment. Reprinted with permission from [16]

itself partially to form a helix. The high-temperature heat treatments can induce structural transformations in the inner layers changing from cone-helix to stacked-cone structure. Further, structural variations such as branched CNFs by the formation of a Y-junction have been reported frequently (Fig. 7.4b). Zhao et al. have reported the formation of branched CNFs through treelike growth due to the fragmentation of catalyst particles during CVD growth [14].

Another structural variation found in fibrous carbon nanostructures is the segmented structure, termed as bamboo-CNT, CNF or filament. The structure consists of a combination of CNT-like seamless graphitic outer and capped CNT-like inner compartment structures. Figure 7.5 shows HRTEM of a wall–compartment joint with numbers representing (1) the joint between the wall and the bamboo compartment, (2) graphite sheets, (3) the defect-free joint between the compartment and the wall and (4) the disappearance of graphite sheets at outer side [16]. The compartment–wall joint consists of defect-free graphitic carbon sheets.

7.2 Synthesis

CNFs are known for long as a byproduct of the catalytic conversion of hydrocarbons. Vapor growth, electrospinning, pyrolysis of biomaterials, etc., are the frequently used methods to synthesize CNFs. Principally, the methods are based either on a direct growth of CNFs from elemental carbon or on a two-step process, where pre-synthesized nanofibers are carbonized to obtain CNFs.

7.2.1 Chemical Vapor Deposition (CVD)

CVD is a versatile technique to synthesize carbon nanomaterials including CNFs, CNTs and graphene. The process involves the decomposition of a carbon precursor to produce elemental carbon, which, in the presence of catalysts, forms the desired carbon allotrope. A schematic representation of the thermal CVD setup is shown in Fig. 7.6, where carbon precursor is decomposed by thermal energy to form various carbon nanomaterials. Generally, a catalyst-coated substrate is kept in a high-temperature furnace, where an inert atmosphere is created to avoid the oxidation or combustion of the carbon precursor. The presence of suitable catalysts is essential to synthesize the desired carbon allotrope. For example, when no catalyst is present, amorphous carbon becomes the dominating product, whereas, in the presence of a transition metal (Cr, Fe, Co, Ni or their alloy), 1-D and 2-D carbon nanomaterials are deposited on the catalyst-coated substrate. Moreover, the fractional amounts of various carbon allotropes obtained during the CVD process are governed by the parameters such as reaction temperature, composition, size and shape of the catalyst particles, pressure, nature, partial pressure and feed rate of the carbon precursor. To obtain the desired allotrope preferably, the parameters must be optimized accordingly. Larger diameter catalyst particles, lower temperature, high precursor feeding rates, etc., are some of the parameters to obtain CNFs over MWCNTs. For the growth of CNTs and CNFs, catalysts in the form of nanoparticles are used, where each nanoparticle acts as the nucleation center for CNT/CNF growth. With increasing growth time, the length and thickness of the CNT/CNF increase and slowly reach toward saturation due to the decreasing activity of the catalyst particles combined with the diffusion limitations [17–19]. This prevents the use of CVD as a continuous synthesis process, since with increasing reaction time, no more nucleation can take place due to the absence of active catalyst particles.

This problem can be overcome by using another form of the thermal CVD, where catalyst particles can be synthesized in situ by decomposing a catalyst precursor along with the carbon precursor. In this case, the catalyst particles are formed continuously,

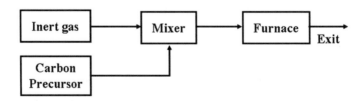

Fig. 7.6 Schematic representation of the setup for vapor growth of CNF

and the carbon nanomaterials are grown on these 'floating' particles. The deposits are collected either by the reactor walls or by a cooled collector. The schematic representation of the setup is shown in Fig. 7.7. Contrary to the substrate deposition method, the 'floating catalyst' method is more suitable for large-scale production. To synthesize CNFs, carbon precursors such as acetylene, benzene and catalyst precursors such as ferrocene, cobaltocene, along with a small amount of sulfur promoter, are used frequently [20, 21].

Plasma-enhanced CVD (PECVD) is the other form of CVD used most commonly to synthesize CNFs and particularly, the vertically aligned CNFs (VACNFs) [22–24]. Being a variation of CVD, the PECVD process includes the decomposition of a carbon precursor in the presence of a catalyst (transition metal nanoparticles)-coated substrate. Unlike thermal CVD, the carbon precursor is decomposed to form a plasma by applying an AC/DC electric discharge between two electrodes (substrate being

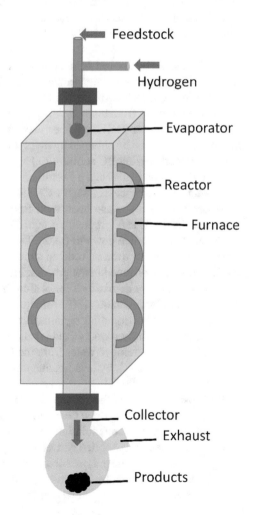

Fig. 7.7 Schematic representation of floating catalyst method. Adopted from [20]

one electrode) with a vacuum level of $\sim 1 \times 10^{-5}$ torr. The plasma is formed between the substrate and another electrode, where the carbon precursor vapor is filled. The purpose of using plasma is to provide the required activation energy to decompose the carbon precursors and to form the new carbon nanostructure at comparatively low temperatures. This makes it possible to synthesized CNFs at room temperature [25–27].

7.2.2 Pyrolysis of Polymer Nanofibers

Pyrolysis of polymer nanofibers (PNFs) is another route to synthesize CNFs. In the carbonization process, the PNFs are heat-treated in an inert atmosphere to form CNFs by converting the polymer into residual carbon while retaining their fibrous shape intact. Numerous methods have been reported to derive PNFs with desired diameter, length and orientations. The PNFs can be derived from biomaterials, electrospinning, ice-segregation-induced self-assembly (ISISA) or template method.

The PNFs can be derived from natural materials such as plants, silk larvae, bacteria. Cellulose and lignin, the primary constituents of plant cells, can be extracted and used to form PNFs. Apart from this, natural polymers in the form of fiber can be obtained directly from biomaterials. For example, nanofibrous biomaterials such as bacterial cellulose (BC) have been used extensively to synthesize CNFs. A 3-D network of CNFs can be obtained by pyrolytic carbonization of BC [28]. CNFs synthesized by carbonizing BC at 600–1200 °C have an internal structure consisting of turbostratic carbon planes and a fiber diameter of 10–20 nm [29].

Among others, processes such as electrospinning and ice-segregation-induced self-assembly (ISISA) have been used frequently to synthesize PNFs from bulk polymers. In the ISISA process, an aqueous solution or suspension of the polymer/monomer is cooled down to freeze. During freezing, the ice crystal phase separates the polymer to form an ordered structure. Under controlled conditions, polymeric fibers can be formed in the process. Finally, the frozen water is removed from the ordered polymer matrix by sublimation. Spender et al. have synthesized CNFs by using a similar technique [30]. An aqueous solution of lignin (0.1–0.3% w/w) was cooled by using liquid nitrogen to form an ice-templated lignin nanofiber network, which was further lyophilized (vacuum dried at low temperature) at -15 °C in a freeze dryer to remove water from the fibrous lignin. Finally, the lignin fibers were converted to CNFs by heat treatment in air at 200 °C for 18 h followed by the carbonization at ~ 1000 °C.

Electrospinning is another technique to produce micro and nanofibers. The technique uses electric field to draw fibers from a polymer solution. In this process, PNFs are prepared from a viscous solution/precursor of the polymer. As shown in Fig. 7.8, a DC or AC potential is applied between the syringe and the grounded target, which charges the surface of the polymer solution droplet at the syringe tip (spinneret). The charges on the drop repulse each other, and the repulsion force works against the surface tension. Above the critical condition, the repulsive force dominates, which

Fig. 7.8 Schematic
representation of
electrospinning setup.
Reprinted with permission
from [31]

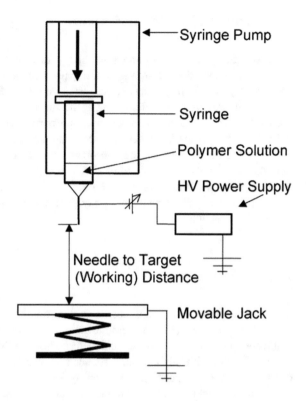

leads to a jet formation from the drop. The jet is accelerated toward the grounded plat, and the diameter is reduced due to acceleration. Hence, the jet diameter can be controlled by adjusting the applied voltage. The jet is converted into a solid polymer fiber with the solvent evaporation in the process. The electrospun polymer fibers are further carbonized at high temperatures >1000 °C) to attain the CNFs [32].

Another process to synthesize a well-defined array of CNFs is the template method. In this process, a template is used to synthesize polymer nanofibers/3-D network. Polymer precursor or liquid solution is infiltrated in a template having pores or 3-D network of desired size and shape. Finally, the polymer-infiltrated template is annealed at high temperatures, where the carbonization of the polymer precursor takes place. The template is dissolved in a suitable solvent to obtain the CNF network. For example, Wang et al. have synthesized mesoporous CNFs by using triblock copolymers F127 with phenolic resol infiltrated within anodic alumina templates [33]. The synthesis of CNFs by carbonization of pre-synthesized PNFs offers a certain advantage over the vapor growth technique. Using electrospinning, it is possible to synthesize relatively long CNFs. The advantages of the template method include the ability to control the orientation, density, length and diameter of the CNFs more accurately.

7.3 Structural Characterizations

The properties of the materials are governed by their bulk and surface structures. The bulk structure includes the atomic bonding and arrangement, presence of impurities and defects, etc. Similarly, the surface structure encompasses the surface morphology, atomic arrangement and bonding at the surface, presence of functional groups and adatoms, etc. Assessment of structure–property correlations is essential to attain the desired properties by structural modifications. Structural characterizations are performed to derive such structure–property correlations. Being potentially important, the surface and bulk structures of CNFs have been studied significantly. Key techniques for the structural characterizations of CNFs include microscopy, specific surface area (SSA) analysis, X-ray diffraction (XRD), Raman spectroscopy, electrochemical analysis, etc.

7.3.1 Microscopy

Microscopy techniques such as scanning electron microscopy (SEM), transmission electron microscopy (TEM) and atomic force microscopy (AFM) are the most important techniques used for structural characterization of CNFs. While the surface morphology, along with other parameters such as length, diameter, presence of other carbon allotropes, impurities, is studied by SEM (Fig. 7.9a), high-resolution TEM reveals the arrangement of graphitic layers and hence the internal structure of CNFs (Fig. 7.9b). Moreover, the selected area diffraction (SAD) pattern reveals valuable information regarding the crystal structure and the order of crystallinity.

Study of TEM images not only provides information about the structure of CNFs, but also helps to realize the growth mechanism. Similarly, AFM can be used to study the surface topography of CNFs with very high resolution. However, as the average diameter of CNFs may reach few tens to hundreds of nanometers, AFM may not be suitable for very rough samples.

Apart from the study of the surface morphology, AFM is used to measure the mechanical properties such as strength and modulus of CNFs. AFM technology can measure extremely small forces required to bend or stretch a single CNF. Figure 7.10 shows such a measurement of bending modulus of an individual CNF. The CNF mounted on a grid is scanned by the AFM tip to obtain the AFM topographic image. Force–displacement measurements are performed by placing the AFM tip in the middle of the fiber as shown in the inset of Fig. 7.10a. The force–displacement curve is obtained by applying a load (F) on the fiber and measuring the midspan displacement (δ) (Fig. 7.10b). The bending modulus is calculated for fixed end beams given by (7.1) [15].

$$F = \frac{192EI}{L^3}\delta \qquad (7.1)$$

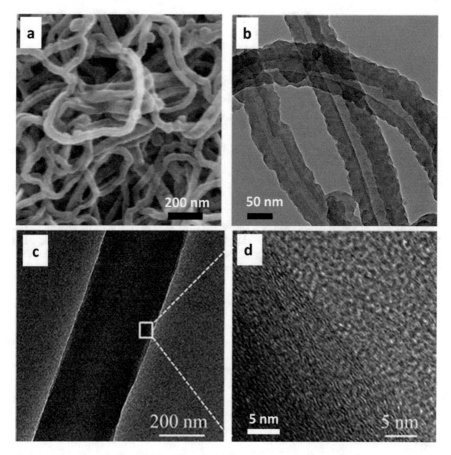

Fig. 7.9 SEM **a** and TEM **b–d** images of CNFs. The lattice fringes of **d** correspond to the graphitic fiber and are characteristic of high crystallinity of the CNFs. Reprinted with permission from [34] **a, b** and [35] **c, d**

where E is the bending modulus, I is the second area moment of the cross-section and L is the beam span.

7.3.2 X-Ray Diffraction

A great deal of information regarding the crystal structure, crystallographic orientation, degree of crystallinity, presence of impurities and defects, etc. of CNFs can be inferred by studying their X-ray diffraction (XRD) pattern. The broadness and intensity of a diffraction peak carries the information of average grain size, degree of crystallinity and orientation of planes. Typical XRD patterns of CNFs formed by ethanol flame synthesis are shown in Fig. 7.11. The XRD patterns exhibit two

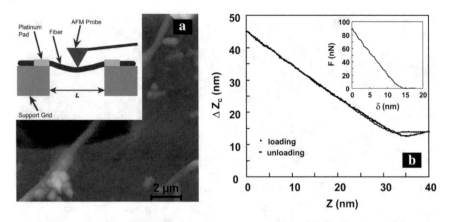

Fig. 7.10 **a** AFM topographic image of a suspended CNF fixed at both ends by platinum pads. Inset shows a schematic of the 3-point bending by using AFM tip and **b** deflection of the AFM cantilever (ΔZ_c) versus vertical coordinate (Z) of piezo during the loading–unloading cycle of the bending test. Inset shows the force (F)–deflection (δ) curve. Reprinted with permission from [15]

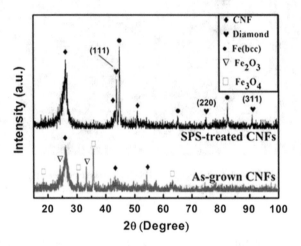

Fig. 7.11 XRD pattern of the CNFs synthesized by ethanol flame (as-grown as well as spark plasma sintered). Reprinted with permission from [37]

distinct peaks for 2θ ranging between 20° and 25° and 40°–45°, corresponding to the graphitic (002) and (101) planes, respectively. At lower carbonization temperatures below 1200 °C, the XRD peaks are generally broad due to the lower crystallinity of CNFs. With increasing carbonization temperature, the peaks become sharper due to the increased graphitic nature as well as crystallite (grain) size. For a carbonization temperature of 2200 °C, the fibers are graphitized, exhibiting a sharp peak at $2\theta = 26°$, the characteristic peak of graphite (002) plane. A spark plasma sintering (SPS) treatment has been found to decrease the XRD peak widths. The effect of synthesis technique on the structure of CNFs can be further understood by the XRD pattern of VGCNFs [36]. The VGCNFs synthesized by using CH_4 as carbon precursor in a microwave-CVD at 1050–1150 °C, exhibit graphite (002) peak at 2θ of 26°. The comparatively broad peak suggests the presence of graphitic crystallites of small size. Moreover, the absence of (101) peak suggests the orientation of fibers along the c-axis without a lateral extension.

7.3.3 Raman Spectroscopy

Raman spectroscopy is another tool to probe the structure of CNFs. The presence of graphitic structure and defects can be detected by studying the Raman spectrum. The Raman spectrum of graphite is well understood [38]. Principally it exhibits three peaks corresponding to the D, G and D* bands (Fig. 7.12). The G band represents the graphitic order as it originates from the in-plane vibrations of covalently bonded sp^2 hybridized carbon atoms. On the other hand, the D band is associated with the breathing mode of the atoms located at the edge of the basal plane (scattering of electrons from a state close to K point to another one of K' points). [39, 40]. Finally, the D* band originates from the backscattering of electron by a second phonon instead of a defect. The integrated Raman intensities I_G and I_D correspond to the number of graphitic (sp^2 bonded) and disordered carbon atoms, respectively. A high I_G/I_D ratio corresponds to higher graphitic nature. The effect of synthesis technique on the structure of CNFs can be understood again by studying their Raman spectra. Figure 7.12a shows the Raman spectrum of electrospun CNFs [41]. The varying degree of graphitization has been observed for different carbonization temperatures [36]. The samples carbonized at 2200 °C exhibit characteristic graphitic peaks. Figure 7.12b shows the Raman spectrum of CNFs synthesized by ethanol flame [37]. Peaks present at 1343

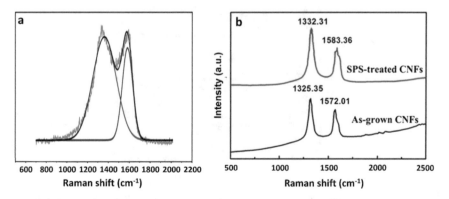

Fig. 7.12 Raman spectrum of **a** the electrospun CNFs. Reprinted with permission from [41]) and **b** the CNFs synthesized by ethanol flame (as-grown as well as spark plasma sintered. Reprinted with permission from [37]

and 1584 cm^{-1} correspond to the D and G bands, respectively. High intensity of D-band suggests the presence of large number of defects in the graphitic structure of CNF.

7.3.4 Electrochemical

Evaluation of electrochemical performance of CNFs is performed either by the performance evaluation of the device fabricated using CNFs as a component or by measuring the relevant parameters separately. In the latter choice, the required characterization techniques depend on the parameters to be evaluated. Among others, cyclic voltammetry (CV) is a versatile technique used to study the electrochemical performance. The technique uses a three-electrode setup consisting of working, counter and reference electrodes. The cyclic voltammograms are recorded by applying a cyclic voltage sweep between the working and reference electrodes while measuring the resulting current between the working and counter electrodes. A peak in the cyclic voltammogram indicates the occurrence of an electrochemical reaction [42–44]. For example, cyclic voltammograms of N-doped CNFs working electrode are shown in Fig. 7.13 [45]. The CV was performed in an oxygen saturated 0.1 KOH electrolyte and the N-doped CNFs as the active material on a glassy carbon electrode. N-doped CNFs synthesized at different carbonization temperatures are analyzed. The current peak at a potential close to -0.1 V (vs, saturated calomel reference electrode) represents the reduction of oxygen present in the electrolyte solution. The electrocatalytic activity for this oxygen reduction reaction (ORR) is of significant importance in fuel cells (FCs). The effect of carbonization temperature on the catalytic activity

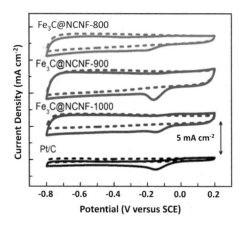

Fig. 7.13 Cyclic voltammograms of porous core–shell Fe$_3$C-embedded N-doped CNFs carbonized at different temperatures in N$_2$ (dashed line) and O$_2$ (solid lines) saturated 0.1 M KOH at a scan rate of 50 mV s^{-1}. For comparison, cyclic voltammogram of a Pt/C catalyst also has been shown. Reprinted with permission from [45]

can be observed clearly from the CV measurements shown here. With increasing carbonization temperature, the ORR peak reduces significantly, suggesting a decreased ORR activity.

7.4 Properties

The properties of CNFs depend strongly on their internal structure and hence, on the synthesis method, heat treatment, purity, etc. Presence of defects, relative orientation of graphitic plane with respect to the fiber axis, etc., affect the properties significantly. As the higher bound, with increasing graphitic nature and decreasing defects, the properties tend to reach to those of MWCNTs.

7.4.1 Specific Surface Area (SSA)

High specific area of nanomaterials is one of the characteristic properties. Unlike CNTs, CNFs have a large range of diameters ranging from a few tens to hundreds of nanometers, depending on their synthesis techniques and parameters. The SSA of CNFs depends significantly on their diameter and hence, on the synthesis process. For electrospun CNFs, the key parameters affecting the final fiber diameter include the polymer precursor concentration and the carbonization parameters such as heating rate, temperature and time. Wang et al. have reported the Brunauer–Emmett–Teller (BET) surface area of electrospun CNFs ranging between ~60 and 800 m^2/g, depending on the processing conditions [46]. BET surface area ranging between 150 and 250 m^2/g has been reported by Zhou et al. for CVD-synthesized CNFs by using various carbon precursors [47]. A similar range of SSA is reported by others as well [48, 49]. The SSA can be further increased by a factor of two or more by activating the electrospun CNFs. Kim et al. have reported the activated CNFs having a BET surface area as high as 1230 m^2/g [50]. Further improvements in the SSA can be obtained by introducing a multiscale length hierarchy. Recently, Chen et al. have synthesized hierarchical structures consisting of CNTs coated on CNFs by CVD and reported an SSA of as high as 1840 m^2/g [51].

7.4.2 Mechanical

The mechanical properties of CNFs are intermediate to those of microsized carbon fibers and CNTs owing to their highly ordered structure consisting of stacked truncated cones of graphene [52]. Theoretical calculations for the conic graphitic-layered structure have suggested that for a tilting angle of θ, the Young's modulus for a single-shell nanocone is $\cos^4 \theta$ of that of an equivalent single-walled CNT

(SWCNT) [53]. The orientation of graphitic layers in CNTs remains along the tube axis, whereas for CNFs, the graphitic layers/cones make an angle θ with the fiber axis. For axial loading, the modulus of a fiber having graphitic layers oriented at an angle θ, the modulus along the loading direction (E) is given by (7.2) [54].

$$\frac{1}{E} = \frac{1}{E_2} + \left(\frac{1}{G_{12}} + \frac{2_{12}}{E_1} - \frac{2}{E_2}\right)\cos^2 + \left(\frac{1}{E_1} + \frac{1}{E_2} - \frac{1}{G_{12}} + \frac{2_{12}}{E_1}\right)\cos^4 \quad (7.2)$$

where E_1, E_2, G_{12} and v_{12} are the longitudinal modulus (modulus along graphitic planes), transverse modulus (modulus perpendicular to graphitic planes), shear modulus and Poisson's ratio, respectively. Uchida et al. have calculated the modulus of one-layered and double-layered CNFs and observed a large variation of modulus based on the graphitic layer orientation [55]. For a single-layered structure, where the graphitic sheets make an angle of ~15° with the fiber axis, the lower bound of modulus was found to be 50 GPa. For double-layered fibers, the modulus lies between 110 and 775 GPa for the respective θ values of 10° and 0°.

Ozkan et al. have measured the mechanical strength and modulus of individual CNFs having different microstructures by using microelectromechanical system (MEMS)-based nanoscale tension measurements [56]. For VACNTs having a turbostratic microstructure, the average values of elastic modulus and strength are found to be 180 GPa and 3.35 GPa, respectively. On the other hand, VGCNFs heat-treated at 2800 °C exhibit much higher modulus values of 245 GPa and a decreased strength of 2.35 GPa. The modulus enhancement is attributed to the increased graphitic order by heat treatment, whereas the strength reduced due to the disappearance of the outermost turbostratic carbon layer. This outer layer restricts the creak initiation to increase the strength. In a similar study, Beese et al. have used MEMS technique to study the tensile behavior of the CNFs synthesized by electrospinning a dispersion of double-walled CNT (DWCNT) and PAN (10% PAN/0.12% DWCNT) in DMF [57]. Smaller diameter fibers exhibit higher modulus and strength as compared to those of larger diameter. Arshad et al. have also reported similar diameter dependence [58].

Another study performed by using a combination of AFM, TEM and focused ion beam has revealed the elastic modulus of individual CNFs to be ranging between 6 and 207 GPa [15]. The study correlates the two-layered structure of the concerned CNFs with the measured modulus values. The modulus of CNFs is governed by the relative thicknesses of the inner and outer walls. Higher modulus is obtained with a thicker outer wall and low overall thickness. This indicates that the outermost layers have less load-bearing capacity. The maximum load is carried by the outer wall layers in the proximity of the inner wall. For CNFs having diameters larger than 80 nm, the modulus was found to be ~25 GPa, independent of the diameter. Being structure sensitive, the mechanical properties of CNFs depend largely on the synthesis techniques. Electrospun CNFs have relatively low strength and modulus as compared to those of VGCNFs. By using the mechanical resonance method, Zussman et al. have reported the banding modulus of electrospun PAN-derived CNFs ranging between 53 and 75 GPa with an average modulus of 63 GPa [59].

7.4.3 Electronic

The electronic properties of CNFs largely depend on their structure. Highly graphitic VGCNFs exhibit electronic conductivity comparable to that of graphite. Less ordered structures have a large number of defects, which may scatter the charge carriers to reduce electronic conductivity. The electron transport in CNFs takes place by 2-D hopping mechanism due to their disordered graphitic structure [60]. Wang et al. have reported a strong dependence of electronic conductivity of PAN-derived CNFs on the pyrolysis temperature [61]. With an increase in the pyrolysis temperature from 600 to 1000 °C, the conductivity increases by an order of four ($\sim 10^0$ to $\sim 10^4$ S/m). A similar value of electronic conductivity (1.3×10^4 S/m) has been reported by Sharma et al. for individual, PAN-derived CNFs pyrolyzed at 900 °C [62]. The presence of graphitic layers in high temperature pyrolyzed CNFs is attributed to their good electronic conductivity, yet smaller by one order or more than that of the graphite crystal and the CNTs [63].

7.4.4 Thermal

Similar to other properties, the thermal conductivity of CNFs depends significantly on their structure and hence, on the processing parameters. Mahanta et al. have used a thermal flash technique to determine the thermal conductivity of individual VGC-NFs [64, 65]. The technique is based on the measurement of the thermal diffusivity (α), which is related to the thermal conductivity (κ) by the relation $\kappa = \alpha \rho C_p$, where ρ and C_p are the specific density and specific heat of the nanofiber, respectively. To determine α, the transient heat flow response of a nanofiber is monitored by heating it at one end while keeping the other end at a constant temperature. When the nanofiber end is contacted to a heater-thermometer (heat source with provision to measure the temperature constantly at very fast ($\sim 1,000,000$ samples/s) rates), its temperature decreases as the heat flows through the nanofiber and reaches to a steady-state. The time required to attain the steady-state of heat flow is further used to calculate the thermal diffusivity and hence, the thermal conductivity. Axial thermal conductivities of VGCNFs, heat-treated at 1100 and 3000 °C, have been found to be 1130 W/m K and 1715 W/m K, respectively, with the corresponding phonon mean free paths of ~ 128 and ~ 178 nm [65]. Low thermal conductivity of CNFs as compared to those of high purity SWCNTs is attributed to the reduced phonon mean free path due to the high phonon scattering by their discontinuous graphitic structure [66].

7.5 Applications

Owing to their excellent properties, CNFs find a significant application potential in the areas including structural composites, electronics, electrochemistry, environmental science, etc. The following sections review a few advanced applications of CNFs in brief.

7.5.1 Composites

The use of CNFs as reinforcement in composites is one of their early applications. CNFs, owing to their high strength, electronic and thermal conductivities, have been used as reinforcements for various matrix materials such as polymers [67–71], rubbers [72], metals [73, 74], ceramics [75–77]. Apart from enhancing the mechanical properties such as strength, modulus, hardness, the incorporation of CNFs affects the thermal [78] and electronic [79] conductivities as well [80]. The performance of composites depends on the degree of dispersion of CNFs in the matrix and the matrix-CNF interface bonding [81]. Recent studies in the area focus on the effects of fiber internal structure and fractional volumes on the properties such as strength, electric conductivity, thermal stability [78, 79, 82, 83] and the hierarchical structured nanocomposites [84–88]. Specific applications of these composites, such as highly stretchable foams and aerogels, have been discussed in the following sections.

7.5.2 Aerogels

Carbon-based aerogels are interesting materials owing to their specific properties such as low density, large SSA, high thermal and electronic conductivities for various applications including catalyst supports, electrodes for supercapacitors, absorbents, sensors, etc. Carbon aerogels consisting of 3-D carbon network are conventionally synthesized either by the pyrolysis of organic aerogels or by using carbon nanostructures such as CNTs, graphene and CNFs. Although the CNT/graphene aerogels exhibit superior properties over the organic-aerogel based counterparts, their complex synthesis and high cost make the large-scale production unsuitable for industrial applications. CNF-based carbon aerogels offer promising advantages in terms of cost, easy processing along with properties comparable to those of CNT/graphene aerogels. Wu et al. have reported low-density, flexible and fire-resistant carbon aerogel synthesized by the pyrolysis of BC [89]. The 3-D network of cellulose nanofibers in BC makes it suitable for CNF aerogel formation. The synthesized aerogel has been found to have excellent fire resistance, high compressibility and absorption capacity of 106–312 times of its weight.

7.5.3 Pollutant Adsorbents

Large SSA carbon nanomaterials are used as adsorbents to remove impurities and pollutants from wastewater, industrial exhausts, etc. Industrial pollution of water and air can be controlled by using such adsorbents before releasing the exhaust gases and wastewater to the external environment. Among others, activated carbon has been used extensively for removal of such pollutants. Recently, other high SSA carbon nanomaterials such as graphene, carbon nanoparticles, CNTs and CNFs have been investigated as adsorbents for removal of liquid and gaseous pollutants such as lead, arsenic, cadmium, fluoride ions, NO, SO_2 [90–92].

Synthesis of CNFs by cost-effective and scalable techniques offers a special advantage over the other carbon nanomaterials. Both electrospun, as well as, CVD-synthesized CNFs have been explored toward their adsorption capacities for pollutants such as heavy metal ions, toxic gases and organic substances [92–97]. The adsorption capacity of CNFs depends on their surface structure and presence of functional groups as well. The functional groups present of the surface may increase or decrease the adsorption performance based on factors such as their interaction with adsorbents, steric hindrance. For organic substances having nucleophile groups, the adsorption is essentially unaffected as the steric hindrance is compensated by the interaction of surface groups with nucleophile groups [98]. Hence, the presence of surface groups may enhance or suppress the adsorption capacities, depending on the nature of pollutants. For example, nitric acid-modified CNFs (presence of surface oxygen group) have been found more effective adsorbents for cadmium [92]. Hierarchal structures including micro- as well as nanoscale carbon materials, exhibit superior properties. Adsorption capacities of hierarchical web of CNFs coated on activated carbon fibers (ACFs) for various water pollutants such as lead, arsenic and fluoride ions have been found superior to that of ACFs [95–97].

Apart from pristine CNFs, their composites with other nanoparticulate adsorbents/chemicals can be used to enhance the absorption capacity. Along with their adsorption properties, the CNFs may act as a support material to provide high surface area to other adsorbents/reactants. High SSA and inert nature of CNFs propose a suitable material to them for such applications. Iron nanoparticle-coated CNFs (Fe–CNFs) offer effective adsorption capacities for volatile organic pollutants [99]. Performance of CNFs as adsorbents for pollutant removal is promising. The cost-effective synthesis of CNFs offers them a superior potential over CNTs and graphene. However, as reported by Diaz et al., CNFs exhibit lower performance compared to graphene and CNTs [100]. Based on the surface energy and enthalpy of adsorption values, the adsorption capacities of graphene, CNTs and CNFs for various organic compounds are found to be in the order of graphene > CNTs > CNFs. Again, the reusability and the lifetime remain a challenge for the large-scale applications of the carbon nanomaterial adsorbents.

7.5.4 Gas Sensors

Apart from pollutant removal, the adsorption capacity of CNFs has been explored for their possible application as chemical sensors. The sensing response depends on the quantitate changes in a measurable parameter that is affected by the presence of the material to be sensed. Ding et al. have reviewed the recent progress in the gas sensor applications of electrospun nanofibers and listed their advantages such as high sensitivity, short response time, reversibility, reproducibility, over the conventional solid flat film sensors [101]. CNFs, owing to their high SSA, may offer similar advantages. Hsieh et al. have used an aligned CNF array synthesized by template synthesis method to fabricate a sensor for *n*-hexane and shown high adsorption capacity [102]. The regular pore structure of such arrays provides high monolayer adsorption capacities. Moreover, CNFs can be used as a support to provide high SSA for other sensor materials to enhance their performance. Based on a similar concept, Jang et al. have fabricated a polypyrrole-coated CNF (CNF/polypyrrole) nanocable and evaluated the sensing capabilities for toxic gases such as NH_3 and HCl [103]. The CNF/polypyrrole sensor exhibits significant enhancement of the sensor response as compared to that of bulk polypyrrole.

Apart from gas sensors, Yarns developed from CNFs and other carbon nano-material composites may be used as strain sensors owing to their high flexibility [104–106]. Highly sensitive strain sensors may be developed using CNF-based yarns [107]. Recently, Wu et al. have developed multifunctional polydimethylsiloxane (PDMS)/CNF composites having high strain sensitivity and electronic conductivity [108].

7.5.5 Power Generation and Storage

Carbon nanomaterials have widespread applications in the fields of power generation as well as storage. Various components of modern power generation devices such as FCs and solar cells (SCs) use carbon nanomaterials. In low-temperature FCs, components such as gas diffusion layer (GDL) and catalyst layer (CL) are composed of high SSA, corrosion-resistant, electrically conductive carbon nanomaterials. Similarly, in dye-sensitized solar cells (DSSCs), carbon nanomaterials find applications both in photoanode as well as counter electrode [109–112]. Similar to other carbon nanomaterials, high SSA, electronic conductivity and chemical inertness of CNFs propose suitable electrode materials for power generation and storage applications [113–115]. Other important and more recent applications include their use as the electrode materials for supercapacitors and batteries [114, 116–123]. Moreover, a few structurally modified carbons exhibit electrocatalytic activities that can be explored to replace the expensive conventional catalysts used in FCs, DSSCs and other applications.

7.5.5.1 Supercapacitors

Supercapacitors form a new class of energy storage devices ideal for applications requiring high power output for short durations. While their power densities remain higher than those for batteries do, supercapacitors exhibit energy densities superior as compared to conventional capacitors. Based on the principle of storing charge, supercapacitors are classified as electric double-layer (EDL), pseudo- and hybrid-capacitors. While charge storage in pseudocapacitors takes place in the form of faradaic charge transfer between electrodes by redox reactions, Helmholtz double layer is formed at the electrode–electrolyte interface in EDL capacitors. The third type of supercapacitors is based on a hybrid mechanism of EDL capacitors and pseudocapacitors [124]. The performance of supercapacitors (specific capacitance (capacitance/mass), energy and power densities, lifetime, etc.) depends largely on the structure of electrodes. High SSA, electronic conductivity and electrochemical stability are some of the basic requisites for the electrode materials.

Nanostructured carbons, having graphitic layers as the basic structural units, exhibit a perfect combination of these properties and are widely acceptable electrode materials for supercapacitors [125–130]. Various carbon materials such as activated carbon, CNTs, graphene, CNFs have been studied as the electrode materials. Among these, activated carbon is the conventionally used material. High SSA-activated carbon-based supercapacitors offer a capacitance in the range of 50–200 F/g. The pore size becomes an important parameter to achieve high energy and power densities. Larger pore size is suitable for high power density, whereas, for higher energy density, it should be small. Hence, a proper optimization of pore size is required. The use of 3-D networks of fibrous carbon nanomaterials such as CNTs and CNFs can provide an optimized pore size distribution along with high SSA, chemical inertness and excellent mechanical strength. Moreover, the fibrous structure of CNF electrodes reduces the diffusion distance for electrolyte ions. Mesoporous CNFs have shown more than 100% enhanced specific capacitance as compared to that of mesoporous carbon [33]. Compared to CNTs, relatively easy and scalable synthesis methods offer CNFs an additional advantage toward large-scale applications. Both vapor-grown and electrospun CNFs have been studied for supercapacitor electrode materials. Niu et al. have reported interbonded CNFs by pyrolyzing electrospun polyvinylpyrrolidone (PVP) and polyacrylonitrile (PAN) fibers kept side by side. The specific capacitance of ~220 F/g is observed CNF network synthesized by using a 9:1 (w/w) PVP/PAN side-by-side network [131]. Fan et al. have synthesized an asymmetric supercapacitor having graphene/MnO_2 and activated CNF as positive and negative electrodes, respectively, and reported an energy density as high as 51.1 Wh/kg and a specific capacitance of 113.5 F/g [28]. The high areal capacitance is required in small-scale electronic and stationary energy storage devices. Although CNF electrodes exhibit high mass capacitance, the low loading makes it difficult to obtain high areal capacitances. To increase the areal capacitance, 3-D foam-like metallic substrates have been used as a substrate to load/grow CNFs [132]. High areal capacitance of 1.2 F/cm^2 has been reported by using the 3-D network CNF electrode.

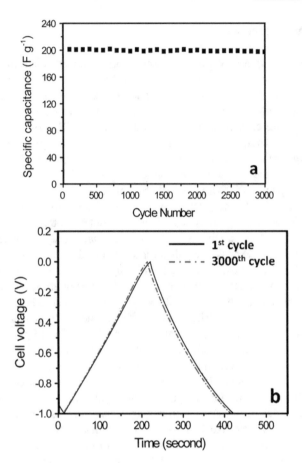

Fig. 7.14 Cycling performance of the supercapacitor based on *N*-doped porous CNFs as the electrode material showing high stability. Reprinted with permission from [136]

Heteroatom-doped CNFs, containing one or more types of heteroatoms such as B, N, O, P, S in the carbon framework have been recently synthesized and explored for their application as supercapacitor electrodes [133–135]. Figure 7.14 shows the performance of a typical supercapacitor based on *N*-doped porous CNFs as the electrode material [136].

7.5.5.2   Lithium-Ion Batteries

Lithium-ion batteries (LIBs) have attracted a great deal of attraction due to their high energy density and lifetime. Graphitic carbon is used conventionally as the negative electrode of LIBs, where lithium is intercalated between the graphitic layers. The unsatisfactory performance of natural graphite triggers the need of using the costly synthetic graphite. The high cost of synthetic graphite stems from the complex synthesis process requiring high temperatures (>2800 °C). The drawback can be overcome by the use of other graphitic carbons, which can be synthesized at lower temperatures. More recently, allotropes of graphitic carbon, including graphene, CNTs, CNFs, etc., have been investigated as a replacement of graphite. CNF-based electrodes can be prepared by mixing the CNFs with a polymer binder [137] or by using CNFs as a filler along with the other electrode materials [138]. During their earlier developments, Abe et al. have reported a GCNF/LiCoO$_2$ battery having energy density and specific capacity of 763 mAh and 107 Wh/kg, respectively [137]. Using CVD-synthesized CNFs as anode material, Yoon et al. have reported the capacity of 297–431 mAh/g with a comparatively lower columbic efficiency of 60–70% [139]. As the lower power density is one of the principal limitations of LIBs, electrode materials having fast charging–discharging capability are desirable. Batteries consisting of lithium metal and (CNF + PVDF 10% + carbon black 10%) electrodes have shown high columbic efficiency (98%) and reversible specific capacities of 461 and 170 mAh/g at the charging/discharging rates of 0.1 and 10 C, respectively [140]. To improve the specific capacities of CNF-based LIBs, CNFs with modified structure have been studied. Rather than the CVD synthesized, the electrospun CNFs are used for such structural modifications. Using electrospun porous CNFs, Ji et al. have obtained specific reversible capacities of ~566 mAh/g [141, 142]. The lifetime of electrodes in terms of capacity retention, rate capability and large charge–discharge cycles can be improved by structural modifications of CNF electrodes. Incorporation of transition metal oxides in the CNF-based electrodes [143], encapsulation or filling of the hollow core of CNFs by metals such as tin (Sn) and non-metals such as sulfur (S) are some of the approaches employed to reduce the capacity fading with charging/discharging cycles. Hollow CNFs synthesized by anodized alumina template method and encapsulated with sulfur have been reported with a high specific capacity of 730 mAh/g after 150 cycles of charging/discharging [36]. A high reversible capacity of 737 mAh/h at a rate of 0.5 C has been reported by Yu et al. for a Sn-encapsulated hollow CNF (Sn–CNF) electrode LIB [144]. Chen et al. have recently reported a hybrid hierarchical material having high SSA, and pore volume (BET surface area 1840 m^2/g, pore volume 1.21 m^3/g) consisting of CNF supported CNTs synthesized by CVD [51]. When used as LIB electrodes (CNT/CNF 80 wt%, PVDF 10 wt%, CB 10 wt%), it exhibits a reversible capacity of ~1150 mAh/g at 0.27 C after 70 cycles. The specific capacity variation with charging/discharging cycling of the LIB is shown in Fig. 7.15. With changing–discharging, the capacity decreases gradually and reaches to ~250–300 mAh/g for cycle number >500. The columbic efficiency of the battery remains close to 100% throughout the cycling. The advancements to date have enhanced the specific capacity of CNF-based LIBs significantly. However, the capacity fading with increasing charging/discharging cycles and the large-scale synthesis of the high capacity electrode materials are the issues to be solved.

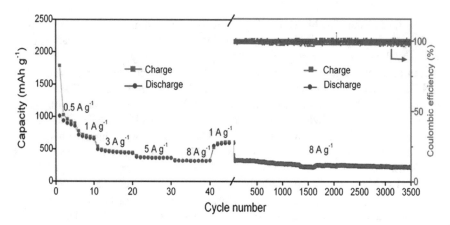

Fig. 7.15 Charge–discharge cycling performance of CNF-based electrode LIB. Reprinted with permission from [51]

7.5.5.3 Catalyst Supports

Excellent corrosion resistance, electronic conductivity, chemical inertness and high SSA of carbon nanomaterials have proposed them a suitable catalyst support material in electrochemical devices such as FCs and DSSCs [110]. Both electrospun as well as CVD-grown CNFs, along with other forms of carbon such as carbon-black, CNTs, graphene, have been studied extensively as the catalyst support in FCs/DSSCs electrodes [145–147]. Similar to FCs, where large SSA carbon nanomaterials are required to support the catalyst particles, DSSCs require materials to support the dye molecules at photoanode and the catalyst particles at counter electrode.

7.5.5.4 Electrocatalysts

The cathode reaction (ORR) in polymer electrolyte membrane (PEM) FCs (PEM-FCs) requires the presence of electrocatalysts. Platinum (Pt) and Pt-alloys are the conventionally used catalyst materials, which are unsuitable for commercial applications due to their high cost and scarce nature. Similar catalysts are required at DSSC counter electrodes, where reduction of electrolyte takes place by electron transfer from the electrode to the oxidized electrolyte molecules. Pt-coated conducting glass is generally used as the DSSC counter electrode, where Pt acts as the electrocatalyst. Carbon nanomaterials, particularly nitrogen (N)- and boron (B)-doped CNTs, have been investigated as a new class of Pt-free electrocatalysts [148–155]. The catalytic activity of carbon nanomaterials stems from the presence of defect sites in the otherwise regular graphitic structure. CNFs, being composed of defective graphitic layers, are being considered for their potential as electrocatalysts in PEMFCs as well as DSSC counter electrodes.

Similar to N-doped CNTs, N-doped CNFs have been reported having significant electrocatalytic activity toward the ORR at the cathode of PEMFCs [156]. Qiu et al. have reported the synthesis of N-doped CNFs by carbonizing electrospun PAN fibers in the presence of NH_3 [157]. The presence of NH_3 during the carbonization introduces N-atoms in the CNFs. The N-atoms in otherwise pure carbon network of CNFs act as electrocatalytic sites to impart high ORR activity to N-doped CNFs. Similar electrocatalysts have been reported for DSSC counter electrodes. DSSCs fabricated by using VGNNFs-based counter electrode having a power conversion efficiency (η) comparable to that of a Pt-based counter electrode (~5.6% for VGCNT electrode compared to ~6.5% for Pt-electrode) has been reported recently [158]. The carbon-based electrocatalysts are superior to Pt-based catalysts in terms of cost, stability, immunity to CO poisoning, etc.

7.5.6 Field Emitters

The high aspect ratio of 1-D nanomaterials offers high field enhancement factor, which enables field electron emission at considerably low applied voltages as compared to metal-tip electron emitters. Carbon nanomaterials, particularly CNTs, have been proposed promising field electron emitters for various applications [159–163]. Being composed of similar graphitic structure, CNFs exhibit good field emission characteristics. Sugimoto et al. have fabricated an X-ray source using a CNF-based field electron emitter as an electron source by coating CNFs on a metal tip [164]. High current density (5×10^9 A/m^2) and smaller focal size of an electron beam (0.4 μm) have been obtained by using the CNF field emitter. Such highly focused beams have been used to produce fine-focused X-ray sources with an X-ray image resolution of ~0.5 μm. Single CNF field emitters can be used to produce gated cathode structures for flat-panel displays [165–167]. Such a structure, fabricated by the combination of microfabrication techniques such as lithography and PECVD, is shown in Fig. 7.16. The structure consists of VACNF field electron emitter inside a SiO_2 cavity positioned vertically on a silicon wafer.

Arrays of VACNFs synthesized by CVD or PECVD exhibit excellent field emission properties. PECVD-synthesized closely packed array (10^9–10^{10}/cm^2) of VAC-NTs has been reported to produce field emission currents sufficient for run the flat-panel displays at an applied field value of 2.5 V/μm [168]. Factors such as the diameter and density of aligned CNFs play a vital role to determine their field emission properties. A closely packed array of CNFs is not ideal field emitter due to the electric field shielding effect. Teo et al. have studied the effect of the CNF density on the field emission properties of VACNF arrays [169]. Patterned VACNT array with a low areal density of 10^6 CNF/cm^2 has shown the highest field emission characteristics compared to the other closely packed (10^9–10^{10}/cm^2) arrays. Furthermore, the field enhancement factor can be improved by structural changes in the CNF structures. Weng et al. have used argon plasma treatment of VACNF arrays to obtain cone-shaped structures having enhanced field emission properties [170]. The enhancement is also

Fig. 7.16 SEM of gated
cathode structure using
VACNF (at the center) as
field emitter. Reprinted with
permission from [165]

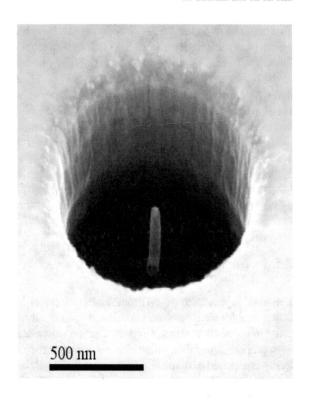

Fig. 7.16 SEM of gated cathode structure using VACNF (at the center) as field emitter. Reprinted with permission from [165]

attributed to the removal of catalyst particles from the CNF tips. Similarly, Tanemura et al. have also synthesized cone-shaped amorphous CNFs by bombarding Ar^+ ions on various surfaces such as graphite, carbon-coated silicon and carbon-coated nickel and observed promising field emission properties with a field enhancement factor of as high as 1951 [171].

7.6 Concluding Remarks

Graphitic structure of CNFs provides them properties comparable to other sp^2-carbon nanomaterials. The structure and properties of CNFs vary significantly with the synthesis techniques and parameters. Both vapor-grown and electrospun CNFs exhibit promising properties, which propose them suitable for a range of applications including composites, power generation and storage, electronic devices, pollutant absorbents, sensors, etc. Dedicated studies have proven the application potential of CNFs comparable to that of MWCNTs. Their easy and scalable synthesis is the obvious advantage of CNFs over CNTs, making CNFs attractive for large-scale applications. The dependence of properties on the synthesis techniques and parameters requires more optimizations to improve their reliability. Among others, their applications in power generation and storage devices turn out to be more promising. CNF-based electrocatalysts/catalyst supporters for FCs/DSSCs and electrode materials for batteries/supercapacitors have considerable application potential in the near future.

Acknowledgements The authors acknowledge the financial support provided by Indian Space Research Organization (ISRO), India, for carrying out this research work.

References

1. H.W. Kroto, J.R. Heath, S.C. O'Brien, R.F. Curl, R.E. Smalley, Nature **318**, 162 (1985)
2. S. Iijima, Nature **354**, 56 (1991)
3. A.K. Geim, K.S. Novoselov, Nat. Mater. **6**, 183 (2007)
4. R.H. Baughman, A.A. Zakhidov, W.A. de Heer, Science **297**, 787 (2002)
5. R. Kumar, S. Sahoo, E. Joanni, R.K. Singh, W.K. Tan, K.K. Kar, A. Matsuda, Prog. Energy Combust. Sci. **75**, 100786 (2019)
6. N.M. Rodriguez, A. Chambers, R.T.K. Baker, Langmuir **11**, 3862 (1995)
7. A.V. Melechko, V.I. Merkulov, T.E. McKnight, M.A. Guillorn, K.L. Klein, D.H. Lowndes, M.L. Simpson, J. Appl. Phys. **97**, 041301 (2005)
8. A. Chambers, C. Park, R.T.K. Baker, N.M. Rodriguez, J. Phys. Chem. B **102**, 4253 (1998)
9. S.-H. Yoon, S. Lim, S.-H. Hong, I. Mochida, B. An, K. Yokogawa, Carbon **42**, 3087 (2004)
10. S.-H. Yoon, S. Lim, S.-H. Hong, W. Qiao, D.D. Whitehurst, I. Mochida, B. An, K. Yokogawa, Carbon **43**, 1828 (2005)
11. M. Tsuji et al., Langmuir **23**, 387 (2007)
12. B. Ekşioğlu, A. Nadarajah, Carbon **44**, 360 (2006)
13. J. Lawrence, L. Berhan, A. Nadarajah, J. Nanopart. Res. **10**, 1155 (2008)
14. T. Zhao, I. Kvande, Y. Yu, M. Ronning, A. Holmen, D. Chen, J. Phys. Chem. C **115**, 1123 (2011)
15. J.G. Lawrence, L.M. Berhan, A. Nadarajah, ACS Nano **2**, 1230 (2008)
16. C.J. Lee, J. Park, Appl. Phys. Lett. **77**, 3397 (2000)
17. R.F. Wood, S. Pannala, J.C. Wells, A.A. Puretzky, D.B. Geohegan, Phys. Rev. B **75**, 235446 (2007)
18. R. Xiang, Z. Yang, Q. Zhang, G. Luo, W. Qian, F. Wei, M. Kadowaki, E. Einarsson, S. Maruyama, J. Phys. Chem. C **112**, 4892 (2008)
19. R. Seidel, G.S. Duesberg, E. Unger, A.P. Graham, M. Liebau, F. Kreupl, J. Phys. Chem. B **108**, 1888 (2004)
20. L. Ci, Y. Li, B. Wei, J. Liang, C. Xu, D. Wu, Carbon **38**, 1933 (2000)
21. L. Ci, J. Wei, B. Wei, J. Liang, C. Xu, D. Wu, Carbon **39**, 329 (2001)
22. V.I. Merkulov, A.V. Melechko, M.A. Guillorn, D.H. Lowndes, M.L. Simpson, Appl. Phys. Lett. **79**, 2970 (2001)
23. V.I. Merkulov, D.H. Lowndes, Y.Y. Wei, G. Eres, E. Voelkl, Appl. Phys. Lett. **76**, 3555 (2000)
24. J.B.O. Caughman, L.R. Baylor, M.A. Guillorn, V.I. Merkulov, D.H. Lowndes, L.F. Allard, Appl. Phys. Lett. **83**, 1207 (2003)
25. B.O. Boskovic, V. Stolojan, R.U.A. Khan, S. Haq, S.R.P. Silva, Nat. Mater. **1**, 165 (2002)
26. S. Hofmann, C. Ducati, J. Robertson, B. Kleinsorge, Appl. Phys. Lett. **83**, 135 (2003)
27. T.M. Minea, S. Point, A. Granier, M. Touzeau, Appl. Phys. Lett. **85**, 1244 (2004)
28. Z. Fan, J. Yan, T. Wei, L. Zhi, G. Ning, T. Li, F. Wei, Adv. Funct. Mater. **21**, 2366 (2011)
29. Y. Wan, G. Zuo, F. Yu, Y. Huang, K. Ren, H. Luo, Surf. Coat. Technol. **205**, 2938 (2011)
30. J. Spender, A.L. Demers, X. Xie, A.E. Cline, M.A. Earle, L.D. Ellis, D.J. Neivandt, Nano Lett. **12**, 3857 (2012)

31. S. Megelski, J.S. Stephens, D.B. Chase, J.F. Rabolt, Macromolecules **35**, 8456 (2002)
32. M. Inagaki, Y. Yang, F. Kang, Adv. Mater. **24**, 2547 (2012)
33. K. Wang, Y. Wang, Y. Wang, E. Hosono, H. Zhou, J. Phys. Chem. C **113**, 1093 (2008)
34. Y. Xu, C. Zhang, M. Zhou, Q. Fu, C. Zhao, M. Wu, Y. Lei, Nat. Commun. **9**, 1720 (2018)
35. X. Zhang, D. Liu, L. Li, T. You, Sci. Rep. **5**, 9885 (2015)
36. G. Zheng, Y. Yang, J.J. Cha, S.S. Hong, Y. Cui, Nano Lett. **11**, 4462 (2011)
37. C. Luo, X. Qi, C. Pan, W. Yang, Sci. Rep. **5**, 13879 (2015)
38. Y.-H. Qin, H.-H. Yang, X.-S. Zhang, P. Li, C.-A. Ma, Int. J. Hydrogen Energy **35**, 7667 (2010)
39. D. Sebastian, J.C. Calderon, J.A. Gonzalez-Exposito, E. Pastor, M.V. Martinez-Huerta, I. Suelves, R. Moliner, M.J. Lazaro, Int. J. Hydrogen Energy **35**, 9934 (2010)
40. H. Raghubanshi, M.S.L. Hudson, O.N. Srivastava, Int. J. Hydrogen Energy **36**, 4482 (2011)
41. Y. Liu, J. Ma, T. Lu, L. Pan, Sci. Rep. **6**, 32784 (2016)
42. R. Sharma, K.K. Kar, RSC Adv. **5**, 66518 (2015)
43. R. Sharma, K.K. Kar, J. Mater. Chem. A **3**, 11948 (2015)
44. R. Sharma, K.K. Kar, Electrochim. Acta **191**, 876 (2016)
45. G. Ren, X. Lu, Y. Li, Y. Zhu, L. Dai, L. Jiang, A.C.S. Appl. Mater. Interfaces **8**, 4118 (2016)
46. P. Wang, D. Zhang, F. Ma, Y. Ou, Q.N. Chen, S. Xie, J. Li, Nanoscale **4**, 7199 (2012)
47. J.-H. Zhou, Z.-J. Sui, P. Li, D. Chen, Y.-C. Dai, W.-K. Yuan, Carbon **44**, 3255 (2006)
48. R. Ma, B. Wei, C. Xu, J. Liang, D. Wu, J. Mater. Sci. Lett. **19**, 1929 (2000)
49. N.M. Rodriguez, M.-S. Kim, R.T.K. Baker, J. Phys. Chem. **98**, 13108 (1994)
50. C. Kim, K.S. Yang, Appl. Phys. Lett. **83**, 1216 (2003)
51. Y. Chen, X. Li, K. Park, J. Song, J. Hong, L. Zhou, Y.-W. Mai, H. Huang, J.B. Goodenough, JACS **135**, 16280 (2013)
52. V.Z. Mordkovich, Theor. Found. Chem. Eng. **37**, 429 (2003)
53. C. Wei, D. Srivastava, Appl. Phys. Lett. **85**, 2208 (2004)
54. T. Liu, S. Kumar, Nano Lett. **3**, 647 (2003)
55. T. Uchida, D. Anderson, M. Minus, S. Kumar, J. Mater. Sci. **41**, 5851 (2006)
56. T. Ozkan, M. Naraghi, I. Chasiotis, Carbon **48**, 239 (2010)
57. A.M. Beese, D. Papkov, S. Li, Y. Dzenis, H.D. Espinosa, Carbon **60**, 246 (2013)
58. S.N. Arshad, M. Naraghi, I. Chasiotis, Carbon **49**, 1710 (2011)
59. E. Zussman, X. Chen, W. Ding, L. Calabri, D.A. Dikin, J.P. Quintana, R.S. Ruoff, Carbon **43**, 2175 (2005)
60. Q. Ngo, T. Yamada, M. Suzuki, Y. Ominami, A.M. Cassell, J. Li, M. Meyyappan, C.Y. Yang, Nanotechnology. IEEE Transactions on **6**, 688 (2007)
61. Y. Wang, S. Serrano, J.J. Santiago-Aviles, J. Mater. Sci. Lett. **21**, 1055 (2002)
62. C.S. Sharma, H. Katepalli, A. Sharma, M. Madou, Carbon **49**, 1727 (2011)
63. T.W. Ebbesen, H.J. Lezec, H. Hiura, J.W. Bennett, H.F. Ghaemi, T. Thio, Nature **382**, 54 (1996)
64. N.K. Mahanta, A.R. Abramson, Rev. Sci. Instrum. **83**, 054904 (2012)
65. N.K. Mahanta, A.R. Abramson, J.Y. Howe, J. Appl. Phys. **114**, 163528 (2013)
66. S. Berber, Y.-K. Kwon, D. Tománek, Phys. Rev. Lett. **84**, 4613 (2000)
67. G. Morales, M.I. Barrena, J.M.G.d. Salazar, C. Merino, D. Rodríguez, Compos. Struct. **92**, 1416 (2010)
68. P. Richard, T. Prasse, J.Y. Cavaille, L. Chazeau, C. Gauthier, J. Duchet, Mater. Sci. Eng. A **352**, 344 (2003)
69. J. Sandler, P. Werner, M.S.P. Shaffer, V. Demchuk, V. Altstädt, A.H. Windle, Compos. Part A **33**, 1033 (2002)
70. M. Cadek et al., AIP Conf. Proc. **633**, 562 (2002)
71. K.K. Kar, S. Rana, J. Pandey, *Handbook of Polymer Nanocomposites Processing, Performance and Application* (Springer, 2015)
72. H.-X. Jiang, Q.-Q. Ni, T. Natsuki, Polym. Compos. **31**, 1099 (2010)
73. Y. Jang, S. Kim, S. Lee, D. Kim, M. Um, Compos. Sci. Technol. **65**, 781 (2005)
74. J.M. Ullbrand, J.M. Córdoba, J. Tamayo-Ariztondo, M.R. Elizalde, M. Nygren, J.M. Molina-Aldareguia, M. Odén, Compos. Sci. Technol. **70**, 2263 (2010)

75. A. Duszová, J. Dusza, K. Tomášek, J. Morgiel, G. Blugan, J. Kuebler, Scripta Mater. **58**, 520 (2008)
76. S. Maensiri, P. Laokul, J. Klinkaewnarong, V. Amornkitbamrung, Mater. Sci. Eng. A **447**, 44 (2007)
77. J. Dusza, G. Blugan, J. Morgiel, J. Kuebler, F. Inam, T. Peijs, M.J. Reece, V. Puchy, J. Eur. Ceram. Soc. **29**, 3177 (2009)
78. J. Liang, M.C. Saha, M.C. Altan, Procedia Eng. **56**, 814 (2013)
79. N. Sabetzadeh, S.S. Najar, S.H. Bahrami, J. Appl. Polym. Sci. **130**, 3009 (2013)
80. K.K. Kar, *Composite Materials: Processing, Applications, Characterizations* (Springer, 2016)
81. P. Agnihotri, S. Basu, K.K. Kar, Carbon **49**, 3098 (2011)
82. S. Bal, Mater. Des. **31**, 2406 (2010)
83. A.J. Paleo, V. Sencadas, F.W.J. van Hattum, S. Lanceros-Méndez, A. Ares, Polym. Eng. Sci. **54**, 117 (2014)
84. M. Colloca, N. Gupta, M. Porfiri, Compos. Part B **44**, 584 (2013)
85. R.L. Poveda, S. Achar, N. Gupta, Compos. Part B **58**, 208 (2014)
86. R.L. Poveda, N. Gupta, Mater. Des. **56**, 416 (2014)
87. M.J. Palmeri, K.W. Putz, T. Ramanathan, L.C. Brinson, Compos. Sci. Technol. **71**, 79 (2011)
88. R. Sharma, A.K. Yadav, V. Panwar, K.K. Kar, J. Reinf. Plast. Compos. **34**, 941 (2015)
89. Z.-Y. Wu, C. Li, H.-W. Liang, J.-F. Chen, S.-H. Yu, Angew. Chem. Int. Ed. **52**, 2925 (2013)
90. S. Chen, X. Zhan, D. Lu, C. Liu, L. Zhu, Anal. Chim. Acta **634**, 192 (2009)
91. X. Ren, C. Chen, M. Nagatsu, X. Wang, Chem. Eng. J. **170**, 395 (2011)
92. G. Andrade-Espinosa, E. Muñoz-Sandoval, M. Terrones, M. Endo, H. Terrones, J.R. Rangel-Mendez, J. Chem. Technol. Biotechnol. **84**, 519 (2009)
93. W.G. Shim, C. Kim, J.W. Lee, J.J. Yun, Y.I. Jeong, H. Moon, K.S. Yang, J. Appl. Polym. Sci. **102**, 2454 (2006)
94. K.J. Lee, N. Shiratori, G.H. Lee, J. Miyawaki, I. Mochida, S.-H. Yoon, J. Jang, Carbon **48**, 4248 (2010)
95. A. Chakraborty, D. Deva, A. Sharma, N. Verma, J. Colloid Interface Sci. **359**, 228 (2011)
96. A.K. Gupta, D. Deva, A. Sharma, N. Verma, Ind. Eng. Chem. Res. **48**, 9697 (2009)
97. A.K. Gupta, D. Deva, A. Sharma, N. Verma, Ind. Eng. Chem. Res. **49**, 7074 (2010)
98. M.R. Cuervo, E. Asedegbega-Nieto, E. Díaz, A. Vega, S. Ordóñez, E. Castillejos-López, I. Rodríguez-Ramos, J. Chromatogr. A **1188**, 264 (2008)
99. M. Bikshapathi, S. Singh, B. Bhaduri, G.N. Mathur, A. Sharma, N. Verma, Colloids Surf. A **399**, 46 (2012)
100. E. Díaz, S. Ordóñez, A. Vega, J. Colloid Interface Sci. **305**, 7 (2007)
101. B. Ding, M. Wang, J. Yu, G. Sun, Sensors **9**, 1609 (2009)
102. C.-T. Hsieh, W.-Y. Chen, F.-L. Wu, Carbon **46**, 1218 (2008)
103. J. Jang, J. Bae, Sens. Actuators, B **122**, 7 (2007)
104. J. Zhu, S. Wei, J. Ryu, Z. Guo, J. Phys. Chem. C **115**, 13215 (2011)
105. F.-Y. Chang, R.-H. Wang, H. Yang, Y.-H. Lin, T.-M. Chen, S.-J. Huang, Thin Solid Films **518**, 7343 (2010)
106. C. Pang, G.-Y. Lee, T.-I. Kim, S.M. Kim, H.N. Kim, S.-H. Ahn, K.-Y. Suh, Nat. Mater. **11**, 795 (2012)
107. T. Yan, Z. Wang, Y.-Q. Wang, Z.-J. Pan, Mater. Des. **143**, 214 (2018)
108. S. Wu, J. Zhang, R.B. Ladani, A.R. Ravindran, A.P. Mouritz, A.J. Kinloch, C.H. Wang, A.C.S. Appl. Mater. Interfaces **9**, 14207 (2017)
109. P. Joshi, L. Zhang, Q. Chen, D. Galipeau, H. Fong, Q. Qiao, A.C.S. Appl. Mater. Interfaces **2**, 3572 (2010)
110. L. Li, P. Zhu, S. Peng, M. Srinivasan, Q. Yan, A.S. Nair, B. Liu, S. Samakrishna, J. Phys. Chem. C **118**, 16526 (2014)
111. S.-H. Park, B.-K. Kim, W.-J. Lee, J. Power Sources **239**, 122 (2013)
112. P. Poudel, L. Zhang, P. Joshi, S. Venkatesan, H. Fong, Q. Qiao, Nanoscale **4**, 4726 (2012)
113. P. Bosch-Jimenez, S. Martinez-Crespiera, D. Amantia, M. Della Pirriera, I. Forns, R. Shechter, E. Borràs, Electrochim. Acta **228**, 380 (2017)

114. D. Ji, S. Peng, J. Lu, L. Li, S. Yang, G. Yang, X. Qin, M. Srinivasan, S. Ramakrishna, J. Mater. Chem. A **5**, 7507 (2017)
115. C.-T. Hsieh, J.-M. Chen, R.-R. Kuo, Y.-H. Huang, Appl. Phys. Lett. **84**, 1186 (2004)
116. Y. Liu, L.-Z. Fan, L. Jiao, J. Mater. Chem. A **5**, 1698 (2017)
117. Y. Liu, J. Zhou, W. Fu, P. Zhang, X. Pan, E. Xie, Carbon **114**, 187 (2017)
118. Q. Xia, H. Yang, M. Wang, M. Yang, Q. Guo, L. Wan, H. Xia, Y. Yu, Adv. Energy Mater. **7**(1701336), 1701336 (2017)
119. Y. Yang, F. Yang, H. Hu, S. Lee, Y. Wang, H. Zhao, D. Zeng, B. Zhou, S. Hao, Chem. Eng. J. **307**, 583 (2017)
120. A. Di Blasi, C. Busaccaa, O. Di Blasia, N. Briguglioa, G. Squadritoa, V. Antonuccia, Appl. Energy **190**, 165 (2017)
121. J. Guo, J. Liu, H. Dai, R. Zhou, T. Wang, C. Zhang, S. Ding, H.-G. Wang, J. Colloid Interface Sci. **507**, 154 (2017)
122. J.S. Lee, W. Kim, J. Jang, A. Manthiram, Adv. Energy Mater. **7**(1601943), 1601943 (2017)
123. D.W. Porter et al., NanoImpact **6**, 1 (2017)
124. A. Burke, J. Power Sources **91**, 37 (2000)
125. P. Simon, Y. Gogotsi, Nat. Mater. **7**, 845 (2008)
126. Y. Zhai, Y. Dou, D. Zhao, P.F. Fulvio, R.T. Mayes, S. Dai, Adv. Mater. **23**, 4828 (2011)
127. J. Cherusseri, K.K. Kar, J. Mater. Chem. A **3**, 21586 (2015)
128. J. Cherusseri, K.K. Kar, Phys. Chem. Chem. Phys. **18**, 8587 (2016)
129. J. Cherusseri, K.K. Kar, RSC Adv. **6**, 60454 (2016)
130. J. Cherusseri, R. Sharma, K.K. Kar, Carbon **105**, 113 (2016)
131. H. Niu, J. Zhang, Z. Xie, X. Wang, T. Lin, Carbon **49**, 2380 (2011)
132. J.R. McDonough, J.W. Choi, Y. Yang, F. La Mantia, Y. Zhang, Y. Cui, Appl. Phys. Lett. **95**, 243109 (2009)
133. L.-F. Chen, Y. Lu, Y. Lu, X.W. Lou, Energy Environ. Sci. **10**, 1777 (2017)
134. W. Na, J. Jun, J.W. Park, G. Lee, J. Jang, J. Mater. Chem. A **5**, 17379 (2017)
135. Y. Cheng et al., Nano Energy **15**, 66 (2015)
136. L.-F. Chen, X.-D. Zhang, H.-W. Liang, M. Kong, Q.-F. Guan, P. Chen, Z.-Y. Wu, S.-H. Yu, ACS Nano **6**, 7092 (2012)
137. H. Abe, T. Murai, K. Zaghib, J. Power Sources **77**, 110 (1999)
138. M. Endo, Y.A. Kim, T. Hayashi, K. Nishimura, T. Matusita, K. Miyashita, M.S. Dresselhaus, Carbon **39**, 1287 (2001)
139. S.-H. Yoon, C.-W. Park, H. Yang, Y. Korai, I. Mochida, R.T.K. Baker, N.M. Rodriguez, Carbon **42**, 21 (2004)
140. V. Subramanian, H. Zhu, B. Wei, J. Phys. Chem. B **110**, 7178 (2006)
141. L. Ji, X. Zhang, Nanotechnology **20**, 155705 (2009)
142. L. Ji, X. Zhang, Electrochem. Commun. **11**, 684 (2009)
143. L. Ji, X. Zhang, Electrochem. Commun. **11**, 795 (2009)
144. Y. Yu, L. Gu, C. Wang, A. Dhanabalan, P.A. van Aken, J. Maier, Angew. Chem. Int. Ed. **48**, 6485 (2009)
145. Y. Shao, J. Liu, Y. Wang, Y. Lin, J. Mater. Chem. **19**, 46 (2009)
146. B.C.H. Steele, A. Heinzel, Nature **414**, 345 (2001)
147. Y. Li, W. Zhou, H. Wang, L. Xie, Y. Liang, F. Wei, J.-C. Idrobo, S.J. Pennycook, H. Dai, Nat. Nano **7**, 394 (2012)
148. Z. Lin, G.H. Waller, Y. Liu, M. Liu, C.-P. Wong, Nano Energy **2**, 241 (2013)
149. Z. Chen, D. Higgins, Z. Chen, Carbon **48**, 3057 (2010)
150. Y. Shao, J. Sui, G. Yin, Y. Gao, Appl. Catal. B Environ. **79**, 89 (2008)
151. K. Gong, F. Du, Z. Xia, M. Durstock, L. Dai, Science **323**, 760 (2009)
152. T.C. Nagaiah, S. Kundu, M. Bron, M. Muhler, W. Schuhmann, Electrochem. Commun. **12**, 338 (2010)
153. K. Parvez, S. Yang, Y. Hernandez, A. Winter, A. Turchanin, X. Feng, K. Müllen, ACS Nano **6**, 9541 (2012)
154. S. Ci, Y. Wu, J. Zou, L. Tang, S. Luo, J. Li, Z. Wen, Chin. Sci. Bull. **57**, 3065 (2012)

155. N.P. Subramanian, X. Li, V. Nallathambi, S.P. Kumaraguru, H. Colon-Mercado, G. Wu, J.-W. Lee, B.N. Popov, J. Power Sources **188**, 38 (2009)
156. J. Yin, Y. Qiu, J. Yu, ECS Solid State Letters **2**, M37 (2013)
157. Y. Qiu, J. Yu, T. Shi, X. Zhou, X. Bai, J.Y. Huang, J. Power Sources **196**, 9862 (2011)
158. S.M. Mahpeykar, M.K. Tabatabaei, H. Ghafoori-fard, H. Habibiyan, J. Koohsorkhi, Nanotechnology **24**, 435402 (2013)
159. W.A. de Heer, A. Châtelain, D. Ugarte, Science **270**, 1179 (1995)
160. Y. Saito, S. Uemura, Carbon **38**, 169 (2000)
161. S. Fan, M.G. Chapline, N.R. Franklin, T.W. Tombler, A.M. Cassell, H. Dai, Science **283**, 512 (1999)
162. Q.H. Wang, A.A. Setlur, J.M. Lauerhaas, J.Y. Dai, E.W. Seelig, R.P.H. Chang, Appl. Phys. Lett. **72**, 2912 (1998)
163. L. Nilsson, O. Groening, C. Emmenegger, O. Kuettel, E. Schaller, L. Schlapbach, H. Kind, J.-M. Bonard, K. Kern, Appl. Phys. Lett. **76**, 2071 (2000)
164. W. Sugimoto, S. Sugita, Y. Sakai, H. Goto, Y. Watanabe, Y. Ohga, S. Kita, T. Ohara, J. Appl. Phys. **108**, 044507 (1 (2010)
165. M.A. Guillorn, M.L. Simpson, G.J. Bordonaro, V.I. Merkulov, L.R. Baylor, D.H. Lowndes, J. Vac. Sci. Technol. B **19**, 573 (2001)
166. M.A. Guillorn, A.V. Melechko, V.I. Merkulov, D.K. Hensley, M.L. Simpson, D.H. Lowndes, Appl. Phys. Lett. **81**, 3660 (2002)
167. M.A. Guillorn, A.V. Melechko, V.I. Merkulov, E.D. Ellis, M.L. Simpson, D.H. Lowndes, L.R. Baylor, G.J. Bordonaro, J. Vac. Sci. Technol. B **19**, 2598 (2001)
168. Y. Chen, S. Patel, Y. Ye, D.T. Shaw, L. Guo, Appl. Phys. Lett. **73**, 2119 (1998)
169. K.B.K. Teo, M. Chhowalla, G.A.J. Amaratunga, W.I. Milne, G. Pirio, P. Legagneux, F. Wczisk, D. Pribat, D.G. Hasko, Appl. Phys. Lett. **80**, 2011 (2002)
170. C.H. Weng, K.C. Leou, H.W. Wei, Z.Y. Juang, M.T. Wei, C.H. Tung, C.H. Tsai, Appl. Phys. Lett. **85**, 4732 (2004)
171. M. Tanemura, J. Tanaka, K. Itoh, Y. Fujimoto, Y. Agawa, L. Miao, S. Tanemura, Appl. Phys. Lett. **86**, 113107 (2005)

Chapter 8
Characteristics of Conducting Polymers

Tanvi Pal, Soma Banerjee, P. K. Manna, and Kamal K. Kar

Abstract Conducting polymers (CPs) have gained recent attention from the research community due to the extraordinary combination of properties including tunable electrical conductivity, easy route of preparation, light in weight, easy to process, etc. These exclusive properties enable the use of CPs in many intriguing applications including modern electrochemical devices. The commonly adopted synthesis strategies utilized for the synthesis of CPs include chemical and electrochemical polymerization routes. The electrical conductivity of CPs can be tuned easily by altering the types of doping and concentration of doping. This chapter also discusses the efficacy and benefits of dopants to improve the conduction properties of common CPs. The new application areas of CPs comprise supercapacitors, sensors, solar cells, corrosion inhibitors, light-emitting diodes, EMI shielding, electrochromic devices, transistors, and many more. This review provides concise yet comprehensive introductory information on the CPs including the types of common CPS, their mechanism of conduction, attractive properties, and diverse application areas to have a broad overview of this field of research.

T. Pal · S. Banerjee · K. K. Kar (✉)
Advanced Nanoengineering Materials Laboratory, Material Science Programme, Indian Institute of Technology Kanpur, Kanpur 208016, India
e-mail: kamalkk@iitk.ac.in

T. Pal
e-mail: findingprerna09@gmail.com

S. Banerjee
e-mail: somabanerjee27@gmail.com

T. Pal
A.P.J. Abdul Kalam Technical University, Lucknow 226031, India

P. K. Manna
Indus Institute of Technology and Management, Kanpur 209202, India
e-mail: pkmanna8161@yahoo.co.uk

K. K. Kar
Advanced Nanoengineering Materials Laboratory, Department of Mechanical Engineering, Indian Institute of Technology Kanpur, Kanpur 208016, India

© Springer Nature Switzerland AG 2020
K. K. Kar (ed.), *Handbook of Nanocomposite Supercapacitor Materials I*,
Springer Series in Materials Science 300,
https://doi.org/10.1007/978-3-030-43009-2_8

8.1   Introduction

Conducting polymers (CPs) are of huge research interest over the last 30 years due to the excellent stability, lightweight, workability, corrosion resistivity, sufficient electrical conduction properties, and many more. Some of the common areas of applications are batteries, display devices, sensors, EMI shielding, printed electronic circuits, etc. [1–3]. The historical development of these materials has been started since the discovery of polysulfur nitride following which the field takes to progress in polyacetylene, nitrogen, and sulfur-containing heterocycles such as polypyrrole and polythiophene and its derivatives, polyaniline [4–6]. Classification of common CPs is in nine families as proposed by Otero [7]. These are substituted CPs (poly(3,4-ethylene dioxythiophene)), salts of pure CPs (polythiophene/ClO$_4$ doped), self-ionized conjugated CPs (PEDOT-3-butanesulfonic acid), CP and organic compound in blend forms (PEDOT:PSS), copolymer-type CPs (poly(EDOT-pyrrole-EDOT), CPs blended with inorganic compounds (polypyrrole (PPy)-polyoxometalates), CPs blended with organic compounds (PEDOT:PSS), organic- and carbon-based electroactive composites (fullerene charged poly(3-hexylthiophene)), etc. [8, 9]. CPs have been established as a material of choice in a wide field of electrochemical research that opens up their utility in the diverse fields of application areas. By the term CP, mainly we mean to say about electrically conducting polymers; however, redox polymers having redox site or polymer electrolytes or polyelectrolytes also belong to this class in a wider sense. Although the redox polymers and CPs differ from each other in terms of mechanism of electron transport due to the dissimilarity in their structural arrangements. A redox polymer is structurally defined as one with electrostatic or especially localized redox sites that is capable to be oxidized or reduced. In this case, the electrons are transported by hopping between the neighboring sites [10]. For electronically conducting polymers also commonly known as intrinsically conducting polymers or ICPs, the conduction process proceeds by the motion of the electrons through conjugation. The hopping mechanism may take place between the chains and defects present [11]. The details of the conduction mechanism with examples will be discussed in the succeeding section of this chapter.

8.2   Structure of Conducting Polymers

The inherent conducting nature of CPs arises from the nature of bonding along the main polymer chain, composed of single and double bonds. Suppose an electron has been added to the backbone of CP by reduction also called n-type doping or via removal of an electron popularly named as p-type doping is capable to generate structural arrangement leading to ease in traveling of charges when kept under a potential field. During the process of chemical and electrochemical doping, the doping degree decides the range of electrical conductivity from insulator to semiconductor to metal. The electrical conduction is very much dependent on the type of dopants, polymeric

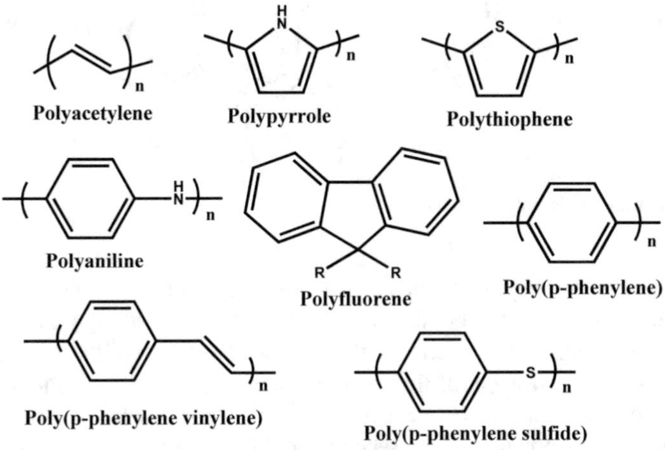

Fig. 8.1   Main chain repeating units of common conducting polymers

nature, and also on the processing technology used for the preparation of the polymers since the anisotropic alignment of the doped polymer chains enhances the conductivity by two orders of magnitude [12]. The CPs are of an excellent combination of properties, e.g., optical, electrical and mechanical when decorated with special doping characteristics, reversible doping and dedoping process, redox doping, etc. These structural characteristics of CPs open up their possibility to be utilized in different technological application areas. CPs having semiconducting behavior are preferably used in electronic devices, e.g., light-emitting diodes, solar cells, field-effect resistors, and many more. When the conductivity of CPs falls in the region of metal, they are used in EMI shielding and microwave absorption. The metallic behaviors of CPs are usefully combined with reversible redox behavior to be extensively used in the field of supercapacitors and rechargeable batteries. The electrochemical doping or dedoping induces a change in color of CPs, which opens up their utilization as displays and electrochromic windows. A judicial choice of the introduction of conductivity in a sensitive way makes them suitable to fabricate chemical sensors, drug-releasing agents, and gas separation membranes. The structure of some common CPs has been represented in Fig. 8.1.

8.3   Mechanism of Conduction in Conducting Polymer

The mechanism of conduction in conducting polymers has been explained in brief in this section to understand the transport mechanism. Here, we refer to polyacetylene as an example of a common conducting polymer to explain the conduction process. Polyacetylene possesses a simple chemical structure and of excellent electrical conductivity as shown in Fig. 8.2. The common features in most conducting polymers,

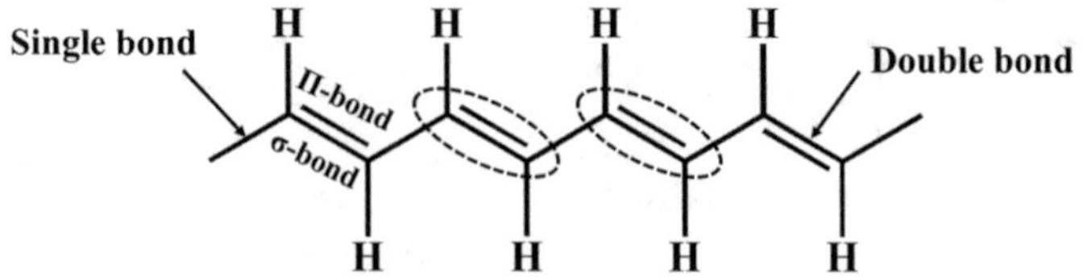

Fig. 8.2 Conjugation in polyacetylene

e.g., conjugated ones are due to the presence of alternating single and double bonds in the main chain. Both the bonds are composed of localized sigma bonds, which lead to the formation of strong chemical bonds, whereas each of the double bonds is composed of localized π-bonds that are comparatively weaker in nature [13].

The π-bond remains in between the first and second carbon atom has been exchanged with the position of the second and third carbon atom and so on. Again, the π-bond in between the third and fourth carbon travels to the next carbon and this process is repeated again and again. This, in turn, leads to the movement of the electrons across the double bonds and the conjugated double bonds allow the flow of the electrons. This is the basis of the conduction in common conducting polymers with the conjugation in the polymer skeleton. However, the conjugation in polymer does not lead to a highly conductive nature.

The conductivity of CPs has been further improved by doping [14]. Depending on the extent of doping, the conductivity may range from insulator to semiconductors to metallic. In CPs, the conductivity raised by doping is governed by redox reactions, where charges are transferred with the formation of the charge carriers [15]. The dopants play the role of electron withdrawer with the supply of additional electrons into the system. As per the concept of electronic structure, the doping induces the extraction of electrons from the highest occupied molecular orbital (HOMO) of the valence band to the lowest unoccupied molecular orbital (LUMO) of the conduction band. The process of oxidation and reduction in the polymer backbone generates different types of charge carriers such as polarons, bipolarons, and solitons. Again, CPs may have both degenerate and non-degenerate forms based on their ground-state configurations. Solitons are referred to the charge carriers of degenerate state in polyacetylenes for example. On the other hand, polarons and bipolarons are the charge carriers present in both the degenerate and non-degenerate forms, e.g., in polypyrrole and polythiophene. Polymers in degenerate states are found to have two identical geometric configurations in their ground state, while the polymers in the non-degenerate state retain two different structures of different energies in its ground state. The movements of the charge carriers are the responsible ones to promote conduction in the polymer structures [16].

CPs can be doped both p- and n-types as shown in Figs. 8.3 and 8.4. Positive and negative polarons and bipolarons are produced by doping. The charge carriers are in delocalized states in the entire polymer backbone promoting the conduction process. In general, n-doped polymers are not that stable as compared to the p-doped ones,

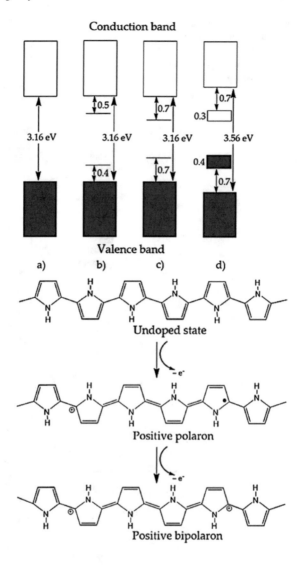

Fig. 8.3 Band and chemical structures of PPY in doped and undoped state. Redrawn and reprinted with permission from [16]

which extend the research interest of the p-doped polymers to the scientific community. The common CP, polypyrrole shows conductivity as a result of p-type doping. In the undoped form, it acts as an insulator having a large band gap of 3.16 eV. By oxidation of undoped PPY, π-electrons have been removed from the polymer chain and the local disturbance in the polymer backbone leads to the formation of polaron [17]. Again, as the oxidation proceeds further, another electron has been removed from the polymer chain leading to the formation of doubly charged bipolaron. As the process of oxidation proceeds further, the polymer is more and more oxidized and finally forms two narrow bands of bipolarons resulting in the reduction of band gap from 3.16 to 1.4 eV (Fig. 8.3).

Fig. 8.4 Band and chemical structure for n- and p-types doping in polythiophene. Redrawn and reprinted with permission from [16]

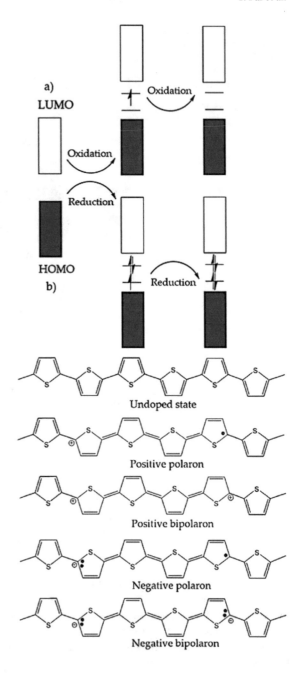

8.4 Characteristics of Common Conducting Polymers

CPs have drawn a huge interest in the field of nanoscience and nanotechnology specifically due to the excellent combination of properties such as tunable chemical and electrochemical properties, electrical conductivity, simple processing conditions, possibility to dope and dedope as per requirement [18]. A wide range of CPs is utilized in different field of applications such as polyacetylene (PA), Polyaniline (PANI), polypyrrole (PPY), polythiophene (PTh), and their derivatives. [19, 20].

8.4.1 Polyacetylene

Polyacetylene (PA) is accidentally invented by Shirakawa in the mid-seventies [21]. PA is an organic polymer having $(C_2H_2)_n$ as the repeating unit that is capable to conduct electricity. This invention has propelled the wheel of research trends in the field of CPs with the possible discovery of new polymers capable to conduct electricity. These materials, later on, have grabbed special attention in microelectronics. Later on, it has been observed that the electrical conductivity of PA can be enhanced at least six orders of magnitude when combined with iodine. This important observation has explained the possible utilization of charge carriers to improve the electrical conductivity in CPs. Polyacetylene in both cis and trans forms (Fig. 8.5) has been prepared as flexible films that are free standing on substrates like glass or metals. The trans form of the polymer is thermodynamically more stable. The cis–trans ratio of the synthesized polymer can be maintained at low temperatures; however, the complete transformation of cis form to trans can be possible once the polymer formed after the synthesis is heat treated [22]. The chemical and electrochemical doping of the

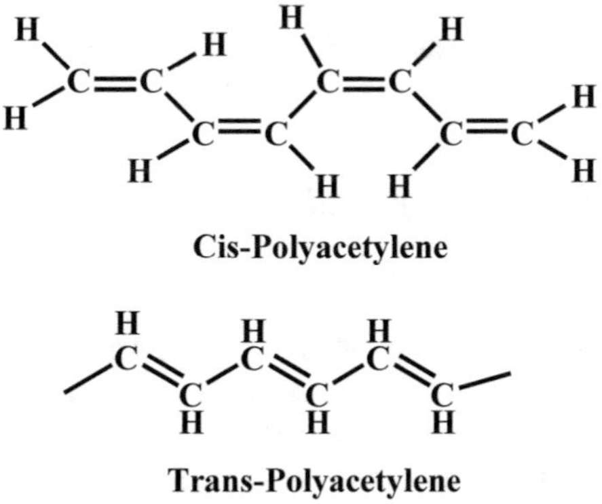

Fig. 8.5 Chemical structure and conductivity values of cis and trans polyacetylene [25]

Cis-Polyacetylene

Trans-Polyacetylene

Cis-polyacetylene	$\sigma = 1.7 \times 10^{-9}$ S cm^{-1}
Trans-polyacetylene	$\sigma = 4.4 \times 10^{-5}$ S cm^{-1}

polymer makes it possible to vary the conductivity in the range of over 12 orders of magnitude [23]. The conductivity values of 10^5 S cm^{-1} have been reported [24].

8.4.2 Polyaniline

Polyaniline remains as one of the conducting polymers having heteroatoms in its backbone structure considered as one of the important members of this class of materials. Polyaniline is characterized by controllable electrical conduction property along with redox properties and environmental stability. In the chemical structure of PANI, nitrogen has been present as the heteroatom in between the constituent phenyl rings of the polymer backbone. PANI may exist in a number of forms depending on the level of oxidation. Around the year 1835, PANI is commonly known as aniline black a product obtained due to oxidation of aniline. Later stage, it has been extensively studied and prepared from oxidation over aromatic amine. Polyaniline can be synthesized both by chemical and electrochemical routes. Polyaniline is crystalline in nature and a simple protonation leading to the formation of highly conducting emeraldine state and the conductivity depends on the pH of the solution. Polyaniline exhibits conjugated structure with wide charge delocalization, which leads to the development of an organic–polymer salt rather than conventional oxidation that remains a common phenomenon in all the other CPs. Polyaniline has the unique property of variation in electronic states both by changing the number of electrons and number of protons in the repeating unit of the main chain. As discussed earlier, polyaniline can be synthesized by chemical oxidative polymerization routes producing conducting emeraldine salt (ES) typically in the form of hydrochloride having a conductivity of 10^0 S cm^{-1} [26]. The treatment of polyaniline with the alkaline solution leads to the formation of a non-conducting form also called emeraldine base (EB) having conductivity in the range of 10^{-8}–10^{-10} S cm^{-1} (Fig. 8.6). EB can produce salts with different acids producing ES of different conductivity, density, and hydrophobicity [27].

8.4.3 Polypyrrole

Polypyrrole is an interesting conducting polymer due to the high conductivity, flexibility in preparation strategies, and good mechanical properties. Polypyrrole has been material of choice in the field of electronics and electrochromic devices, sensors, batteries, electrodes, etc. [30, 31]. Polypyrrole can be synthesized by both chemical and electrochemical methods [32, 33]. Polypyrrole is produced in bulk by oxidative polymerization of the monomer pyrrole in presence of chemical oxidants in both aqueous and non-aqueous media. In this respect, iron (III) chloride is found to be most extensively used as the chemical oxidant and water remains the best media to carry out the chemical polymerization of pyrrole [34]. During polymerization of pyrrole to

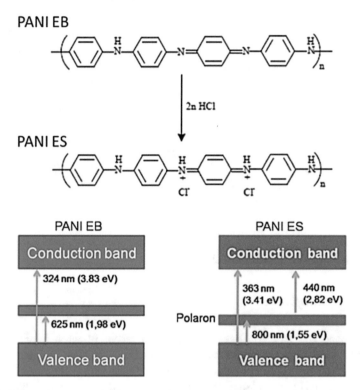

Fig. 8.6 Energy diagrams and PANI structure before (emeraldine base or EB) and after (emeraldine salt or ES) doping. Redrawn and reprinted with permission from [28], numerical values used in the figure are taken from [29]

form polypyrrole, at first, pyrrole gets oxidized by Fe^{3+} to generate pyrrole radical cation as shown in Fig. 8.7. This process is followed by formation of the dimer using two radical cations with the liberation of two protons. Finally, the chain propagation proceeds via the formation of new radical cation and so on. Again, good solubility of the monomer makes it easy to synthesize the polymer by electropolymerization in different aqueous and non-aqueous solvents [35]. Electrical conductivity remains one of the most important properties of this CP. The conductivity of polypyrrole is governed by two important factors, e.g., the number of charge carriers may be electrons or holes and second is the mobility of the charge carrier. Again, an increment in the doping level leads to the enhancement in the density of the charge carrier as well. The conductivity is found to decrease with a reduction in temperature similar to conventional semiconductors. This CP exhibits conductivity greater than $500\,S\,cm^{-1}$ [36].

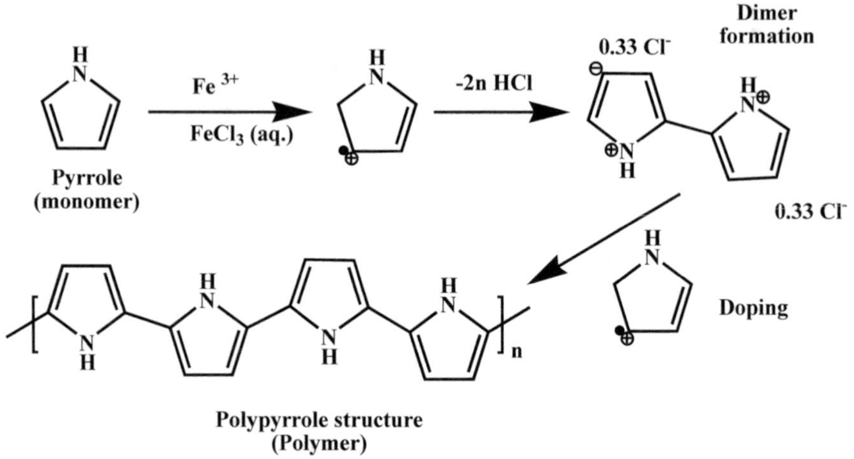

Fig. 8.7 Mechanism of polymerization of pyrrole [37]

8.4.4 Polythiophene

Polythiophene and its derivatives have been in recent research interest for the researchers working in the field of conducting polymers [38–41]. The general methods used for the synthesis of polythiophenes are chemical and electrochemical procedures. Thiophene, furan, and selenphene are polymerized using initiators like sulfuric acid, Zeigler catalyst, and iron chloride. The polyheterocycles are doped with both donor and acceptor types dopants [42, 43]. However, the high air sensitivity of the product allows the extensive use of polymers with acceptor types only. Polythiophenes are commonly doped with halogens like iodine and bromine to generate huge conductivity in the polymer; however, the material becomes unstable due to the slow evaporation of dopants [44, 45]. Other dopants utilized are different types of organic acids such as propionic, sulfonic, and trifluoroacetic acid that generate conductivity although a little bit lower side compared to the iodine doped one; however, these doped materials are of good environmental stability [46]. Polythiophene and derivatives are insoluble and infusible in nature. Polythiophenes are conductive in nature when oxidized. The delocalization of the electrons leads to the development of electrical conductivity in polythiophenes. Electron delocalization also induces optical properties in polythiophenes. They may undergo a remarkable color change in presence of solvent, applied potential, attachment with different molecules, and finally change in temperature. Changes in color of the polymer are also accompanied by the change in electrical properties of this CP. The mechanism of curving the polymer backbone and distraction of the configurations makes them attractive for applications in sensors and electrical responses [47, 48]. These CPs have found applications in display devices. The reversible doping and dedoping characteristics of the polymers make them a promising material for secondary battery electrode as well [49]. They have also been used in photovoltaic devices. Polythiophene has been used as photoactive materials for the conversion of solar energy into the electricity using undoped form [50]. They have been also utilized for protection against corrosion by grafting with thin films on the surface.

8.4.5 Others

Other conducting polymers in recent progress are polyphenylene vinylene, poly(p-phenylene sulfide), poly(p-phenylene), etc. Polyphenylene is a composite polymer made of polyacetylene and polyphenylene and is also conductive in nature when doped with AsF_5 [51]. Compared to the individual polymers, the composite polymer exhibits much lower conductivity of about 3 S cm^{-1}. The bandwidth of the HOMO of the composite polymer is reported to be 2.8 eV [5]. Poly (p-phenylene sulfide) is a highly conducting polymer when doped and is considered as the first non-rigid CP. The chemical modification of the polymer makes it possible to attain conductivity in the range of 0.01–3 S cm^{-1} [52]. Poly (p-phenylene) is one of the other CPs having conductivity as high as 500 S cm^{-1} when doped with AsF_5 [53]. Poly (p-phenylene) has found its application in the field of batteries [54]. The band gap of this CP is 3.4 eV, which is twice that of trans form of polyacetylene [53].

8.5 General Method of Synthesis of Conducting Polymers

CPs are commonly synthesized by two main procedures, electrochemical oxidation or chemical oxidation of monomer [5, 55]. However, other less common methods such as photochemical way of polymerization and enzyme-assisted polymerization are also utilized [56–58]. In practice, polymerization takes place in starting material, i.e., the monomer, and leads to the formation of the oligomer of low molecular weight. These low molecular weight oligomers lead to the formation of high molecular weight polymers by further process of oxidation at a potential lower than that of the monomer [55]. Chemical polymerization proceeds via chemical oxidation of monomers to polymer using chemical oxidants such as ammonium persulfate [$(NH_4)_2S_2O_8$], ferric chloride ($FeCl_3$), and ferric nitrate [$Fe(NO_3)_3$] [59]. Again, for the electrochemical polymerization routes, polymers get electrodeposited on to the electrode when a potential is applied in electrode dipped in a monomer solution [60–62].

8.5.1 Electrochemical Polymerization

Many techniques are utilized for the synthesis of CPs by electrochemical polymerization technique such as potentiostats, where a constant voltage is applied, galvanostatic method, where a constant current is applied at electrode dipped in the monomer solution subjected to be polymerized, and potentiodynamic, where a flexible voltage and current are applied [58, 63–65]. Electrochemical polymerization is carried out in a typical three-electrode setup composed of counter, reference, and working electrode. The polymer deposits in the working electrode during the process of polymerization. As working electrodes, platinum, indium tin oxide, or glassy carbons are used. The requirement of the electrolytes depends on the types of CPs to be prepared, e.g., inorganic salts or ionic liquids are required to polymerize aniline; however, they are not vital for the preparation of PEDOT or polypyrrole during the process of synthesis. The protons present maintain acidic pH range to avoid the formation of branched structures, undesired products, and generation of doped CPs [66, 67].

8.5.2 Chemical Polymerization

In chemical polymerization, monomers are oxidized with the help of oxidizing agents that initiate the process of polymerization. The oxidizing agents commonly used are ammonium persulfate and ferric chloride [68]. For the polymerization of polyaniline, ferric chloride has been extensively used to produce high molecular weight polyaniline although the oxidation potential of ferric chloride remains much lower compared to other oxidizing materials [69]. Polymerization of aniline-based CPs requires excess protons; hence, the reaction mixture has to be kept at around pH < 3 for successful polymerization to proceed [70].

The polymerization of PANI and PEDOT takes place by two general routes, first formation of an undoped polymer via polymerization of the respective monomers and second doping of the as-formed neutral polymer due to the presence of excess oxidant present in the reaction mixture (Fig. 8.8). The chemistry of oxidation can be explained by the steps such as oxidation of monomer into the radical cation, formation of dimers by radical coupling, and finally, by propagation of the polymeric chains to produce a final polymer of desired molecular weight.

8.6 Properties of Conducting Polymers

8.6.1 Electrical Conductivity

The electrical conductivity is the most important property of a conducting polymer. Conductivity is found to be in the range of 10^{-3}–10^3 S cm^{-1} in case of doped polymers, whereas for intrinsically CPs with conjugation, the conductivity lies in the range of 10^{-9}–10^{-6} S cm^{-1} [72]. Hence, the study reveals that doping has a pronounced effect on the conductivity of CPs. CPs are of amorphous nature in general with the exception of some ordered domains. Hence, the transport mechanism lies

Fig. 8.8 Mechanism of
polymerization of PANI and
PEDOT. Reprinted with
permission from [71]

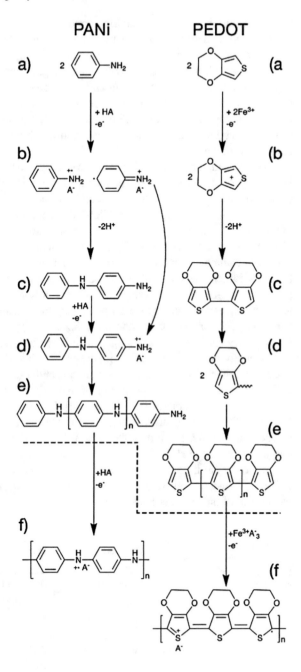

to be different than that of crystalline conductive materials. For a complete under-standing of the mechanism of electrical conduction, refer to the mechanism portion of this chapter. In CPs, the conductivity remains temperature-dependent and varies enormously with the change in temperature. Again, the conductivity is very much dependent on the degree of doping and the degree of ordering in solid films. The degree of doping is related to the concentration of the charge carrier on the main chain. The doping at a low doping degree, i.e., for doping much lower than the saturation, conductivity follows a linear trend with a doping degree in the case of conjugated polymers.

As discussed, the conductivity of CPs can be varied by changing the dopant types. Depending on the molecular size, dopants are classified as a large polymeric class or small ions (cations or anions). The examples of polymeric species are polyvinyl and polystyrene sulfonates and Cl^- and ClO_4^- are that for the small ions. Some examples of conductivity of CPs with the types of dopants are enlisted in Table 8.1. The nature of dopants alters the conductivity of the conducting polymers with the additional benefit of change in surface and structural characteristics. As the size of the dopants becomes larger, the physical properties, surface morphology, and density of the polymer change. Another advantage of using large dopants is that they can be strongly anchored with the polymer matrix minimizing the leaching tendency of the dopants from the polymer matrix. On the other hand, as the size of the dopants remains smaller, they can be easily exchanged with other ions present in the medium [73].

Conductivity in CPs increases with the progressive addition of dopants in the con-ducting polymers. The increment in electrical conduction resulted from enhancement in dopant concentration has been studied in detail by Tsukamoto et al. where poly-acetylene has been doped with iodine [24]. The electrical conductivity increases due to the formation of stack structures with the successive addition of dopants and reaches saturation after doping for several hours. The extensive doping inside the polymer structure leads to slow diffusion and the structure of the polymer becomes progressively hard, dense, and ordered in nature [74]. The process of doping and dedoping is usually reversible and dedoping forms original undoped state without weakening of the polymer backbone.

Table 8.1 Conductivity with types of dopants of some common conducting polymers	Conducting polymer	Dopants	Conductivity ($S\ cm^{-1}$)
	Polyacetylene	AsF_5, I_2, Na	10^4
	Polypyrrole	BF_4^-, ClO_4^-	500–7.5×10^3
	Polythiophene	BF_4^-, ClO_4^-, $FeCl_4^-$	10^3
	Polyphenylene	AsF_5, Li, K	10^3
	Polyaniline	HCl	200
	Poly phenylene vinylene	AsF_5	10^4

Data are reproduced with permission from [57]

8.6.2 Absorption Characteristics

The absorption spectrum of CPs is a clear-cut evidence of the presence of doping in it. A strong absorption peak has been evidenced near-infrared region for doped CPs due to the presence of polarons and bipolarons, which disappears after dedoping. For example, the absorption spectrum of PPY in doped state exhibits an absorption peak at 700–ca. 2000 nm due to the presence of polaron and bipolaron energy levels, whereas for the dedoped state, there exists an absorption peak at 400 nm due to the π-π* transition in PPY main chain [72]. In the case of PANI, the case is a little more complicated due to the presence of different structural transformations in PANI. The absorption spectra of PANI show that a strong peak is present at around 950 nm in doped state due to the presence of the polaron and bipolaron. Another absorption peak has been evidenced at around 320 nm due to the π-π* transition in PANI. With further reduction in PANI, the near-infrared absorption peak disappears in completely reduced state. The conducting PANI can be further proton acid dedoped to be in the state of pernigraniline, where it shows two absorption peaks (s) at ca. 320 and 630 nm. The absorption peak at about 320 nm corresponds to π-π* transition and 630 nm is due to the electronic transition between HOMO and LUMO of the PANI structure [72]. Figure 8.9 shows the optical absorption spectrum of PANI thin film indicating three different peak positions corresponding to the transitions due to polarons, bipolarons, and π-π*.

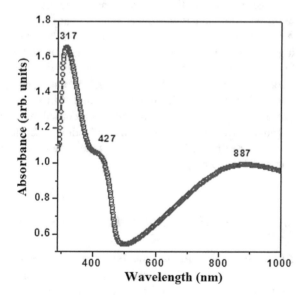

Fig. 8.9 Optical absorbance spectra for PANI showing peaks at 317, 427, and 887 nm due to the π-π* electronic, polaron, and bipolaron transitions in PANI. Redrawn and reprinted with permission from [75]

8.6.3 Electrochemical Characteristics

CPs have found applications in the area of sensors and capacitors depending on the reversible doping and dedoping characteristics. This reversible doping–dedoping in CPs corresponds to the charging and discharging process. During p-doping, the electrons from the π-bonds are removed and flow through the polymer backbone while, on the other hand, counter ions from the electrolyte go inside the polymer chain to maintain the electrical charge balance. In the case of n-doping, electrons are transported to the main chain of the polymer and the counter ions again placed inside the polymer matrix from the electrolyte to maintain the charge balance. Cyclic voltammetry is the most common process by which electrochemical characterization of CP has been studied. During doping of the polymer chain, the redox reactions proceed leading to the formation of negative or positive charges as it is being reduced or oxidized. Recently, CPs have attracted much attention in the field of electrochemical capacitors. Electrochemical capacitors are mainly two types depending on the mechanism of charge storage, namely electrochemical double-layer capacitors and pseudocapacitors. A number of materials ranging from carbon materials, metal oxides, CPs, and their mutual combinations have been studied. The charge storage mechanism in EDLCs occurs by adsorption of the electrolyte ions over the electrode surface, whereas the pseudocapacitors act like a battery and provide much better specific capacitance and energy density compared to EDLCs. CPs due to the reversible charge storage mechanism follow pseudocapacitive behavior. During the charging process, anions from the electrolyte go inside the polymer matrix and electrons got removed from the CP backbone. Hence, charging always proceeds with oxidation. Again, during discharging process, reduction becomes the phenomenon. Conducting polymers are often used as matrix materials with carbon nanotubes to produce high-performance polymer nanocomposites [76, 77]. A cyclic voltammogram of a common conducting polymer, polyaniline, and its composite with multiwalled carbon nanotube (MWCNT) has been included in Fig. 8.10 to understand the redox process in CP. The oxidation potential is recorded in positive current window and the negative current window shows the reduction potential of the material. The curves also indicate that the current density of the composite electrode remains much higher compared to the polymer.

8.6.4 Solubility Characteristics

CPs are in general insoluble and infusible in nature due to the presence of rigid conjugated polymeric chains, which somehow limit its application in this form. Several strategies have been adopted to overcome this particular issue and make it in a usable form. Cao et al. have prepared soluble PANI and resolved the inherent problem of solubility of PANI, which opens up the possibility of large-scale production of PANI [79]. The solubility of CPs having conjugation has been further improved by the

Fig. 8.10 Cyclic voltammograms of polyaniline and composite with MWCNT. Redrawn and reprinted with permission from [78]

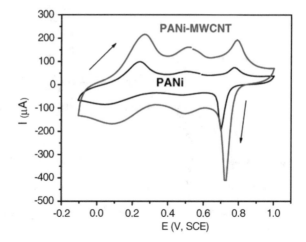

introduction of flexible side-chain substitution, which remains a general approach in this respect. For example, polythiophene without side-chain functionalization is found to be insoluble in organic solvents; however, an introduction of hexyl groups to form the derivative P3HT gains its solubility in several organic solvents such as toluene, chlorobenzene, etc. In this respect, it is worthy to mention that in addition to improve the solubility in CPs, the introduction of side chains has found to be advantageous to tune the band gaps and the electronic states in CPs that are of huge importance to be used in the field of photovoltaics and electroluminescence.

8.6.5 Swelling and De-Swelling

CPs when subjected to redox process during doping or dedoping face excessive volumetric changes inside it [80, 81]. The complete process of swelling or de-swelling can be divided into two components. First, the intrinsic one that happens due to the alteration in bond lengths and conformational changes in the polymer backbone and the second one is the osmotic expansion in the polymer matrix [82]. This process is usable in the field of actuators. CP-based actuators have the advantage of tunable properties by chemical or electrochemical ways. CP when acts as an actuator, the polymeric chains develop a positive charge when electrons are removed by the oxidation process. Small anions are often incorporated to maintain charge balance inside the polymer structure. In the next process, ionic crosslinks are generated between the anions and polymeric chains leading to an expansion in polymer volume. The actuation process guided by anions causes swelling and de-swelling in the polymer matrix by oxidation and reduction [83].

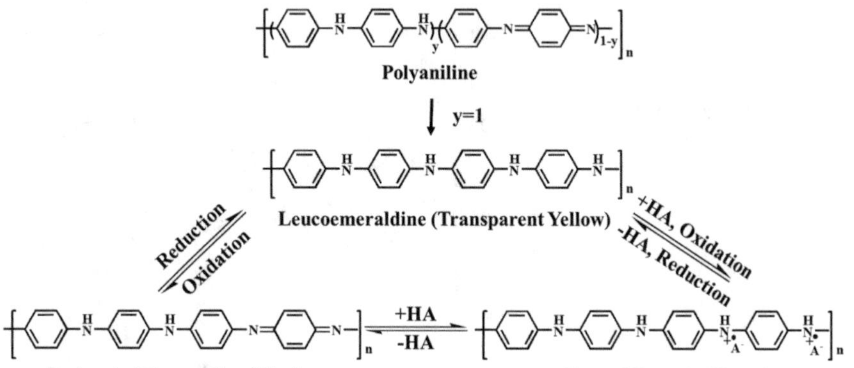

Fig. 8.11 Redox states and corresponding colors of polyaniline. Redrawn and reprinted with permission from [16]

8.6.6 Electrochromism

Again, CPs possess electrochromism since the reversible redox process leads to the development of color change in the polymer. This particular property makes the CPs exploitable in the field of electrochromic applications such as displays, rearview mirrors, and smart windows. [84, 85]. The color change depends on the energy gap in the polymers and also the presence of dopants. The doping in the CP leads to reformation of the electronic state of the polymer. This addition of the dopant lowers the energy gap of π-π^* transition. The generation of new sub-bands due to the insertion of the dopant molecules leads to alteration in the absorbance of CPs, which finally changes the color of the polymer. In general, the undoped CPs have an energy gap higher than 3 eV and the films are transparent and colorless in the undoped state (Fig. 8.11). However, if the pristine CP has an energy gap of around 1.5 eV, they absorb the visible light strongly leading to the formation of high color contrast in the undoped state as well. Doping shifts the absorption to near-infrared region [84]. Hence, the color contrast in CPs is due to difference in the redox states.

8.7 Applications

The applications of CPs include supercapacitors, solar cells, sensors, corrosion inhibitors, light-emitting diodes, EMI shielding, electrochromic devices, antistatic coating, transistors, and many more [86]. The research in the field of supercapacitors has been one of the main focuses of the CPs. CPs in combination with different inorganic materials, such as graphene, carbon nanoparticles, nanofibers, and graphene oxide, have been widely explored by several research groups [87–90]. A combination of these materials in CPs along with metal oxides of high surface area and mechanical properties is sound promising [91, 92]. CPs are of low cost and capable to exhibit good pseudocapacitive behavior; however, they often lack in mechanical strength. Metal oxide-based supercapacitors are traditional and well established; however, many of them are quite expensive, e.g., RuO_2. Hence, a proper combination of these materials with CPs can exhibit a synergistic effect to be used as a new emerging electrode material for supercapacitor devices. CPs also play a dominating role in the field of batteries [93, 94]. Various reports, such as rechargeable sodium metal/CP batteries, paper-based storage devices, flexible batteries, and ultralight polymers, have also been studied [93, 94]. CPs are found to be a promising material in the field of biosensors as well [95, 96]. Biomimetic membranes made of CPs are used for the detection of taurine [97]. Other examples are CP/CNT modified electrode materials used for the preparation of enzyme biosensors for detection of superoxides that have been reported as well [98]. CPs and their composites find extensive applications in photoelectrochemical and photocatalytic processes [99, 100]. Polyanilne is one of the well-known examples of photoelectrochemical reducer for the reduction of CO_2 to alcohols [101]. Another application area for CPs includes electrocatalysts where they have been used without any other materials [102, 103], corrosion resistance [104, 105], microwave absorption [106], etc.

8.8 Concluding Remarks

Soon after the discovery of conducting polyacetylene, CPs have gained the attention of research community to a great extent. The unique combination of attractive properties of CPs such as electrical conductivity, mechanical, and thermal stability, has made them a material of choice for a number of applications. The electrical conductivity of the CPs can be easily tailored by judicially selecting the types and amount of doping. CPs can be synthesized by two common methods of polymerizations such as chemical and electrochemical methods. This chapter elucidates the types of CPs, their conduction mechanism, properties, and applications to understand the recent technological advancement in the field of conducting polymers.

Acknowledgements The authors acknowledge the financial support provided by Department of Science and Technology, India (DST/TMD/MES/2K16/37(G)), for carrying out this research work.

References

1. G. Inzelt, J. Solid State Electrochem. **21**, 1965 (2017)
2. K.K. Kar, *Composite Materials: Processing, Applications, Characterizations* (Springer, 2016)
3. S.K. Singh, H. Prakash, M. Akhtar, K.K. Kar, A.C.S. Sustain, Chem. Eng. **6**, 5381 (2018)
4. H. Shirakawa, A. McDiarmid, A. Heeger, Chem. Commun. **2003**, 1 (2003)
5. T.A. Skotheim, J. Reynolds, *Handbook of Conducting Polymers, 2 Volume Set* (CRC press, 2007)

6. T.M. Swager, Macromolecules **50**, 4867 (2017)
7. T. Otero, J. Martinez, J. Mater. Chem. B **4**, 2069 (2016)
8. J. Cherusseri, K.K. Kar, RSC Adv. **5**, 34335 (2015)
9. K.K. Kar, S. Rana, J. Pandey, *Handbook of Polymer Nanocomposites Processing, Performance and Application* (Springer, 2015)
10. S. Banerjee, K.K. Kar, M.K. Das, Recent Pat. Mater. Sci. **7**, 173 (2014)
11. S. Banerjee, K.K. Kar, Recent Pat. Mater. Sci. **7**, 131 (2014)
12. M. Hundley, P. Adams, B. Mattes, Synth. Met. **129**, 291 (2002)
13. R. Ravichandran, S. Sundarrajan, J.R. Venugopal, S. Mukherjee, S. Ramakrishna, J. Royal Soc. Interface **7**, S559 (2010)
14. A.G. MacDiarmid, R. Mammone, R. Kaner, L. Porter, Philos. Trans. Royal Soc. A **314**, 3 (1985)
15. D. Trivedi, H. Nalwa, Wiley, New York **2**, 505 (1997)
16. T.-H. Le, Y. Kim, H. Yoon, Polymers **9**, 150 (2017)
17. A. Patil, A. Heeger, F. Wudl, Chem. Rev. **88**, 183 (1988)
18. K.K. Kar, A. Hodzic, *Developments in Nanocomposites* (2014)
19. X. Lu, W. Zhang, C. Wang, T.-C. Wen, Y. Wei, Prog. Polym. Sci. **36**, 671 (2011)
20. J. Stejskal, I. Sapurina, M. Trchová, Prog. Polym. Sci. **35**, 1420 (2010)
21. H. Shirakawa, Angew. Chem. Int. Ed. **40**, 2574 (2001)
22. H. Shirakawa, T. Ito, S. Ikeda, Macromol. Chem. Phys. **179**, 1565 (1978)
23. P.J. Nigrey, D. MacInnes, D.P. Nairns, A.G. MacDiarmid, A.J. Heeger, J. Electrochem. Soc. **128**, 1651 (1981)
24. J. Tsukamoto, A. Takahashi, K. Kawasaki, Jpn. J. Appl. Phys. **29**, 125 (1990)
25. A. Upadhyay, S. Karpagam, Rev. Chem. Eng. **35**, 351 (2019)
26. J. Stejskal, R. Gilbert, Pure Appl. Chem. **74**, 857 (2002)
27. P. Bober, T. Lindfors, M. Pesonen, J. Stejskal, Synth. Met. **178**, 52 (2013)
28. R. Castagna, R. Momentè, G. Pariani, G. Zerbi, A. Bianco, C. Bertarelli, Polym. Chem. **5**, 6779 (2014)
29. A. Eftekhari, *Nanostructured Conductive Polymers* (Wiley, 2011)
30. J. Cherusseri, K.K. Kar, J. Mater. Chem. A **4**, 9910 (2016)
31. V.V. Tat'yana, O.N. Efimov, Russ. Chem. Rev. **66**, 443 (1997)
32. R. Ansari, J. Chem. **3**, 186 (2006)
33. J. Cherusseri, K.K. Kar, PCCP **18**, 8587 (2016)
34. S. Banerjee, K.K. Kar, J. Environ. Chem. Eng. **4**, 299 (2016)
35. J. Cherusseri, K.K. Kar, RSC Adv. **6**, 60454 (2016)
36. R.A. Khalkhali, Iran. Polym. J. **53** (2004)
37. B. Yeole, T. Sem, D. Hansora, S. Mishra, Am. J. Sens. Technol. **4**, 10 (2017)
38. H. Kokubo, T. Sato, T. Yamamoto, Macromolecules **39**, 3959 (2006)
39. H.-F. Lu, H.S. Chan, S.-C. Ng, Macromolecules **36**, 1543 (2003)
40. I. McCulloch et al., Nat. Mater. **5**, 328 (2006)
41. T. Yamamoto et al., J. Am. Chem. Soc. **120**, 2047 (1998)
42. G. Tourillon, F. Garnier, J. Electroanal. Chem. Interfac. Electrochem. **135**, 173 (1982)
43. T. Yamamoto, K.-I. Sanechika, A. Yamamoto, Bull. Chem. Soc. Jpn **56**, 1503 (1983)
44. M. Loponen, T. Taka, J. Laakso, K. Väkiparta, K. Suuronen, P. Valkeinen, J.-E. Österholm, Synth. Met. **41**, 479 (1991)
45. R.D. McCullough, S. Tristram-Nagle, S.P. Williams, R.D. Lowe, M. Jayaraman, J. Am. Chem. Soc. **115**, 4910 (1993)
46. J. Bartuš, J. Macromol. Sci. Chem. **28**, 917 (1991)
47. D.T. McQuade, A.E. Pullen, T.M. Swager, Chem. Rev. **100**, 2537 (2000)
48. C.B. Nielsen, I. McCulloch, Prog. Polym. Sci. **38**, 2053 (2013)
49. K. Kaneto, K. Yoshino, Y. Inuishi, Jpn. J. Appl. Phys. **22**, L567 (1983)
50. S. Glenis, G. Horowitz, G. Tourillon, F. Garnier, Thin Solid Films **111**, 93 (1984)
51. G.E. Wnek, J.C. Chien, F.E. Karasz, C.P. Lillya, Polymer **20**, 1441 (1979)
52. R. Chance, L. Shacklette, G. Miller, J. Chem. Soc. Chem. Commun. **348** (1980)

53. D. Ivory, G. Miller, J. Sowa, L. Shacklette, R. Chance, R. Baughman, J. Chem. Phys. **71**, 1506 (1979)
54. L.W. Shacklette, R.L. Elsenbaumer, R.R. Chance, J.M. Sowa, D.M. Ivory, G.G. Miller, R.H. Baughman, J. Chem. Soc. Chem. Commun. **361** (1982)
55. D.J. Sandman, Mol. Cryst. Liq. Cryst. **325**, 260 (1998)
56. S. Annapoorni, N. Sundaresan, S. Pandey, B. Malhotra, J. Appl. Phys. **74**, 2109 (1993)
57. D. Kumar, R. Sharma, Eur. Polym. J. **34**, 1053 (1998)
58. G.G. Wallace, P.R. Teasdale, G.M. Spinks, L.A. Kane-Maguire, *Conductive Electroactive Polymers: Intelligent Materials Systems* (CRC press, 2002)
59. G.A. Snook, A.I. Bhatt, M.E. Abdelhamid, A.S. Best, Aust. J. Chem. **65**, 1513 (2012)
60. L. Groenendaal, G. Zotti, P.H. Aubert, S.M. Waybright, J.R. Reynolds, Adv. Mater. **15**, 855 (2003)
61. G.A. Snook, A.S. Best, J. Mater. Chem. **19**, 4248 (2009)
62. I. Villarreal, E. Morales, T. Otero, J. Acosta, Synth. Met. **123**, 487 (2001)
63. R. Gracia, D. Mecerreyes, Polym. Chem. **4**, 2206 (2013)
64. T. Otero, J. Martinez, J. Arias-Pardilla, Electrochim. Acta **84**, 112 (2012)
65. G.A. Snook, G.Z. Chen, D.J. Fray, M. Hughes, M. Shaffer, J. Electroanal. Chem. **568**, 135 (2004)
66. G.A. Snook, T.L. Greaves, A.S. Best, J. Mater. Chem. **21**, 7622 (2011)
67. G. Zotti, S. Cattarin, N. Comisso, J. Electroanal. Chem. Interfac. Electrochem. **239**, 387 (1988)
68. W.K. Maser, R. Sainz, M.T. Martínez, A.M. Benito, Contrib. sci. **187** (2008)
69. A.G. Green, A.E. Woodhead, J. Chem. Soc. Transac. **101**, 1117 (1912)
70. M. Wu, G.A. Snook, V. Gupta, M. Shaffer, D.J. Fray, G.Z. Chen, J. Mater. Chem. **15**, 2297 (2005)
71. M.E. Abdelhamid, A.P. O'Mullane, G.A. Snook, RSC Adv. **5**, 11611 (2015)
72. Y. Li, *Organic Optoelectronic Materials*, vol 91 (Springer, 2015)
73. N.K. Guimard, N. Gomez, C.E. Schmidt, Prog. Polym. Sci. **32**, 876 (2007)
74. J. Tsukamoto, Adv. Phys. **41**, 509 (1992)
75. B. Ashwini, P.U. Rohom, N. Chaure, Nanosci. Nanotechnol. **6**, 83 (2016)
76. K. Kar, A. Hodzic, *Carbon Nanotube based Nanocomposites: Recent Development* (Research Publishing, Singapore, 2011)
77. R. Kumar, S. Sahoo, E. Joanni, R.K. Singh, W.K. Tan, K.K. Kar, A. Matsuda, Prog. Energy Combust. Sci. **75**, 100786 (2019)
78. T.H. Le, N.T. Trinh, L.H. Nguyen, H.B. Nguyen, V.A. Nguyen, T.D. Nguyen, Adv. Natural Sci. Nanosci. Nanotechnol. **4**, 025014 (2013)
79. Y. Cao, P. Smith, A.J. Heeger, Synth. Met. **48**, 91 (1992)
80. L. Lizarraga, E.M. Andrade, F.V. Molina, J. Electroanal. Chem. **561**, 127 (2004)
81. Q. Pei, O. Inganäs, J. Phys. Chem. **96**, 10507 (1992)
82. L. Bay, T. Jacobsen, S. Skaarup, K. West, J. Phys. Chem. B **105**, 8492 (2001)
83. W. Takashima, S.S. Pandey, M. Fuchiwaki, K. Kaneto, Jpn. J. Appl. Phys. **41**, 7532 (2002)
84. A. Barnes, A. Despotakis, T. Wong, A. Anderson, B. Chambers, P. Wright, Smart Mater. Struct. **7**, 752 (1998)
85. H. Pages, P. Topart, D. Lemordant, Electrochim. Acta **46**, 2137 (2001)
86. C. Zhan, G. Yu, Y. Lu, L. Wang, E. Wujcik, S. Wei, J. Mater. Chem. C **5**, 1569 (2017)
87. R.A. Davoglio, S.R. Biaggio, N. Bocchi, R.C. Rocha-Filho, Electrochim. Acta **93**, 93 (2013)
88. M. Hao et al., J. Solid State Electrochem. **20**, 665 (2016)
89. S. Wang, T. Gao, Y. Li, S. Li, G. Zhou, J. Solid State Electrochem. **21**, 705 (2017)
90. Z. Zhou, X.-F. Wu, J. Power Sources **262**, 44 (2014)
91. S. Hashmi, J. Solid State Electrochem. **18**, 465 (2014)
92. J. Kim, Y. Jung, S. Kim, J. Nanosci. Nanotechnol. **15**, 1443 (2015)
93. A. Guerfi, J. Trottier, C. Gagnon, F. Barray, K. Zaghib, J. Power Sources **335**, 131 (2016)
94. L. Nyholm, G. Nyström, A. Mihranyan, M. Strømme, Adv. Mater. **23**, 3751 (2011)
95. H. Hu, L. Cao, Q. Li, K. Ma, P. Yan, D.W. Kirk, RSC Adv. **5**, 55209 (2015)

96. M.A. Komkova, E.A. Andreyev, V.N. Nikitina, V.A. Krupenin, D.E. Presnov, E.E. Karyakina, A.K. Yatsimirsky, A.A. Karyakin, Electroanalysis **27**, 2055 (2015)
97. J. Kupis-Rozmysłowicz, M. Wagner, J. Bobacka, A. Lewenstam, J. Migdalski, Electrochim. Acta **188**, 537 (2016)
98. M. Braik, M.M. Barsan, C. Dridi, M.B. Ali, C.M. Brett, Sensors Actuators B: Chem. **236**, 574 (2016)
99. A. Kormányos, B. Endrődi, R. Ondok, A. Sápi, C. Janáky, Materials **9**, 201 (2016)
100. E.P. Krivan, D. Ungor, C. Janáky, Z. Németh, C. Visy, J. Solid State Electrochem. **19**, 37 (2015)
101. D. Hursán, A. Kormányos, K. Rajeshwar, C. Janáky, Chem. Commun. **52**, 8858 (2016)
102. D. Hursán, G. London, B. Olasz, C. Janáky, Electrochim. Acta **217**, 92 (2016)
103. T.V. Magdesieva, O.M. Nikitin, E.V. Zolotukhina, M.A. Vorotyntsev, Electrochim. Acta **122**, 289 (2014)
104. P.P. Deshpande, D. Sazou, *Corrosion Protection of Metals by Intrinsically Conducting Polymers* (CRC Press, 2016)
105. U. Riaz, C. Nwaoha, S. Ashraf, Prog. Org. Coat. **77**, 743 (2014)
106. J. Yan, Y. Huang, X. Chen, C. Wei, Synth. Met. **221**, 291 (2016)

Chapter 9
Characteristics of Electrode Materials for Supercapacitors

Kapil Dev Verma, Prerna Sinha, Soma Banerjee and Kamal K. Kar

Abstract Device performance is based on the individual properties of the materials and performance of the components in the working environments. For example, the fabrication of high-performance supercapacitors and the electrode material should have a high specific surface area and high electrical conductivity along electrical and thermal stability. For different charge storage mechanisms in supercapacitors like electrical double-layer capacitors, pseudocapacitors and hybrid capacitors, and different types of electrode materials are proposed. Electrode materials of a supercapacitor decide the storage of charge in the device and thereby the capacitance of the final device. The effective surface area including electrical conductivity remains the parameter of importance to produce high capacitance. Carbon materials are proposed as the electrode material by storage of the charge at the surface of the material via electrical double-layer capacitance. High surface area, appropriate pore size, pore size distribution and the presence of functional groups complement the capacitance of the device. Some commonly used carbon-based materials of interest are graphite, graphene, carbon nanotube, activated carbon, etc. Other materials of importance remain metal oxides, conducting polymers, metal–organic frameworks, MXenes, black phosphorus, metal nitrides, etc. This chapter provides a short yet comprehensive overview of the characteristics of suitable electrode material for supercapacitor

K. D. Verma · P. Sinha · S. Banerjee · K. K. Kar (✉)
Advanced Nanoengineering Materials Laboratory, Materials Science Programme, Indian Institute of Technology Kanpur, Kanpur 208016, India
e-mail: kamalkk@iitk.ac.in

K. D. Verma
e-mail: kdev@iitk.ac.in

P. Sinha
e-mail: findingprerna09@gmail.com

S. Banerjee
e-mail: somabanerjee27@gmail.com

K. K. Kar
Advanced Nanoengineering Materials Laboratory, Department of Mechanical Engineering, Indian Institute of Technology Kanpur, Kanpur 208016, India

© Springer Nature Switzerland AG 2020
K. K. Kar (ed.), *Handbook of Nanocomposite Supercapacitor Materials I*,
Springer Series in Materials Science 300,
https://doi.org/10.1007/978-3-030-43009-2_9

devices. Activated carbon, carbon nanotubes, graphene, polyaniline (PANI), polypyrrole (PPY) and polythiophene (PTH) are examples of some of the suitable electrode materials.

9.1 Introduction

Every material has its specific property and limitation. Properties of a material govern the specific performance and its potential applications. It is rare that the performance of material depends only on one property; in contrast, it is always the combination of different properties. For better performance of a device, it is necessary to know the variation of one property against other property to understand the limitations and correlations among the material properties.

Nowadays, energy storage devices are an important area for research so that the various smart devices can be made to overcome the energy crises of the present age. A new technology, known as supercapacitor, has emerged with an excellent ability to store energy. It utilizes the properties of high surface area electrodes and thinner dielectrics to attain higher capacitance value. Supercapacitor follows the same governing equations, which are used for conventional capacitors. It is an electrical storage device having high energy storage density simultaneously with a limited-power density (compared to capacitor). As a result, supercapacitors may become an attractive power solution for an increasing number of applications. The supercapacitors have the capability of delivering high power as compared to the batteries. It has capacitance much higher than the conventional capacitors that bridges the gap between electrolyte capacitors and rechargeable batteries. It provides a hundred to thousand times of high power as compared to the conventional batteries having the same volume and delivers charges at must faster rate. Hence, supercapacitors are suitable materials for those application areas, where high power is needed. Applications where rapid charge/discharge cycles are required rather than long term energy storage, supercapacitors are the material of interest. There are two major developments in the area of supercapacitors: The first device is electrochemical double-layer capacitors (EDLCs), and the other is the pseudocapacitors. EDLCs store charges by the formation of a electrochemical double layer at the electrode/electrolyte interfaces, whereas pseudocapacitors use electron transfer mechanisms such as Faradaic reactions to store charges [1–3].

In the context of high-performance supercapacitor, for the material selection of supercapacitors, it is necessary to take care of four components. These are current collector, electrode, electrolyte and separator. Supercapacitor performance depends on the selection of materials of these four elements. But this chapter discusses the characteristics of electrode materials used in supercapacitors.

9.2 Historical Background of Supercapacitors

The history of supercapacitors is not very old; however, within a small span of time it has received a lot of attention. In 1853, Hermann von Helmholtz has introduced electric double-layer capacitor (EDLC), which stores charge electrostatically at the electrode and electrolyte interface. Practical use of electrostatic double-layer capacitance has been accomplished after filing a patent on the porous carbon-based capacitor in the year of 1957 by Becker at the General Electric Corporation (Becker 1957). For commercial purpose, the first supercapacitor based on porous carbon and tetra alkaline ammonium salt has been manufactured by the Sohio Corporation (now called BP), Cleveland, in the year of 1969. Sohio Corporation abandons its further marketing, and the license of manufacturing has been given to NEC Corporation (Japanese multinational provider of information technology services and products). NEC Corporation has successfully marketed a supercapacitor for the memory backup of computers [4]. Pseudocapacitor is introduced in the year of 1971. Storage of the proton on the surface of thin-filmed RuO_2 surface results in a fast Faradaic reaction called pseudocapacitor [5]. Till the year of the 1990s, many companies have been introduced in the field of supercapacitor, which provides improved performance and low-cost supercapacitor for hybrid vehicles [4]. For energy density improvement of supercapacitor, a hybrid capacitor comes into the picture with one capacitor-type electrode and other battery-type electrode [6–8].

9.3 Types of Supercapacitors

Electrochemical capacitor stores charge electrostatically on the surface of the electrode material, unlike the battery, where phase change takes place. Here, charge accumulates at the interface of the high surface area electrode and the electrolyte. Because of the physical storing of the charge, the process is fast and highly reversible. When it is connected with the external source, the electrolyte ions get collected over the electrode surface. The electrodes already have a particular charge either positive or negative, and opposite ions get accumulated on the electrode also called electric double-layer capacitor. Electrolyte ions and electrode surface are separated by the solvent molecules, which act as a dielectric material. Solvent molecules make a single layer over the surface of electrolyte ions called solvation. Solvation occurs because of hydrogen bonding, van der Waals forces and ion–dipole interactions. Some ions pass through the solvent layer and intercalate or adhere to the surface of electrode, and fast Faradaic reaction takes place called pseudocapacitor. Figure 9.1 demonstrates types of supercapacitors like EDLCs, pseudocapacitor, hybrid capacitor with their corresponding electrode materials.

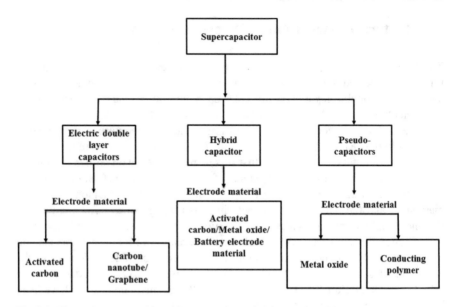

Fig. 9.1 Types of supercapacitor with commonly used electrode material

9.3.1 Electric Double-Layer Capacitors

During charging, EDLC stores ions at the interface of the electrode and electrolyte by fast ion adsorption [9]. In EDLCs, non-Faradaic and highly reversible charge–discharge reactions take place, which manifest mirror image voltammograms [9–11]. The charge storage mechanism in EDLC is based on three models: Helmholtz model, Gouy–Chapman model and Stern model (Fig. 9.2). Helmholtz model considers that two parallel layers of opposite charges form at the interface of the electrode and electrolyte (Fig. 9.2a). It is stated that the potential window varies linearly till the outer Helmholtz layer. Gouy–Chapman model considers the thermal motion of the ions, which forms a diffused double layer at the surface of the electrode material instead of closely packed ions on the surface of electrode material (Fig. 9.2b). Again, the potential varies exponentially in this case. Stern model is the combination of the Helmholtz and Gouy–Chapman models. This model concludes that there will be compact ions on the electrode surface (Helmholtz model) and next to it there will be a diffused layer of ions (Gouy–Chapman model). Here, the potential varies linearly till the outer Helmholtz layer and then exponentially in the diffused layer (Fig. 9.2c). Equivalent capacitance in the Stern model is a series combination of the capacitance of the charges held inside the outer Helmholtz layer and capacitance of the diffused layer [12]. EDLC is known for its high power capability and cycle life, while pseudocapacitor provides high capacitance and high energy density [9]. For high EDLC, the porosity of the electrode material and electrolyte ion size plays an important role [13].

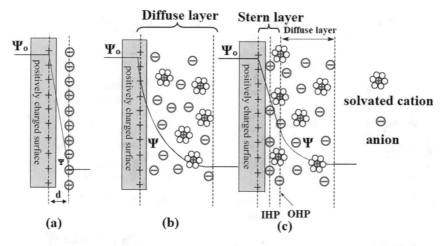

Fig. 9.2 Charge and potential distribution in Helmholtz, Gouy–Chapman and stern model. Redrawn and reprinted with permission from [14]

Accumulation of ions at the surface of the electrode also depends on pore size of the electrode material. When the pore size of electrode material remains smaller than diameter of the hydrated ions, then ions will not be able to participate in double-layer formation. Electro-adsorption of the ions into different pore size materials not necessarily provides the same capacitance value even if they are of same surface area.

9.3.2 Pseudocapacitors

Although EDLC and pseudocapacitor coexist together in the supercapacitor device, for convenience these are explained separately for better understanding. In pseudo-capacitor, charge stores in a Faradaic way, which allows the charges to pass across the double layer as batteries. However, capacitance arises because of the passing of charges across the interface, Δq, and potential change, ΔV. Hence, the derivative, dq/dV, is equivalent to capacitance [15]. There are many charge storage mechanisms for pseudocapacitor like adsorption pseudocapacitance, redox pseudocapacitance and intercalation pseudocapacitance as demonstrated in Fig. 9.3. In adsorption pseudocapacitance, electrolyte ions adsorb and desorb on the surface of the electrode material. Redox pseudocapacitance originates from fast reversible Faradaic reactions by electro-adsorption or intercalation of electrolyte ions on the surface of electrode. In intercalation pseudocapacitance, electrolyte ions intercalate on the van der Waals gaps or lattice of redox-active electrode materials [16, 17]. Although pseudocapacitive electrode material also takes place in the redox reaction shown ever, it is different

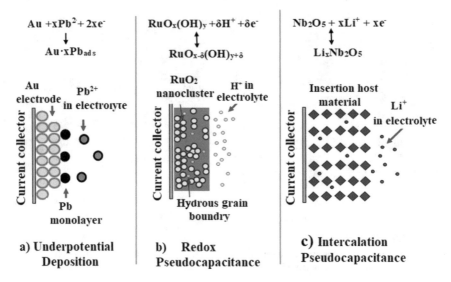

Fig. 9.3 Charge storage mechanisms in pseudocapacitor. Redrawn and reprinted with permission from [5]

from battery-type electrode material. Some examples of commonly used pseudocapacitive electrode materials are metal oxides (RuO_2, MnO_2, Fe_3O_4, Fe_2O_4, PbO_2, MoO_3), metal hydroxides ($Ni(OH)_2$) and conducting polymers (PANI and PPY) [4].

9.3.3 Hybrid Capacitors

Hybrid capacitors come into the existence because of provisions of selection of different types of electrode materials in the same device. There are some confusions in the hybrid and asymmetric capacitor. In asymmetric capacitor, both electrode materials will be capacitive in nature but of different behaviors like one will be EDLC-type electrode material made of carbon-based materials and others are based on pseudocapacitive electrode materials like metal oxides, metal hydroxides and conducting polymers, while in hybrid capacitor one electrode material is of capacitive nature and other made of battery-type material. The goal behind the development of hybrid capacitor is to increase the energy density. In a symmetric capacitor, there is a problem of low potential window in the aqueous media because of gas evolution reactions and electrochemical oxidation of the electrode materials at high potential [16]. In the place of positive electrode, any Faradaic material with high oxygen evolution reaction over potential, e.g., Pb/PbO_2 and $Ni(OH)_2$, may be used. Positive redox electrode and negative carbon electrode can operate in the complementary electrochemical window [16]. This increases not only potential window and energy density but also the capacitance of the cell because of the presence of pseudocapacitive electrode materials. Low self-discharge, high reliability, high operating temperature, high potential window and high energy density are the main advantages of the hybrid electrochemical capacitors.

9.4 Characteristics of Electrode Materials Used in Supercapacitors

Supercapacitor has four major elements/components, e.g., current collector, electrode, electrolyte and separator. Its performance depends on the materials used in all these four elements. The contribution of electrode and electrolyte remains high in deciding the performance (capacitance and energy density). The requirement of the electrode is high electrical conductivity. It should have a large surface area and mesoporous structure. The EDLC electrode does not undergo any chemical change during storing the charges, whereas electrodes used in pseudocapacitors are electroactive in nature and take part in Faradaic reactions. Also, electrodes in EDLCs are rigid; however, conducting polymer-based electrodes are much flexible.

For maintaining higher performance of SCs, proper selection of electrode materials is of uttermost importance. The performance of both types of SCs, EDLC and pseudocapacitors depends on the electrode material. The tuneable features of electrode can be achieved by using advanced materials, which show different properties from the bulk materials. By using nanoscience and nanotechnology, various interesting phenomena can be used for designing the materials suitable for electrodes. Since nanomaterial has a high strength-to-weight ratio and high surface area-to-volume ratio, they exhibit various attractive properties such as high conductivity, large surface area, good temperature stability and high corrosion resistance. The porosity of electrode can also be modified by using nanostructured materials. By using large surface area materials like carbon nanomaterial, the electrode-specific capacitance can be enhanced. Other important issues for electrode materials are the safety concern. The electrodes of SCs are safer to use for a long period, whereas are major safety issues are arising with batteries. Depending on the utility of the materials in different types of SCs, the materials can be classified as EDLC electrode material, pseudocapacitance electrode material and hybrid electrode material as demonstrated in Fig. 9.4. EDLC electrode materials are based on carbon materials. On the other hand, pseudocapacitor materials are composed of transition metal oxides (TMOs) and conducting polymers (CPs). The hybrid capacitors are made of carbon material, pseudocapacitor electrode material and battery electrode materials in combination.

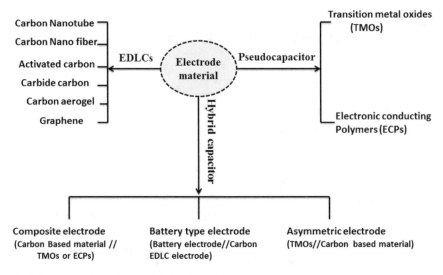

Fig. 9.4 Various electrode materials used in SCs

9.4.1 Electrodes for Supercapacitors

Electrode material is responsible for charge storage and capacitance of the super-capacitor device. Design of electrode material, which has a high specific surface area, high electrical conductivity, high chemical stability, lightweight, high thermal stability, corrosion resistance and low manufacturing cost, is the utmost goal for supercapacitor devices. Instead of high surface area, nanostructured pore size plays an important role in achieving high capacitance. An effective surface area is an important parameter for the high capacitance of supercapacitor. In the electrode material, all pores do not take part in charge storage. If the ion size is bigger than the pore size, ions will not accumulate on the surface of the pore; hence, those pore areas will be ineffective. Generally, pore size on the electrode surface should be double of the ion sizes to allow the effective coverage of the pore walls [18]. Pore size should be of around 0.7 nm for high capacitance, and as the pore size increases or decreases over or below 0.7 nm, capacitance starts decreasing [19]. Hence, pore size and pore distributions are the important criteria for the high capacitance of the SC devices. In Fig. 9.5, the specific capacitance of supercapacitor using different electrode materials like carbons for EDLC, and polymers and metal oxides for pseudocapacitor has been demonstrated. The last column demonstrates electrode materials for asymmetric and hybrid capacitors [20, 21].

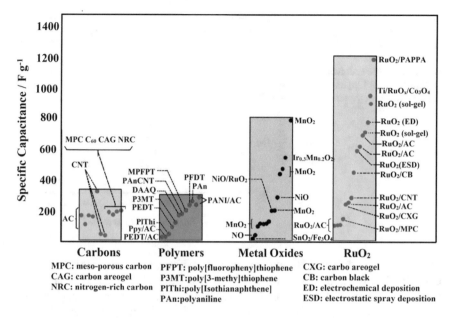

Fig. 9.5 Specific capacitance of supercapacitor using different electrode materials. Reprinted with permission from [22]

9.4.1.1   Emerging Materials: Carbon-Based Materials

Carbon exists in many allotropes like graphite, diamond, fullerene, graphene, carbon nanotube, amorphous carbon, etc. Among these allotropes, graphite, carbon nanotubes and graphene are good conductors of the electricity, which is an important parameter to serve as effective supercapacitor electrode material. Carbon has high conductivity and high chemical stability and is cheap and easily available, and it can achieve a high surface area, which makes it a suitable material to be used as electrode material for supercapacitors [23]. Carbon materials are used for EDLCs because they accumulate charge on the surface of the electrode material. High surface area, desirable pore size, pore size distribution and the presence of functional groups like oxygen, nitrogen and sulfur (responsible for pseudocapacitance) complement capacitance of the device. Commonly used carbon allotropes used as electrode materials are graphite, graphene, carbon nanotube, activated carbon and carbon aerogel [13, 24].

Activated carbon

Activated carbon is mostly used carbon for supercapacitor electrode material because of high electrical conductivity, easy availability and moderate cost. It is derived from carbonaceous materials like waste biomasses (e.g., rice husk, human hair, coconut shell, etc.), coal and wood [13]. Activation of the carbonaceous material is performed via chemical or physical activation processes. Physical activation occurs in two steps,

Fig. 9.6 Specific capacitance versus specific surface area of activated carbons used for the electrode. Figure is drawn using the data of [26] and Table 9.2 of this chapter

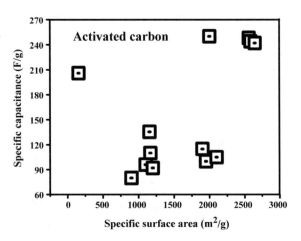

first pyrolysis at 400–900 °C and then partial gasification at 350–1000 °C using an oxidizing gas (e.g., O_2, CO_2, steam and their mixture) [25]. Chemical activation is done by pyrolysis at 440–900 °C using chemical activators (e.g., KOH, NaOH, H_3PO_4, $ZnCl_2$, etc.) [25]. Activation process evolves different pore sizes, which are distributed over the surface. Pore size distribution shows the presence of micropores (<2 nm), mesopores (2–50 nm) and macropores (>50 nm). Microspores evolve inside the mesopores and mesopores inside the macrospores, respectively. It is assumed that capacitance is linearly dependent on the specific surface area of the electrode material. Figure 9.6 demonstrates the relation between BET surface area and specific capacitance, which is contradicting the relation between specific surface area and specific capacitance though it depends on scan rate, current density, pore size of electrodes, pore size distribution of electrodes, etc. [26, 27]. An exclusive overview of the characteristics of activated carbon has been given in Chap. 4.

Graphene

Graphene is the two-dimensional single layer of carbon. It has high electrical conductivity, excellent chemical stability, high surface area and lightweight. It is independent of the pore size distribution of the solid electrode. Many methods have been developed for graphene productions such as chemical vapor deposition, micromechanical exfoliation, arc discharge method, unzipping of the carbon nanotubes, epitaxial growth, and electrochemical and chemical method. An exclusive overview of the characteristics of graphene has been given in Chap. 5.

Carbon nanotubes

CNT is a single-atom graphite sheet, which curls into a cylindrical form. CNTs have high electrical conductivity (10^5 S cm^{-1}), high mechanical strength, high surface area, high flexibility and high chemical and thermal stability [13]. CNT is categorized into single-walled carbon nanotube (SWCNT) and multiwalled carbon nanotube (MWCNT). SWCNT and MWCNT both forms are used as supercapacitor electrode

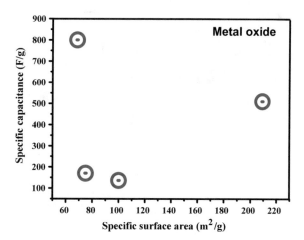

Fig. 9.7 Specific capacitance versus specific surface area of metal oxides used for the electrode. Figure is drawn using the data of Table 9.2 of this chapter

material application. An exclusive overview of the characteristics of CNT has been given in Chap. 6.

9.4.1.2 Emerging Materials: Metal Oxides

Transition metal oxides are also used as electrode materials because of high surface area, high ionic conductivity and mesoporous structure, which make them an effective electrode material for charge storage application. Metal oxides are used for pseudocapacitor, where fast Faradaic reaction takes place on the surface of the electrode material. Some commonly used metal oxide-based electrode materials are ruthenium oxide (RuO_2), manganese dioxide (MnO_2), nickel oxide (NiO), cobalt oxide (Co_2O_3), iridium oxide (IrO_2), iron oxide (Fe_2O_3), vanadium oxide (V_2O_5), indium oxide (In_2O_3) and tin oxide (SnO_2) [13]. Figure 9.7 demonstrates the relation between BET surface area and specific capacitance, which is again contradicting the relation between specific surface area and specific capacitance though it depends on several parameters like scan rate, current density, pore size of electrodes, pore size distribution of electrodes, etc. An exclusive overview of the characteristics of metal oxide has been given in Chap. 3.

9.4.1.3 Emerging Materials: Conducting Polymers

Conducting polymer is widely studied in the field of supercapacitors from the last two decades because of high energy density compared to metal oxides, low cost, easy to fabricate and low equivalent series resistance [13]. Conducting polymers are used for pseudocapacitors and store charges by redox reactions. Oxidation induces doping of electrolyte ions in polymer matrix, while reduction leads to de-doping of ions from the polymer matrix [23]. Examples of some commonly used conducting polymers

Fig. 9.8 Specific
capacitance versus specific
surface area of conducting
polymers used for the
electrode. Figure is drawn
using the data of Table 9.2 of
this chapter

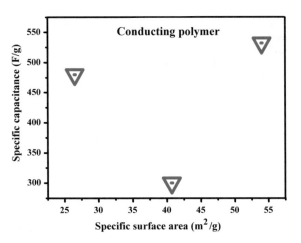

for supercapacitor electrode materials are polyaniline (PANI), polypyrrole (PPY),
polythiophenes, etc. [13]. Figure 9.8 demonstrates the relation between BET surface
area and specific capacitance, which is also contradicting the relation between specific
surface area and specific capacitance though it depends on various parameters like
scan rate, current density, pore size of electrodes, pore size distribution of electrodes,
etc. An exclusive overview of the characteristics of conducting polymers has been
given in Chap. 8.

9.4.1.4 Emerging Materials: Special Types

Metal–organic frameworks

Metal–organic frameworks (MOFs), subclass of coordination polymers, are a class
of compounds consisting of metal ions or clusters coordinated to organic ligands to
form one-, two- or three-dimensional structures. MOFs have received attention in
the field of the supercapacitors because of highly porous and molecular-level tunable
frameworks. These are lightweight and of high surface area, high pore volume and
controllable pore size structures. It is built by coordination bond formation resulting
from weak interactions like van der Waals forces, hydrogen bonding, P-P stacking
between organic and inorganic groups. MOF allows arranging single atomic active
metal, which increases the interaction between electrode and electrolyte. It is the
combination of inorganic node, which creates a corner of the framework, and these
nodes get connected by the organic linker. Pore size of MOFs can be controlled in
the range of 0.6–2 nm and can also incorporate pseudocapacitive electrode material
within the framework, which extends its suitability as electrode material for supercapacitor devices. Low electronic conductivity is a big drawback of most of the MOFs.
Composite of MOFs with carbon material is a solution to overcome this issue. Carbon
materials, e.g., CNTs, rGO, etc., help to improve the electronic conductivity [18].

Covalent organic framework

Covalent organic frameworks (COFs) are linked by strong covalent bonds. These are two-dimensional and three-dimensional organic solids. The basic structure of COFs is similar to MOFs. They have a high surface area, controllable pore size and high flexibility. But instead of inorganic node and organic linkers, the framework is composed by lightweight elements like carbon, boron, oxygen with covalent bonds (B–O, C–N, B–N). COFs are light in weight, and now some of them are chemically and thermally more stable than MOFs [18].

MXenes

MXenes are a few atomic layer thick material composed of transition metal carbide, nitride and carbonitride (Nb_2C, Ti_3C_2) leading to the formation of two-dimensional structure. It consists of $M_{n+1}X_n$, where M represents transition metal and X is related to nitrogen and carbon with X (1, 2, 3, 4, 5 …) [18]. MXenes have high electronic conductivity, good mechanical strength and hydrophilicity that establish them as a suitable material for supercapacitor energy storage applications.

Metal nitrides

Metal nitrides are gaining attention in the energy storage applications because of high electronic conductivity. A variety of metal nitrides are involved in the supercapacitor applications like molybdenum nitride (MoN), titanium nitride (TiN), vanadium nitride (VN), niobium nitride (NbN), ruthenium nitride (RuN), chromium nitride (CrN), which not only serve as the electrode material but also work as the mechanical backbone [28, 29].

Black phosphorus

Black phosphorus has got interest in energy storage application because of its 2D puckered structure with high surface area and electronic mobility with strong mechanical strength. Black phosphorus contains semi-connected nanosheets, which provide an open structure in the sponge [18].

9.4.2 Characteristics of Electrode Materials

Electrode accumulates and intercalates the electrolyte ions on its surface during charging and transfers electrical current to the current collector at the time of discharge. Electrode material is responsible to get high specific capacitance and to decide the potential window of supercapacitor device. As the charge storage capacity of the electrode increases, capacitance of the device also improves. Effective surface area of the electrode material influences the charge storage capacity of the device. The following are the characteristics of suitable electrode material (Table 9.1). Selected characteristics of a few electrode materials are given in Table 9.2. Figure 9.9

Table 9.1 Characteristics of electrode material for electrochemical supercapacitors

Primary characteristics	Secondary characteristics	Tertiary characteristics
High specific surface area	High chemical stability	Low cost
High electrical conductivity	Wide working temperature range	Non-toxicity
High charge storage capacity	High thermal conductivity	Easy to fabricate
Lightweight	High corrosion resistance	Easy availability
	Controllable pore size	

demonstrates the relation between BET surface area and specific capacitance, which is contradicting the relation between specific surface area and specific capacitance. Note that capacitance depends on scan rate, current density, pore size of electrodes, etc. It is desirable to have a high surface area.

9.5   Concluding Remarks

Selection of appropriate material is a critical task for the fabrication of any high-performance device. A supercapacitor is composed of four essential counterparts, e.g., current collector, electrode, electrolyte and separator. The overall performance of the device depends on the properties of the individual component. Electrode material is used for the accumulation of ions on the surface of the electrode material during charging and vice versa. The main characteristics of an electrode material are highly effective specific surface area, extraordinary electrical conductivity, lightweight, chemical and thermal stability and cost-effective. According to the common requirements of electrode material, graphene, carbon nanotube, ruthenium oxide, activated carbon, nickel oxide, polypyrrole, manganese dioxide, polyaniline, etc., are found to be the most appropriate electrode materials to be used for the fabrication of high-performance supercapacitor devices.

Table 9.2 Properties of different electrode materials for supercapacitor performance

Electrode material			Specific surface area (m²/g)	Density (kg/m³)	Specific capacitance (F/g)	Electrolyte	References
Carbon materials	Activated carbon	Shrimp shell	156.4	2000–2100	206 F/g at 0.1 A/g	6M KOH	[30]
		Rice straw	2646		242 F/g at 0.5 A/g	6M KOH	[31]
		Oily sludge	2561		248.1 F/g at 0.5 A/g	6M KOH	[32]
		Coconut shell	2000		250 F/g	1M H₂SO₄	[33]
		Citrus peel	1167		110 F/g at 0.1 A/g	1M NaClO₄	[34]
		Sunflower seed	2585		244 F/g at 0.25 A/g	6M KOH	[35]
		Rotten carrot	1154.9		135.5 F/g at 2.2 A/g	6M KOH	[36]
	Graphene		2630	2266	138 F/g at 10 A/g	Aqueous electrolyte	[37, 38]
	CNT		380–2005	1600	227 F/g at 5 A/g	H₂SO₄	[39]
Metal oxide	Ruthenium oxide (RuO₂)		68.6	6970	800 F/g at 1 A/g	1M H₂SO₄	[40]
	Manganese dioxide (MnO₂)		80–120	7200	72–201 F/g	Aqueous electrolyte	–
	Nickel oxide (NiO)		153.2–265	6810	138.6–982 F/g at 1 A/g	6M KOH	[41]
	Iron oxides		75	5250	170 F/g at 2 mV/s	1M Na₂SO₄	[42]
Composite	RuO₂/CNT		750	–	1340 F/g at 25 mV/s	0.1M H₂SO₄	[43]
	V₂O₅·XH₂O/CNT		–	–	910 F/g at 10 mV/s	1M LiClO₄/PC	[44]
	NiO–Co₃O₄		23.3	–	801 F/g at 1 A/g	3M KOH	[45, 46]
Conducting polymer	Polyaniline (PANI)/α MnO₂		–	–	626 F/g at 2 A/g	1M H₂SO₄	[47]
	Polyaniline (PANI)		30.9–77.1	1245	532 F/g at 1.5 A/g	1M H₂SO₄	[48, 49]
	Polypyrrole (PPY)		26.5	1480	480 F/g at 10 mV/s	1M KCl	[50, 51]
	Polythiophenes (PTH)		40.7	1400	300 F/g at 5 mV/s	0.1M LiClO₄/PC	[52, 53]

Fig. 9.9 Specific capacitance versus specific surface area used for the electrode. Figure is drawn using the data of [26] and Table 9.2 of this chapter

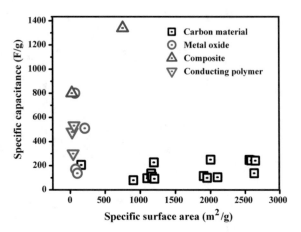

Acknowledgements The authors acknowledge the financial support provided by the Department of Science and Technology, India (DST/TMD/MES/2K16/37(G)), for carrying out this research work. The authors are thankful to Ms. Tanvi Pal for drafting a few figures.

References

1. A. González, E. Goikolea, J.A. Barrena, R. Mysyk, Renew. Sustain. Energy Rev. **58**, 1189 (2016)
2. S. Mohanty, S.K. Nayak, B.S. Kaith, S. Kalia (eds.), *Polymer Nanocomposites Based on Inorganic and Organic Nanomaterials* (Wiley, Hoboken, 2015)
3. K.K. Kar, A. Rahaman, P. Agnihotri, D. Sathiyamoorthy, Fullerenes. Nanotub. Carbon Nanostruct **17**, 209 (2009)
4. A.K. Samantara, S. Ratha, *Materials Development for Active/Passive Components of a Supercapacitor* (Springer Singapore, Singapore, 2018)
5. V. Augustyn, P. Simon, B. Dunn, Energy Environ. Sci. **7**, 1597 (2014)
6. A. Muzaffar, M.B. Ahamed, K. Deshmukh, J. Thirumalai, Renew. Sustain. Energy Rev. **101**, 123 (2019)
7. K. K. Kar, A. Hodzic, *Carbon Nanotube Based Nanocomposites: Recent Developments*, 1st ed. (Research Publishing, 2011)
8. R. Kumar, S. Sahoo, E. Joanni, R.K. Singh, W.K. Tan, K.K. Kar, A. Matsuda, Prog. Energy Combust. Sci. **75**, 100786 (2019)
9. Y. Zhang, H. Feng, X. Wu, L. Wang, A. Zhang, T. Xia, H. Dong, X. Li, L. Zhang, Int. J. Hydrogen Energy **34**, 4889 (2009)
10. A. Yadav, B. De, S.K. Singh, P. Sinha, K.K. Kar, A.C.S. Appl, Mater. Interfaces **11**, 7974 (2019)
11. K.K. Kar, *Composite Materials*, 1st edn. (Springer, Berlin, 2017)
12. E. Gongadze, S. Petersen, U. Beck, U. Van Rienen, COMSOL Conf. (2009)
13. W. Raza, F. Ali, N. Raza, Y. Luo, K.-H. Kim, J. Yang, S. Kumar, A. Mehmood, E.E. Kwon, Nano Energy **52**, 441 (2018)
14. L.L. Zhang, X.S. Zhao, Chem. Soc. Rev. **38**, 2520 (2009)
15. B.E. Conway, *Electrochemical Supercapacitors* (Springer, US, 1999)
16. C. Zhong, Y. Deng, W. Hu, D. Sun, X. Han, J. Qiao, J. Zhang, *Electrolytes for Electrochemical Supercapacitors*, 1st ed. (CRC Press, 2016)

17. J. Cherusseri, K.K. Kar, J. Mater. Chem. A **4**, 9910 (2016)
18. F. Wang, X. Wu, X. Yuan, Z. Liu, Y. Zhang, L. Fu, Y. Zhu, Q. Zhou, Y. Wu, W. Huang, Chem. Soc. Rev. **46**, 6816 (2017)
19. C. Largeot, C. Portet, J. Chmiola, P.-L. Taberna, Y. Gogotsi, P. Simon, J. Am. Chem. Soc. **130**, 2730 (2008)
20. D.S. Achilleos, T.A. Hatton, J. Colloid Interface Sci. **447**, 282 (2015)
21. K.K. Kar, S.D. Sharma, S.K. Behera, P. Kumar, J. Elastomers Plast. **39**, 117 (2007)
22. K. Naoi, P. Simon, Electrochem. Soc. Interface **17**, 34 (2008)
23. Z.S. Iro, Int. J. Electrochem. Sci. **11**, 10628 (2016)
24. K. K. Kar, A. Hodzic, *Developments in Nanocomposites*, 1st ed. (Research publishing, 2014)
25. M. Sevilla, R. Mokaya, Energy Environ. Sci. **7**, 1250 (2014)
26. J. Gamby, P. Taberna, P. Simon, J. Fauvarque, M. Chesneau, J. Power Sources **101**, 109 (2001)
27. K.K. Kar, S.D. Sharma, P. Kumar, J. Ramkumar, R.K. Appaji, K.R.N. Reddy, Polym. Compos. **28**, 637 (2007)
28. S. Ghosh, S.M. Jeong, S.R. Polaki, Korean J. Chem. Eng. **35**, 1389 (2018)
29. K.K. Kar, S.D. Sharma, P. Kumar, Plast. Rubber Compos. **36**, 274 (2007)
30. J. Qu, C. Geng, S. Lv, G. Shao, S. Ma, M. Wu, Electrochim. Acta **176**, 982 (2015)
31. L. Zhu, F. Shen, R.L. Smith, L. Yan, L. Li, X. Qi, Chem. Eng. J. **316**, 770 (2017)
32. X. Li, K. Liu, Z. Liu, Z. Wang, B. Li, D. Zhang, Electrochim. Acta **240**, 43 (2017)
33. K. Yang, J. Peng, C. Srinivasakannan, L. Zhang, H. Xia, X. Duan, Bioresour. Technol. **101**, 6163 (2010)
34. N.R. Kim, Y.S. Yun, M.Y. Song, S.J. Hong, M. Kang, C. Leal, Y.W. Park, H.J. Jin, A.C.S. Appl, Mater. Interfaces **8**, 3175 (2016)
35. X. Li, W. Xing, S. Zhuo, J. Zhou, F. Li, S.Z. Qiao, G.Q. Lu, Bioresour. Technol. **102**, 1118 (2011)
36. S. Ahmed, A. Ahmed, M. Rafat, J. Saudi Chem. Soc. **22**, 993 (2018)
37. A.S. Lemine, M.M. Zagho, T.M. Altahtamouni, N. Bensalah, Int. J. Energy Res. **42**, 4284 (2018)
38. S. Banerjee, P. Benjwal, M. Singh, K.K. Kar, Appl. Surf. Sci. **439**, 560 (2018)
39. R. Sharma, K.K. Kar, RSC Adv. **5**, 66518 (2015)
40. X. Wu, Y. Zeng, H. Gao, J. Su, J. Liu, Z. Zhu, J. Mater. Chem. A **1**, 469 (2013)
41. C.-C. Hu, W.-C. Chen, K.-H. Chang, J. Electrochem. Soc. **151**, A281 (2004)
42. S.-Y. Wang, K.-C. Ho, S.-L. Kuo, N.-L. Wu, J. Electrochem. Soc. **153**, A75 (2006)
43. R. A. Fisher, M. R. Watt, W. Jud Ready, ECS J. Solid State Sci. Technol. **2**, M3170 (2013)
44. I.-H. Kim, J.-H. Kim, B.-W. Cho, Y.-H. Lee, K.-B. Kim, J. Electrochem. Soc. **153**, A989 (2006)
45. X.W. Wang, D.L. Zheng, P.Z. Yang, X.E. Wang, Q.Q. Zhu, P.F. Ma, L.Y. Sun, Chem. Phys. Lett. **667**, 260 (2017)
46. R. Sharma, A.K. Yadav, V. Panwar, K.K. Kar, J. Reinf. Plast. Compos. **34**, 941 (2015)
47. R.I. Jaidev, A.K. Jafri, S.Ramaprabhu Mishra, J. Mater. Chem. **21**, 17601 (2011)
48. N.H. Khdary, M.E. Abdesalam, G. EL Enany, J. Electrochem. Soc. **161**, G63 (2014)
49. K.K. Kar, J.K. Pandey, S. Rana, *Handbook of Polymer Nanocomposites. Processing, Performance and Application* (Springer, Berlin, 2015)
50. K. Nishio, M. Fujimoto, O. Ando, H. Ono, T. Murayama, J. Appl. Electrochem. **26**, 425 (1996)
51. L.-Z. Fan, J. Maier, Electrochem. Commun. **8**, 937 (2006)
52. M.M. Mulunda, Z. Zhang, E. Nies, C. van Goethem, I.F.J. Vankelecom, G. Koeckelberghs, Macromol. Chem. Phys. **219**, 1800024 (2018)
53. Q. Meng, K. Cai, Y. Chen, L. Chen, Nano Energy **36**, 268 (2017)

Chapter 10
Characteristics of Electrolytes

Kapil Dev Verma, Soma Banerjee, and Kamal K. Kar

Abstract Electrolytes play a major role in determining the energy-storage capacity of the electrochemical supercapacitor devices. It consists of solvent, dissociated positive and negative ions, or pure salt, i.e., a solvent-free ionic liquid. The properties of an electrolyte depend on a number of parameters such as size of the ion and concentration, conductivity, interaction between electrolyte, electrode materials, etc. In this chapter, the types of electrolytes used in electrochemical supercapacitors have been discussed in detail. Based on chemical nature, the electrolytes can be classified in various groups, i.e., aqueous, organic, ionic liquid, solid state, redox, etc. This chapter also points out the effect of these electrolytes on the cell voltage, specific capacitance, and other important parameters for electrochemical double layer, asymmetric, and hybrid supercapacitor devices.

10.1 Introduction

An electrochemical supercapacitor (ES) is a modern energy-storage device that provides high power density and long cycle life with a limitation of low energy density. Electrolyte is one of the responsible materials for efficient energy-storage capacity of the device. It consists of solvent, dissociated positive cations and negative anions, or pure salt, i.e., solvent-free ionic liquid. Characteristics of electrolytes such as ion

K. D. Verma · K. K. Kar (✉)
Advanced Nanoengineering Materials Laboratory, Materials Science Programme, Indian Institute of Technology Kanpur, Kanpur 208016, India
e-mail: kamalkk@iitk.ac.in

K. D. Verma
e-mail: kdev@iitk.ac.in

S. Banerjee · K. K. Kar
Advanced Nanoengineering Materials Laboratory, Department of Mechanical Engineering, Indian Institute of Technology Kanpur, Kanpur 208016, India
e-mail: somabanerjee27@gmail.com

© Springer Nature Switzerland AG 2020 287
K. K. Kar (ed.), *Handbook of Nanocomposite Supercapacitor Materials I*,
Springer Series in Materials Science 300,
https://doi.org/10.1007/978-3-030-43009-2_10

size and concentration, conductivity of the electrolyte and interaction between electrolyte and electrode materials influence the performance of supercapacitor cells [1]. The potential window of a supercapacitor cell largely depends on the electrochemical stability of electrolyte, which influences energy density and power supply of the cell. Over potential window of electrolyte, electrolyte and water, in case of aqueous electrolyte decompose in hydrogen and oxygen, which causes leakage in current flow and thereby the poor performance of the supercapacitor is quite expected. Between a supercapacitor and battery, supercapacitor lacks behind in energy-storage because of the low potential window. In electrical double-layer capacitors (EDLCs), energy-storage capacity is less than 10 Wh/kg and in pseudocapacitors and hybrid capacitors, it is less than 50 Wh/kg [2]. Equivalent series resistance (ESR) of supercapacitor cell also depends on the conductivity of the electrolyte, which affects the power density of cell. ESR is also known as the electrolyte resistance that is responsible for IR drops [it is the voltage drop across any resistance, which is the product of current (I) passing through resistance and resistance value (R)] at the start of discharging in the galvanostatic charge–discharge test. As ionic conductivity of the electrolyte increases, ESR decreases. Charge-storage capacity of a supercapacitor depends on the accumulation of ions on the pores of electrode material. Electrolyte ions accumulate on the surface of opposite charge electrode and form EDLC. Better contact between electrolyte and electrode material leads to a highly effective surface area, which causes better capacitance. During the calculation of specific capacitance of the device, we neglect the mass of the electrolyte. Suppose, C_e is the capacitance of single electrode and m is the mass of two electrodes, overall specific capacitance (C_{ell}) of the device is expressed as $C_{\text{cell}} = \frac{c_e}{m}$. Electrolyte volume should be low so that the weight of electrolyte does not come into the picture. Cycle life of supercapacitor is an important parameter that shows the stability of electrode and electrolyte materials. Commonly used electrode material for EDLC is carbon, which is highly stable. In pseudocapacitors, metal oxides are used that interact with the electrolyte and electrode. Pseudocapacitance in supercapacitor also comes into play with fast intercalation of electrolyte ions into the lattice of electrode material along with a Faradaic charge transfer reaction [1]. Cyclic stability of pseudocapacitors is less in comparison to EDLC because of the non-ideal electrochemical reversibility of redox reaction [2]. Reasons behind current leakage in supercapacitor are short circuit between electrodes, charging of the supercapacitor cell over the potential window of electrolyte, no redistribution of charge of the electrolyte ions into pores of electrode materials, and corrosion of the current collector due to electrolyte. The requirements for the electrolyte to optimize the performance of supercapacitors are high potential window, conductivity and chemical stability, matching with the electrode materials, high operating temperature range, nontoxic, optimum viscosity, environmentally friendly, and low cost. In Fig. 10.1, the distribution of electrolyte ions during the charging and discharging of an electrochemical supercapacitor has been shown.

Important criteria for electrolyte selection are electrochemical stability, i.e., it should be of high potential window and good ionic conductivity that influences energy (10.1) and power density (10.2) of the electrochemical supercapacitor cell, respectively. As shown in (10.1) and (10.2), energy and power density both are

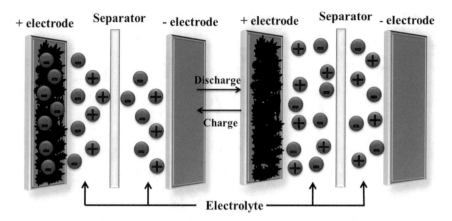

Fig. 10.1 Electrolyte ions distribution behavior during charging and discharging of electrochemical supercapacitor

proportional to the square of potential window (V). Power density also depends on the ESR, which directly depends on the ionic conductivity of electrolyte.

$$E_{\text{cell}} = 1/2 C_{\text{cell}} V^2 \tag{10.1}$$

$$P_{\text{cell}} = \left(\frac{V^2}{4 \times \text{ESR} \times m} \right) \tag{10.2}$$

Different electrolytes are developed and tested on different types of supercapacitors like EDLCs, pseudocapacitor, hybrid capacitor to improve their performance. Till now, no one is able to fullfill all the requirements of ideal electrolyte. For example, an electrolyte of high conductivity say aqueous electrolyte lacks on a potential window. The reverse remains the fact for organic electrolytes. Figure 10.2 demonstrates the fact that as we are moving from aqueous to ionic liquid electrolyte, potential window increases, however, the ionic conductivity still lags behind.

Generally, electrolytes are classified according to the state of electrolyte like liquid and solid electrolytes. However, according to chemical and physical structures, they can be classified into five categories, e.g., aqueous, organic, ionic liquids, solid state or quasi-solid state, and redox-active electrolytes. Behavior of these five electrolytes on EDLC, pseudo and hybrid capacitor will be discussed in detail in the succeeding sections of this chapter.

10.2   Aqueous Electrolytes

In aqueous electrolytes , water is used as a solvent for the salts. An aqueous electrolyte is superior compared to an organic electrolyte due to the environmental effect,

Fig. 10.2 Potential window
and ionic conductivity of
aqueous, organic, and ionic
liquid electrolytes [1, 3]

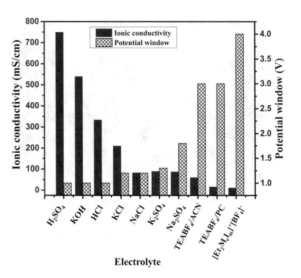

ionic conductivity, electrochemical behavior, easy handling in the open atmosphere, and cost factor. However, the aqueous electrolyte has a low potential window in comparison to organic and ionic liquid (IL) electrolytes. The thermodynamic potential window of water is 1.23; hence, water can be easily electrolyzed into hydrogen and oxygen to form gases at a low potential window [3]. Gases lead to potential damage to the device, which crosses the safety limits. There are two major causes behind the electrolysis of water. First is the ionic equilibrium of water and second is the presence of hydrogen bonds among molecules of water. The constituents of water, hydrogen, and oxygen can split easily in an ionic environment. Again, when O–H bonds in the water get stronger, the hydrogen bonds become comparatively weak. If the water molecules are isolated from each other, the O–H bonds become stronger due to the weakening of hydrogen bonds. The fact is that the O–H bond strength of isolated water in gaseous form remains even greater compared to that in methanol and ethanol [4].

Aqueous electrolyte exhibits high conductivity and capacitance; however, the potential window is small. Generally, potential window of the aqueous electrolyte remains between 1.0 and 1.5 V. This potential window may reach 2.0 V by using neutral aqueous electrolytes like Na_2SO_4 and KCl [1]. Because of the low potential window, the usage of aqueous electrolytes is avoided for commercial applications; however, they are less expensive and easy to fabricate in comparison to organic electrolytes. Electrolyte performance also depends on the ion size, hydrated ion size, corrosive behavior, mobility, and ion conductivity. The conductivity of aqueous electrolyte is one order of magnitude higher compared to organic electrolytes. Chemically, aqueous electrolytes are of three types, i.e., acidic, alkaline, and neutral.

10.2.1 Acid Electrolytes

Sulfuric acid is a typical electrolyte, which is being used not only in supercapacitors but also in batteries because of high conductivity. The conductivity of electrolyte is the main parameter to be optimized using the correct ratio of salt to water. The concentration of electrolyte decides the ionic conductivity. A high ionic conductivity is needed to achieve excellent electrochemical performance. If concentration remains too low, the conductivity of electrolyte decreases. Strong electrolytes completely ionize in water and there will be no neutral atoms, whereas weak electrolytes partially dissolve in water. For reference, water is considered as the weakest electrolyte.

10.2.1.1 Acid Electrolytes for Electrical Double-Layer Capacitors

Figure 10.1 demonstrates the basic charge–discharge principle of electrostatic double-layer capacitor. During charging, ions accumulate on the surface of opposite electrodes separated by solvent molecules. Electrolyte ion distribution during charging and discharging of EDLCs can be demonstrated in the following [1].
On positive electrode

$$E_p + I^- \xrightleftharpoons[Discharging]{Charging} E_p^+ / I^- + e^-$$ (10.3)

and on negative electrode

$$E_n + e^- + I^+ \xrightleftharpoons[Discharging]{Charging} E_n^- / I^+$$ (10.4)

Hence, the net charging and discharging on the device will be

$$E_p + E_n + I^- + I^+ \xrightleftharpoons[Discharging]{Charging} E_p^+ // I^- + E_n^- // I^+$$ (10.5)

where E_p and E_n represent the positive and negative electrodes, respectively. I^+ and I^- are the cations and anions, $//$ depicts the electrostatic double layer on the electrodes.

Accumulation of ions on the surface of electrode also depends on the pore size of electrode material. If the pore size of electrode material is smaller than the diameter of hydrated ions, then those ions will not be able to participate in double-layer formation. Electroadsorption of same ions into different pore size materials not necessarily provides the same capacitance value even if they are of the same surface area (A_{BET}). Examples of some common aqueous electrolytes for carbon-based EDLCs have been summarized in Table 10.1.

Table 10.1 Aqueous electrolytes for carbon-based electrostatic double-layer capacitors (EDLCs)

Electrolyte	Electrode material	Potential window (V)	Specific capacitance (F/g)	References
NaCl (L)	AC	0–0.9	16–26	[5]
1M NaCl	AC	0–1.2	120	[6]
2M KCl	$MnO_2 \cdot nH_2O$	−0.2 to 1	200	[7]
1M Na_2SO_4 (L)	AC	0–1	133.4	[8]
0.5M Na_2SO_4 (L)	AC	0–1.6	135	[9]
2M $NaNO_3$	AC	0–1.6	116	[10]
1M Na_2SO_4	AC-banana fiber	−0.2 to 0.8	74	[11]
1.5M Li_2SO_4	CF/CNT	0–1	227.5	[12]
1M Li_2SO_4	Activated carbon	0–2.2	180	[13]
1M H_2SO_4	AC fiber	0–0.9	280	[14]
1M H_2SO_4	AC-oil palm	0–1	122	[15]
1M H_2SO_4	AC	0–1	130	[16]
1M H_2SO_4	MMPGC	0–0.8	81.05	[17]
1M H_2SO_4	AC fiber	0–0.9	92.80	[18]
1M H_2SO_4	Microporous carbon	0–1	100	[19]
1M H_2SO_4	Graphene quantum dots–3D graphene composites	0–0.8	268	[16]
1M H_2SO_4	Phosphorus-enriched carbon	0–1.3	220	[20]
1M H_2SO_4	Carbon nanotubes	0–1	115.7	[21]
1M H_2SO_4	CNT balls	0–1	80	[22]
1M H_2SO_4	Carbon film	0–0.9	292	[23]
0.1M HCL	AC-sunflower seeds	0–0.9	244	[24]
6M KOH (L)	AC-shrimp shell	0–0.9	206	[25]
6M KOH	AC-corn cob	0–1	120	[25]
6M KOH	3D FHPC	0–1	294	[26]
6M KOH	Highly porous graphene planes	0–1	303	[27]
1M KOH	rGO	0–1	300	[28]
0.5M NaOH	AC-rice husk	−0.5 to 0.5	243	[29]

AC activated carbon, *CNT* carbon nanotube, *MMPGC* macro/mesoporous partially graphitized carbon, *FHPC* flower-like and hierarchical porous carbon material, *rGO* reduced graphene oxide

10.2.1.2   Acid Electrolytes for Asymmetric and Hybrid Capacitors

A combination of two different electrode materials or same materials at different weight ratios with the same charge-storage mechanisms such as EDLC and pseudo-capacitor is called an asymmetric capacitor. In the hybrid capacitor, one electrode will be capacitive in nature and the other one is of battery type. Charge-storage mechanism is totally different in pseudocapacitance from EDLC. In pseudocapacitance, charge crosses the boundary layer and fast Faradaic reaction takes place. It is similar to batteries; however, stores charge like a capacitor. Here, the question is how it works like a capacitor. The extent of charge acceptance (Δq) into the electrode surface across the interference and change in the potential (ΔV) of the device is the thermodynamic reason for capacitance. First derivative ($\Delta q/\Delta V$) provides a charge-storage capacity of pseudocapacitor [30]. EDLC has high cycle stability but low charge-storage capacity in comparison to pseudocapacitor. In acid electrolytes with EDLCs, there is some short of pseudocapacitance as well. Pseudocapacitance accompanies with EDLC in carbon-based supercapacitor. Additional functional groups present in the electrode material like oxygen, nitrogen, sulphr, boron, potassium, etc., enhance pseudocapacitance properties. It is not necessary that these functional groups will be present in the electrode material naturally; however, functional groups can be added in the material by mixing with suitable precursor. The presence of electrolyte has a major influence on pseudocapacitance properties. Different electrolytes provide different surface functionalities on carbon-based materials. Cyclic stability of pseudocapacitors is less in aqueous electrolyte in comparison to the organic electrolyte, because in organic electrolyte, maximum capacitance comes from static capacitance. In some surface, functional groups like phosphorus, addition on carbon surface improves the cyclic stability. For pseudocapacitance, different electrode materials like metal oxide, polymers, and sulfide have been utilized. However, all of these are not stable in a strong acid electrolyte. Ruthenium oxide is stable; however, lack of abundance along with cost factor restricts commercialization. Conducting polymers with carbon material also enhance pseudocapacity of the electrochemical supercapacitor since Faradaic charge storage always dominates the electric double-layer charge storage due to the presence of conducting polymer [31]. Hybrid capacitors are used to increase the potential window using two different electrode materials working at separate potential range [2]. Table 10.2 summarizes different types of aqueous electrolytes used for asymmetric and hybrid capacitors.

10.2.2   Alkaline Electrolytes

Alkaline electrolyte is is also a widely used electrolyte in the supercapacitor applications. Commonly used alkaline materials are KOH, LiOH, and NaOH. Alkaline electrolyte KOH has been extensively studied due to the high conductivity of 0.6 S/cm at a concentration of 6 M at room temperature [1].

Table 10.2 Aqueous electrolytes for asymmetric and hybrid capacitors

Electrolyte	Electrode material	Cell voltage (V)	Specific capacitance (F/g)	References
1M H_2SO_4	RuO_2-Graphene	0–1.2	479	[32]
1M H_2SO_4	Graphene-mPANI	0–0.7	749	[33]
6M KOH	RCGCA	−0.1 to 1.1	704.3	[34]
1M KOH	$Ni(OH)_2$/Graphene	0–1	303	[35]
2M KOH	Porous $NiCo_2O_4$ nanotubes	0–1	647.6	[36]
6M KOH	p-CNTs/ CGBs	0–0.9	202	[13]
1M LiOH	MnO_2 nanoflower	0–0.6	363	[37]
3M KCl	$Ni(OH)_2$	0–0.5	718	[38]
0.5M K_2SO_4	$V_2O_5 \cdot 0.6H_2O$	0–1.5	60	[39]
1M Na_2SO_4	Hydrous RuO_2	0–1.6	56.66	[40]
1M Li_2SO_4	Mesoporous MnO_2	0–1	284	[41]

PANI polyaniline, *RCGCA* RuO_2/g-C_3N_4/rGO aerogel composite, *CNT* carbon nanotube, *CGBs* crumpled graphene balls

10.2.2.1 Alkaline Electrolytes for Electrical Double-Layer Capacitors

Alkaline electrolyte provides the same potential window as acid electrolyte. Selection of proper electrolyte leads to remarkable improvement in electric double-layer capacitance. Small-sized cations can easily enter into the electrode pores and form a double layer. Sometimes, an optimized mixture of two different alkaline materials is used in electrochemical supercapacitor. EDLC generates a square-shaped cyclic voltammetry (CV) curve. For EDLC, the current remains proportional to the scan rate (V/s) in the electrochemical analysis [42].

10.2.2.2 Alkaline Electrolytes for Asymmetric and Hybrid Capacitors

Pseudocapacitance of carbon-based electrode materials is related to the surface functional groups present on the electrode. Nitrogen-doped carbon electrode materials with KOH containing electrolyte is more beneficial. Phosphorus- and nitrogen-doped carbon electrodes offer a wide potential window and stability. Pseudocapacitance provides a quasi-rectangular shape cyclic voltammetry curve with two small peaks. In pseudocapacitor, current is not ideally proportional to scan rate ($I = C \times \frac{dV^n}{dt}$), where n is approximately equal to one.

The common electrode materials utilized in the alkaline electrolytes are metal oxides such as NiO_x, CoO_x, MnO_2, RuO_2, and $NiCo_2O_4$, hydroxides such as $Ni(OH)_2$, $Co(OH)_2$, sulfides, and nitrides, e.g., cobalt sulfide and vanadium nitride

[1]. Properties of the electrolyte affect the performance of supercapacitor. For example, concentration of the electrolyte affects the ESR, capacitance, and ionic conductivity [43]. However, a high concentration of electrolyte corrodes the electrode material [44]. At high temperature, ESR decreases but cyclic stability also gets improved. An optimization in concentration and temperature is of uttermost importance in this respect [45, 46]. MnO_2 gives high capacitance with LiOH electrolyte in comparison to KOH and NaOH since lithium ion is smaller than K^+ and Na^+. In an asymmetric capacitor, generally in pseudocapacitor, positive electrode is of pseudo-capacitive type and the negative electrode is made of carbon, where charge is being stored in EDLC [1].

10.2.3   Neutral Electrolytes

Among the aqueous electrolytes, neutral electrolyte is used most widely due to the wide potential window with less corrosive nature and low cost. Commonly used anions for neutral electrolytes are metal chloride, sulfide, and nitride with small-sized alkaline and alkaline earth metals as cations, e.g., LiCl, Na_2SO_4, KNO_3, $MgSO_4$, and $Ca(NO_3)_2$. The most frequently used neutral electrolyte is Na_2SO_4.

10.2.3.1   Neutral Electrolytes for Electrical Double-Layer Capacitors

Neutral electrolyte provides high energy density because of high electrochemical stability. Acid electrolyte (H_2SO_4) and alkaline electrolyte (KOH) give high capacitance in comparison to the neutral electrolyte [41, 47, 48]. Neutral electrolytes are less corrosive in comparison to acid and alkaline electrolytes because of neutral pH value. Low corrosion of neutral electrolyte is responsible for high cyclic stability. Acid electrolytes have a high concentration of H^+ ions, whereas alkaline electrolytes have OH^- in abundance. Neutral electrolyte has lower concentration of H^+ and OH^- in comparison to acid and alkaline electrolyte; therefore, there is a requirement of high potential for hydrogen and oxygen evolution from the water solvent. In practical supercapacitors, a high concentration electrolyte has been used to get a large number of ions to achieve high performance. However, in some cases, there remain some issues to produce high concentration at lower temperatures [2]. A comparative evaluation of the performance of different neutral electrolytes reveals that bare and size of hydrated ions play an important role herein. Lithium is smaller than sodium and potassium. So for high value of capacitance, smaller size of solute, and high specific surface area of electrode are beneficial. Surface area of activated carbon plays a competent role in enhancing charge-storage capacity [49]. Capacitance of supercapacitor decreases with the hydrated ion size of the electrolyte. Li_2SO_4, Na_2SO_4, and K_2SO_4 provide the decreasing order of capacitance and ESR, respectively [2].

10.2.3.2 Neutral Electrolytes for Asymmetric and Hybrid Capacitors

Using MnO_2 as electrode material, sodium electrolyte shows the highest capacitance compared to lithium and potassium. Again, for alkaline earth metal as electrolyte, capacitance gets doubled compared to alkali metal. Alkali earth metals are divalent cations when intercalated with MnO_2 electrode, and they possess two valence ions of Mn^+ while alkali metals have only one charge on Mn. The reason behind less cyclic stability of the pseudocapacitor compared to EDLC remains the stability of the material at high temperatures. EDLCs can withstand high temperature range because of carbon-based electrode materials while pseudocapacitor electrode materials possess a lower temperature range having a negative impact on cyclic stability. Hybrid capacitors are used to increase the potential window of supercapacitor device for high energy density [50]. Figure 10.3 demonstrates CV and GCD curves for EDLCs and pseudocapacitor. EDCLs have rectangular feature, while pseudocapacitor has a rectangular shape with oxidation and reduction peaks in CV curve. GCD curve of EDLCs is a straight line while the charge– discharge curves of pseudocapacitors also depend on redox reaction. Few examples of neutral electrolytes used in supercapacitors are LiCl, Li_2SO_4, Na_2SO_4, NaCl, KCl, K_2SO_4, etc. Among them, Na_2SO_4 neutral electrolyte has shown promising electrochemical reactions for electrodes especially pseudocapacitance materials.

10.3 Organic Electrolytes

Organic electrolytes are better materials considering the facts, i.e., potential window and operating temperature range than aqueous electrolytes. In organic electrolytes, organic solvents are used with inorganic salts. Potential window of organic electrolyte-based electrochemical supercapacitors ranges from 2.5 to 2.8 V corresponding to high energy and power density [1]. Commonly for EDLCs, inorganic salts ($TEABF_4$) dissolved in organic solvents like acetonitrile (ACN) and propylene

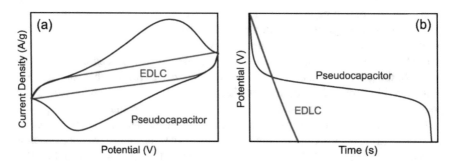

Fig. 10.3 a Cyclic voltammetry curve of EDLCs and pseudocapacitors. **b** Galvanostatic charge–discharge curve of EDLCs and pseudocapacitors. Redrawn and reprinted with permission [51]

carbonate (PC) are used [1]. High viscosity of organic solvents yields low ionic conductivity of the organic electrolyte, which causes high IR drop and low power density. Conductivity of the electrolyte can be optimized by using a solvent with low viscosity and high dielectric constant like dimethylformamide. Very low or very high viscosity of the salt gives low conductivity. In general, there is no relation between dielectric constant and conductivity; however, dielectric constant and conductivity show a linear relationship with frequency. Corrosion resistance of organic electrolyte is high in comparison to aqueous electrolyte. Other drawbacks remained in the organic electrolytes are low capacitance value, high cost, toxic nature, and a need for high purification to optimize the performance. Most important fact is that we cannot assemble the organic electrolyte-derived supercapacitor in an open atmosphere because it reacts with the atmosphere and affects the electrochemical performance. A drop of water or impurities can produce the gases inside the device at high potential. The evolved gas will enter inside the pores of the activated carbon and separator, which will increase the overall resistance of the device. The commonly used solvents for organic electrolytes are acetonitrile (49.6 mS/cm), propylene carbonate (10.6 mS/cm), N-methyloxazolidinone (10.7 mS/cm), N,N-dimethylformamide (22.8 mS/cm), nitromethane (33.8 mS/cm), nitromethane, and 2-nitropropane. Salt for organic electrolytes is also selected according to ion conductivity. Commonly used salts are either ammonium or phosphonium tetrafluoroborate [1]. Interaction between solvent and salt is also a very important aspect in view of capacitance of the supercapacitor cell. Proper dissolution of salt provides good conductivity and double-layer formation at the electrode surface. Poor energy retention is a big issue with electrochemical supercapacitor devices. Organic electrolytes are good in voltage retention, e.g., $TEABF_4/EC$ has good charge transportation with relatively low capacity and hence is able to retain high voltage after many hours of self-discharge.

10.3.1 Organic Electrolytes for Electrical Double-Layer Capacitors

Organic electrolytes have a large ion size of electrolyte and low dielectric constant. So an effective number of pore sizes in electrode material decreases, which is the cause of low electric double-layer capacitance. In EDLCs, capacitance also depends on the specific area of the electrode. Specific area is controlled by pore present in the electrode. Electrolyte ions go inside the pore and generate high capacitance. However, the large size of organic electrolytes restricts entry inside the pore of activated carbon. This creates a negative effect on the performance of the capacitor. Hence, control of the number and sizes of the pore will open up a new era in the field of supercapacitors.

Organic electrolytes can work at high voltage. However, they oxidize the electrode material and evolve some gases. Electrochemical reaction in organic electrolytes causes a decrease in the performance of the supercapacitor. The gases evolved in

Table 10.3 Organic electrolytes for electrical double-layer capacitors

Electrolyte	Electrode materials	Cell voltage (V)	Specific capacitance (F/g)	References
1M TEABF$_4$/ACN	Highly porous interconnected carbon nanosheets	0–2.7	120–150	[53]
TEABF$_4$/PC	Nitrogen-doped carbon	0–2	141	[54]
1M TEABF4/PC	Graphene/activated carbon composite	0–3	103	[55]
1M TEABF$_4$/ACN	Graphene nanomeshes	0–2.7	253	[56]
1M TEABF$_4$/PC	Graphene and CNT composite	0–3	110	[57]
1M TEABF$_4$/HFIP	Activated carbon	–	110	[58]
1.5M TEMA-BF$_4$/PC	AC	0–3.5	103	[59]
1M SBPBF$_4$/ACN	Carbon	0–2.3	109	[60]
1.5M SBPBF$_4$/PC	AC	0–3.5	122	[61]

AC activated carbon, *ACN* acetonitrile, *PC* propylene carbonate, *HFIP* hexafluoroisopropanol

these type of supercapacitors are carbon dioxide and carbon monoxide from positive electrodes and hydrogen, propylene, carbon dioxide, and carbon monoxide from negative electrodes [51]. Low ionic conductivity has also been the main issue with organic electrolytes since this leads to an increase in ESR. Self-discharging of electrochemical supercapacitor for organic electrolyte happens due to the traces of water as well as the type of organic electrolyte being used defines the same [52]. In Table 10.3, we have gathered cell voltage and specific capacitance of some organic electrolytes for EDLCs.

10.3.2   Organic Electrolytes for Asymmetric and Hybrid Capacitors

Activated carbon material shows pseudocapacitance accompanied with EDLC because of the functional groups present therein. However, with organic electrolytes, functional groups do not work effectively due to the pseudocapacitance as a result of proton participation while organic electrolytes are aprotic in nature [1]. Organic electrolytes have no labile H$^+$ to donate to the reagent. With pseudo-capacitive material like metal oxides and hydroxides, organic electrolyte shows fast intercalation process. Ion size of electrolyte salt should be small for high capacitance. Organic electrolyte-derived supercapacitors provide high energy density because of the high

Table 10.4 Organic electrolytes for asymmetric and hybrid capacitors

Electrolyte	Electrode material	Specific capacitance (F/g)	Cell voltage (V)	References
1M LiPF$_6$/(EC–DEC)	Nano porous Co$_3$O$_4$ graphene composite	424 F/g	–	[62]
0.5M LiClO$_4$/PC	PANI/graphite	420 F/g	0–1	[63]
1.5M TEMABF 4/PC	Carbon microbeads/graphitized carbon	–	0–4	[64]
1M LiPF$_6$/EC–DEC–DMC	Fe$_3$O$_4$/graphene	–	0–3	[65]
5M LiClO$_4$/AC	Exfoliated graphite/polypyrrole	0.376 F/cm^2	0–1	[66]

EC ethylene carbonate; *DEC* diethyl carbonate; *DMC* dimethyl carbonate; *ACN* acetonitrile, *PC* propylene carbonate

potential window in comparison to the aqueous electrolyte. Asymmetric and hybrid capacitors both contain high potential window; however, capacitance remains low in comparison to aqueous electrolytes. Table 10.4 shows the examples of organic electrolytes used in asymmetric and hybrid capacitors.

10.3.3   Organic Solvents for Electrochemical Supercapacitors

For high performance of a supercapacitor, a suitable nontoxic and non-flammable organic solvent is needed, which can dissolve salt effectively producing high conductivity, operating temperature range, dielectric constant, and potential window with low viscosity. The common solvents used as organic electrolytes are acetonitrile (49.6 mS/cm) and propylene carbonate (10.6 mS/cm). The working temperature range and dielectric constant of acetonitrile (ACN) are higher than propylene carbonate (PC). Lower conductivity of PC in comparison to ACN causes low power density and efficiency [1]. However, PC is an environmentally friendly solvent compared to ACN.

For supercapacitor applications, organic electrolytes are utilized in both ways, one as a single solvent for electrolyte and other remains a mixture of it. Study reveals that γ-butyrolactone (GBL)-based electrolytes have higher oxidation stability compared to PC- and ACN-based electrolytes. While at atmospheric temperature, GBL has low power performance because of high viscosity and low ionic conductivity in comparison to ACN. Other single organic solvents are also giving high performance because of specific characteristics, e.g., butylene carbonate, hexafluoro-2-propanol, fluorinated organic solvents, etc. However, a single solvent system suffers from limitations such as low operating temperature range, high viscosity, and low dielectric constant. A mixture of solvent is used to avoid the above-mentioned challenges. In this respect, the challenge remains that, which solvent should be mixed and at what optimum ratio should it be used. PC is a nice option for ACN; however, low conductivity and viscosity along with poor thermal stability have a detrimental effect on the performance. Many researches have been going on to improve the use of PC using trimethylene carbonate and ethylene carbonate as solvents.

10.3.4 Conducting Salts in Organic Electrolytes for Electrochemical Supercapacitors

Conducting salts with organic solvents make organic electrolyte and provide cations and ions as charge carriers. Molarity and mobility of charge carriers play an important role in conductivity and power performance. There are some criteria for salt selection like high solubility, electrochemical stability, conductivity, and temperature range, all at low cost. TEABF$_4$ is most commonly used salt. Salt not only affects ionic conductivity but also energy density because of the concentration of ions. Low ionic conductivity corresponds to high equivalent series resistance, hence ion concentration has a pronounced effect on the energy density of a supercapacitor cell.

10.4 Ionic Liquid Electrolytes

Any salt, which is in a liquid state at room temperature or below 100 °C is considered as the ionic liquid (IL). Generally, ILs are neutral at room temperature but the electrolytes have free or paired ions. Water also exists in a liquid state but it does not have free or paired ions like IL electrolytes. There is no need to use additional solvents for IL electrolytes. It has high chemical stability since ionic bonds are stronger than van der Waal forces. IL electrolyte is an organic salt in the liquid state at room temperature. It receives attention because of the high potential window, excellent thermal, electrochemical and chemical stability, nonvolatile, and has low vapor pressure [1]. In nature, first IL has been introduced by the ants. In combat between tawny and fire ants, tawny ants exude selfvenom (formic acid) on own body to detoxify fire ant's venom (alkaloid-based venom) and formed IL [67].

Aqueous electrolytes can work in the range of 1.0–1.5 V and for organic electrolytes reaches from 2.5 to 4.0 V [68][69]. When this limit has been crossed, the aqueous and organic electrolytes get decomposed leading to the formation of gases. Main advantage of IL electrolytes is its high potential (3.5 V) window providing high energy and power density. However, low conductivity, high viscosity, and high cost are the potential drawbacks of IL electrolyte. IL has an ionic conductivity of 14 mS/cm, which is much lower compared to the organic electrolyte.

Viscosity of ionic electrolytes is 41–219 cp, which is higher than aqueous electrolyte [2]. In aqueous electrolytes, viscosity is about 0.3 cp [1]. ESR value of IL-based supercapacitor is high due to high viscosity and low ionic conductivity. This affects the specific power and energy of the supercapacitor. Because of the high viscosity and low conductivity, the specific capacitance of IL remains low compared to aqueous and organic electrolytes. Tetraalkylammonium cation and chloroaluminate anion-based ionic liquids have been studied extensively [68]. Figure 10.4 demonstrates different types of anions used frequently with $[EMIM]^+$ cation. Some salts are not able to maintain its liquid state at room temperature. These salts have to be modified either by continuous heating or by adding some other salts that can bring down the melting point. There is one another class of IL, which has a melting point below room temperature called room-temperature ionic liquids (RTILs) [68]. RTIL cations are usually based on tetralkylammonium $[R_4N]^+$, salfonium $[R_3S]^+$, and phasphonium salt $[R_4P]^+$ cations with cyano group anions $[Ag(CN)_2]^-$ and $[N(CN)_2]^-$. Generally, it consists of large and asymmetrical ions. Small halides are generally not able to form RTIL. Large BF_4^- forms RTIL with imidazolium $[R_1R_2Im]^+$ and pyridinium $[RPi]^+$ cations. For electrochemical supercapacitor application, $[EtMeIm]^+$, $[BuMeIm]^+$, $[BuMePy]^+$, imidazolium $[EMIM]^+$, pyrrolidinium $[PYR]^+$, ammonium $[DEME]^+$, sulfonium $[Me_3S]^+$ cations and tetrafluoroborate $[BF_4]^-$, hexafluourophosphate $[PF_6]^-$, bis(fluorosulphonyl)imide $[FSI]^-$, bis(Trifluouromethanesulphonyl)imide $[TFSI]^-$, teracynoborate $[TCB]^-$, thiocynate $[SCN]^-$, dichloroacitic acid $[DCA]^-$ anions-based ILs are commonly used [68].

ILs can be classified into three categories like aprotic, protic, and zwitterionic IL. Aprotic IL contains substituent in place of proton at the site occupied by the labile H^+. This type of IL is suitable for supercapacitor and lithium ion batteries. Protic IL contains one labile H^+. Protic IL is formed by transfer of proton from bronsted acid (donate proton) to bronsted base. These ILs are suitable for fuel cells. Zwitterionic types are suitable for IL-based membranes [2].

Fig. 10.4 Commonly used $[EMIM]^+$ cations with different types of anions. Redrawn and reprinted with permission [70]

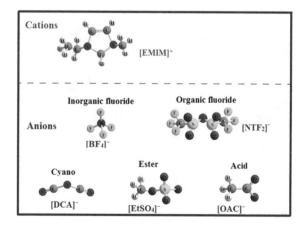

10.4.1 Solvent-Free Ionic Liquid Electrolytes for Electrical Double-Layer Capacitors

To effectively operate as electrolytes for EDLCs excellent conductivity, mobility, working temperature range, and capacitance remain the parameters of uttermost importance. For EDLCs, $[EMIM]^+[BF4]^-$ has been utilized in general due to high conductivity. $[EMIM]^+[SCN]^-$ has low viscosity and high conductivity as compared to $[EMIM]^+[BF4]^-$ [70]. $[EMIM]^+[FSI]^-$ has good rate capability [71]. $[EMIM]^+[TCB]^-$ is of high conductivity and low viscosity, 13 mS/cm and 22 cP, respectively [72]. Again, $[EMIM]^+[PO_2F_2]^-$ has high specific capacitance compared to $[EMIM]^+[BF4]^-$,however, which has less potential window (2.5 V) [73]. $[EMIM]^+[BCN_4]^-$ also provides best performance similar to $[EMIM]^+[BF4]^-$ [73]. Again, $[EMIM]^+[DCA]^-$ delivers high capacitance but a low potential window of about 2.3 V remains the matter of concern. Table 10.5 represents IL behavior on EDLCS as a function of specific capacitance and cell voltage.

MEGO-microwave-expanded graphite oxide

The most commonly used electrolytes are imidazolium-based electrolytes for EDLCs. However, other electrolytes are also in practice, e.g., pyrrolydium, ammonium, sulfonium-based electrolytes. Maximum potential window remains 3.7 V in pyrrolidium-based electrolytes. Pyrrolidinium ($[PYR]^+[TFSI]^-$) ionic electrolyte has a high potential window (3.7 V) and a large working temperature range (-30 to $60\,°C$). Electrode material is asymmetric in nature indicating higher carbon loading at the positive electrode and lower carbon loading at the negative electrode side. Azepanium ($[AZP]^+[TFSI]^-$) electrolytes are cheaper compared to pyrrolidinium; however, they exhibit low energy density. Ammonium ($[DEME]^+[TFSI]^-$) electrolyte has high potential window than $[EMIM]^+[BF_4]^-$ and TEABF$_4$/PC. It also has a high working temperature range. Fluorohydrogenate $[FHILs]^+[TFSI]^-$ has higher specific capacitance as compared to $[EMIM]^+[BF_4]^-$ and TEABF$_4$/PC. Sulphonium

Table 10.5 Ionic electrolytes for electrical double-layer capacitors

Electrolyte	Electrode materials	Specific capacitance (F/g)	Cell voltage (V)	References
$[BMIM]^+[BF4]^-$/AN	Activated MEGO	166	0–3	[74]
$[BMIM]^+[BF4]^-$/AN	Activated MEGO	172	3.5	[75]
$[EMIM]^+$ $[TFSI]^-$/AN	Activated sMEGO	174	3.5	[76]
$[EMIM]^+[BF_4]^-$	Mesoporous graphene nanofiber	231	0–3.5	[55]
$[EMIM]^+[BF_4]^-$	Mesoporous graphene	145	0–4	[77]

[Me$_{3S}$S]$^+$[TFSI]$^-$ electrolytes on the other hand possess a high potential window of up to 5 V with a high specific energy density of 44 Wh/kg [2].

10.4.2   Solvent-Free Ionic Liquids for Asymmetric and Hybrid Capacitors

Capacitance of pseudocapacitors depends on the surface functionalities, hydrophobicity, and free water content of the ionic liquid. The shape of cyclic voltammetry curves and capacitance of RuO_2 are found to be similar to aqueous electrolytes. Charging rate of pseudocapacitors is limited because of high viscosity and slow proton transfer in IL electrolytes. In protic ILs, alkali chain length has a substituent other than the proton and its position affects the capacitance, potential window, and cyclic stability as well. With RuO_2 electrode material, [EMIM]$^+$[BF$_4$]$^-$ electrolyte exhibits 83 F/g specific capacitance and the cyclic voltammetry curve is similar to aqueous electrolyte [78]. MnO_2-based electrodes are less expensive than RuO_2 electrode material. MnO_2 electrode provides 72 F/g specific capacitance with [EMIM]$^+$[DCA]$^-$ electrolyte having 2 V potential window [79]. There are different types of asymmetric electrode configurations such as AC//3-Methilthiophene, AC//MnO$_2$ and AC//graphene supported Fe$_2$O$_3$ [80]. [PYR]$^+$[TFSI]$^-$ can give maximum energy density of 24 Wh/kg and power density of 14 kW/kg after the first 1000 cycle at 60 °C [81]. It can deliver a 30% higher rate of energy density compared to asymmetric EDLC however, it has a very less cyclic stability. Again, capacitance gets decreased to 50% after fifty thousand cycles because of deterioration of the asymmetric electrode. For [EMIM]$^+$[BF$_4$]$^-$ electrolytes, the voltage window will be 0–4 V with energy and power density of 177 Wh/kg and 8 k W/kg, respectively [82].

10.4.3   Mixture of Organic Solvent and Ionic Liquid

Certain drawbacks of IL are high viscosity and low conductivity. To overcome the difficulties, a mixture of organic solvent and ionic liquid electrolyte has been adopted [83]. For example, [EMIM]$^+$[PF$_6$]$^-$/PC can give high specific capacitance and high thermal stability in comparison to TEABF$_4$/PC [84]. [EMIM]$^+$[PF$_6$]$^-$/DMF mixture increases the specific capacitance and decreases internal resistance of the asymmetric electrode, AC//MnO$_2$ [85]. The flammability and volatility of the ACN solvent can be reduced as compared to sulfonium-based ionic liquid electrolyte using a mixture of [TMPA]$^+$[TFSI]$^-$/ACN [86]. Comparative evaluation has also been made to observe the effect on specific capacitance and ionic conductivity. [DEMS][TFSA]/PC has the highest ionic conductivity among the electrolytes, which contain cation [TFSA]. Again, [DEMS][BF$_4$]/PC has higher specific capacitance compared to [EMIM][BF$_4$]/PC [87]. Because of the low potential window, the protic electrolytes have not received as much attention as aprotic electrolytes in a mixture of ionic electrolytes and organic solvents [87].

10.5   Solid State or Quasisolid State Electrolytes

Liquid electrolytes are good in ionic conductivity. However, leakage of the electrolytes remains a major concern. In recent times, we need flexible electronic devices, printed electronic devices, wearable and micro devices. To fullfill these requirements researchers focus on the development of solid-state electrolytes for supercapacitor applications. Solid-state or quasisolid-state electrolyte not only works as an electrolyte but also as separators. Fabrication and packaging are easy in solid electrolytes. As solid electrolytes, synthesis and natural polymer, inorganic solids like ceramics, etc., are generally used [88]. Solid state or quasisolid-state electrolytes contain high electrochemical stability but low ionic conductivity. Solid electrolytes and quasisolid electrolytes can change their volume, which makes the device flexible. These electrolytes have high mechanical strength as well [89]. Polymer electrolytes can be further classified into three categories, solid/dry polymer electrolytes, gel polymer electrolytes, and polyelectrolytes.

10.5.1   Polymer Electrolytes

10.5.1.1   Solid/Dry Polymer Electrolytes

Solid polymer electrolyte (SPE) is solvent-free polymer-based electrolyte [1]. Solid polymer electrolytes (SPEs) have avast fields of energy storage in supercapacitor, batteries, and fuel cells [90]. They provide high energy and cyclic stability. It is free from the toxic solvent as well. Solid polymer matrix holds salt (e.g., NaCl, LiCl) without any solvent (say, water). The main drawback of the SPEs is low ionic conductivity. Conductivity of SPE depends on the ion transfer of the salt in the polymers [91]. In SPEs, ions move in polymer phase in place of solvent (e.g., aqueous or organic electrolytes). Solid electrolyte-derived supercapacitors have high electrochemical stability, good working temperature range, high mechanical strength, and structural stability. Solid electrolytes are preferable for the development of flexible and printable energy storage devices [91, 92].

10.5.1.2   Gel Polymer Electrolytes

In liquid electrolytes, leakage remains the main concern; however, they possess high ionic conductivity (10^{-3} to 10^{-2} S/cm) [93]. On the other hand, SPE is portable and more flexible and possesses no problem of leakage. However, SPEs have low ionic conductivity (10^{-8} to 10^{-5} S/cm) and poor interfacial contact with the electrode material. Gel polymer electrolyte (GPE) has the advantage of both liquid and solid electrolytes. Gel polymer has the highest ionic conductivity among solid electrolytes and polyelectrolytes. It can work as a separator for electrochemical supercapacitors as well. In GPE, the polymer acts as the host matrix. People also use aqueous electrolytes or salts with solvent (Fig. 10.5). Polymer works as a matrix and in the swollen matrix, ion transfer takes place instead of changes in polymeric phases [88, 94].

To get low ESR, the polymer matrix should be conductive. Polymer matrix should have high electrolyte retention capacity with high flexibility and mechanical strength. The commonly used polymers are poly vinyl alcohol (PVA), poly acrylic acid (PAA), potassium poly acrylate (PAAK), poly methylmetha acrylate (PMMA), poly ether-ether ketone (PEEK), and PVDF-HFP [95–97]. Again, based on solvents, GPE can be categorized into three parts as hydrogel polymer electrolyte, organogel electrolyte, and ionic liquid-based solid electrolytes.

Hydrogel polymer electrolytes

For hydrogel polymer electrolytes, polymer is composed of aqueous electrolytes. Most commonly used polymer matrix for hydrogel polymer is polyvinyl alcohol (PVA), a linear polymer [98]. H_2SO_4, KOH, and LiCl are commonly used aqueous electrolytes in PVA matrix. GPEs have the advantage of a high potential windows in comparison to aqueous and organic electrolytes. High potential window of GPEs provides high energy density.

Hydrogel polymer electrolyte for electrical double-layer capacitors

Capacitive performance depends upon the interfacial interaction between the electrolyte and electrode. GPEs exhibit low capacitance and performance because of the high interfacial resistance or limited ion diffusion. An increase in the thickness of the electrode material has found to improve the specific capacitance. A comparison between PVA/H_2SO_4 and PVA/H_3PO_4 reveals that with an increase in the thickness

Fig. 10.5 Schematic diagram of gel polymer electrolyte

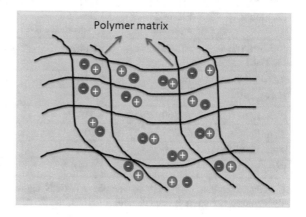

of electrode material, PVA/H$_2$SO$_4$ attains linear increment in specific capacitance compared to PVA/H$_3$PO$_4$ [93]. PVA/H$_2$SO$_4$ attains a faster increment in capacitance with an increase in the thickness of the electrode material. To increase the interface between the electrode and gel electrolyte, different electrodes are used such as CNTs on carbon clothes, activated carbon clothes, 3D graphene, and graphene/porous carbon aerogels [99].

Type of aqueous solution used with polymer matrix also has a potential impact on the capacitive performance of the electrochemical supercapacitor. In comparison, among six electrolytes H$_2$SO$_4$, H$_3$PO$_4$, KOH, NaOH, KCl, and NaCl in PVA matrix, PVA/H$_3$PO$_4$ combination gives the best capacitive performance compared to others [100]. Cross-linked PVA/H$_2$SO$_4$ shows high cyclic stability at high temperature of 70 °C. Various additives are also used such as SiO$_2$, TiO$_2$, graphene oxide (GO) to increase the performance. Study reveals that a small amount of GO in PVA/KOH increases the conductivity and capacitance value to 24% [101]. Environmental stability is another important factor in supercapacitors. In PVA/KOH electrolyte, a decrement in the performance has been evidenced after some time due to dehydration. Tetra ammonium hydride has replaced the KOH in the electrolyte since it has high water retention. Again, to improve the retention of water and minimize the effect of dehydration, SiO$_2$ has been utilized to controls moisture content in the supercapacitor [102]. Examples of hydrogel polymer electrolytes for EDLCs are represented in Table 10.6.

Hydrogel in asymmetric and hybrid capacitors

Aqueous electrolyte hydrogels are of more chemical stability as compared to other hydrogels. Pseudocapacitors acid hydrogel that are commonly used are PVA/H$_2$SO$_4$ and PVA/H$_3$PO$_4$ [107]. Electrode materials for pseudocapacitors are metal oxide, nitride, and selenides. For alkali-based hydrogels, PVA/KOH is the most extensively used electrolyte. Pseudo-capacitive electrode material is made of Ni(OH)$_2$ and MnO$_2$ [108]. For neutral electrolyte-based hydrogels, PVA/LiCl, PVA/Na$_2$SO$_4$,

Table 10.6 Hydrogel polymer electrolytes for electrical double-layer capacitors

Electrolyte	Electrode materials	Specific capacitance (F/g)	Cell voltage (V)	References
EW-GPE (G) + 1M NaCl	AC-rice husk	214.3	0–0.8	[5]
KCl–Fe^{3+}/PAA	AC	87.4	0–1	[15]
H$_3$PO$_4$–PVA	CNT	11	0–0.8	[9]
PVA–KOH (G)	AC-oily sludge	348.1	0–1	[103]
PVA–H$_2$SO$_4$ (G)	CNT	80	0–1	[104]
PVA–KOH–H$_2$O	AC-NiO	73.4	0–1.6	[105]
PVA–KOH–KI	AC	226.9	0–0.8	[106]

AC activated carbon, *CNT* carbon nanotube, *EW-GPE* egg white gel polymer electrolyte, *PAA* polyacrylic acid, *PVA* polyvinyl alcohol and PVA/NaNO$_3$ are used commonly. The pseudo-capacitive electrode material for neutral electrolyte-based hydrogels is MnO$_2$, ZnO, and V$_2$O$_5$ [109]. Like an EDLC, in pseudocapacitors, optimum matching of electrode material and hydrogel electrolyte is needed to achieve high conductivity. For symmetric supercapacitors, hydrogel polymer electrolyte has a low potential window hence of low energy density. Other than PVA, poly(acrylate) and poly(acrylic acid) are the polymer matrix utilized in hydrogel polymer electrolytes [110]. These polymers promote proton conduction. Nafion has also been tested in different solid-state electrochemical supercapacitors. It has high conductivity and supports high potential scan rate [111, 112].

Organogel electrolytes

In gel polymer electrolytes, organic solvents are used as plasticizer. Organic solvent increases the working cell voltage of the electrolyte. For the preparation of organogel electrolytes, various host polymers are tested such as PEEK, PVA, PEO, PMMA, and some copolymers are also tested such as PAN-b-PEG-PAN. PAN-b-PEG-PAN emerges as the most promising polymer matrix in gel polymer electrolyte [113]. By controlling the chain length of acrylonitrile and ethyl glycol, the conductivity of the copolymers can be controlled easily. The copolymer exhibits a high conductivity of 1.1×10^{-2} S/cm, which is quite high compared to organic liquid electrolyte [114]. A high energy and power density of 20 Wh/kg and 10 kW/kg, respectively, have also been reported. Copolymers in gel polymer electrolytes with organic solvents show high conductivity and cell voltage of about 2.5–3.0 V that is reported to be much higher than the hydrogel electrolytes [114]. The ratio of host polymer matrix to salt in the organic solvent has a pronounced effect on the ionic conductivity of the electrolytes.

Ionic liquid-based solid electrolytes

This type of electrolytes is prepared by the incorporation of ionic liquid in a polymer matrix. Ionic liquid-based solid electrolyte gives high thermal stability and potential window for electrochemical supercapacitor. Ionic conductivity and potential window depend upon the nature of the host polymer matrix and ionic electrolyte. PMMA/[BMIM]$^+$[TFSI]$^-$ electrolytes have at least four-fold stretchability. With graphene-based electrode material, it shows capacitance of 83 F/g and high energy and power densities of 26.5 Wh/h and 5 kW/kg, respectively [115]. PVDF/[BMIM]$^+$[BF$_4$]$^-$ exhibits good thermal stability (up to 300 °C), high flexibility, and conductivity (2.42×10^{-3} S/cm). Addition of Li salt in PVDF-HFP/[EMIM]$^+$[FAP]$^-$ electrolytes improves capacitance. For PVDF-HFP/[EMIM]$^+$[FAP]$^-$ ionic electrolyte, capacitance increases from 76 to 126 F/g due to the addition of lithium salts. SiO$_2$ in PEO/EMIHSO$_4$ helps to improve ionic conductivity. Addition of 10% SiO$_2$ could increase the ionic conductivity from 0.85 to 2.15 mS/cm at room temperature. [pDADMA]$^+$[PFSI]$^-$ and [PYR$_{14}$]$^+$[PFSI]$^-$ binary polymer in 4:6 mass ratio gives a high potential window [116]. From an environmental point of view, natural polymers are also being used as gel polymer electrolyte. Renewable sources such as starch, corn, egg white, and chitosan are commonly used in electrochemical supercapacitors. Biodegradable materials like poly (epsilon-caprolactone) are also studied. Natural polymers have low mechanical strength. To overcome this, chitosan and starch have been blended for electrochemical supercapacitor applications [117].

10.5.1.3 Polyelectrolytes

Polyelectrolytes are the polymers having repeating units with electrolyte groups partially dissociate to form poly cations and anions when dissolving in an aqueous solvent. It contains ions of high molecular weight and as polymer, they have high viscosity. Polymer chain contributes to ionic conductivity in these electrolytes [118]. Among three kinds of polymer electrolytes, gel polymer electrolytes have maximum conductivity with a provision to adjust the conductivity as per the requirement. However, the mechanical strength and working temperature range are on the lower side. The three basic requirements are good ionic conductivity, high thermal stability, and enough strength to sustain the mechanical load.

10.5.1.4 Inorganic Solid-State Electrolytes

These electrolytes have no flexibility and are almost nonstretchable. These are mechanically robust and thermally stable. Compared to polymer-based electrolytes, inorganic solid electrolytes receive less attention. Francisco et al. have developed nanostructured solid-state supercapacitors using $Li_2S–P_2S_5$ glass ceramic electrolytes [88]. $Li_2S–P_2S_5$ acts both as a separator and an ion conductor [88, 89]. $Li_2S–P_2S_5$ has high lithium ion conductivity and 10 F/g of specific capacitance has been reported when utilized with CNTs as the electrode material [88].

10.6 Redox-Active Electrolytes

Through the use of redox-active electrolytes, capacitance value can be increased by including pseudo-capacitive behavior of redox-active electrolytes. Here, pseudocapacitance is not only contributed by electrode material but also from the electrolyte. Redox mediator shows reduction and oxidation behaviors [119].

10.6.1 Aqueous Redox-Active Electrolytes

For carbon-based electrochemical supercapacitors, the typically used redox-active electrolyte is iodide/iodine redox pair. In 1.0 M KI, carbon-based positive electrode shows maximum 1840 F/g, which is very high [39]. This happens due to the contribution from redox-active electrolyte in pseudocapacitance. This high capacitance is observed in the positive potential electrode with a low potential window. For two electrodes in EDLCs, specific capacitance of 125 F/g has been observed. As van der Waal radius of the positive electrode increases, specific capacitance is reduced. In 1.0 M electrolyte, specific capacitance follows the order RbI > KI > NaI > LiI. van der Waal's radius is half of the distance between two closest non-bonding atoms [120]. Additional pseudocapacitance contribution can be explained by considering following equation [100].

$$3I^- \underset{Discharging}{\overset{Charging}{\rightleftarrows}} I_3 + 2e^- \tag{10.6}$$

$$2I^- \underset{Discharging}{\overset{Charging}{\rightleftarrows}} I_2 + 2e^- \tag{10.7}$$

$$2I_3^- \underset{Discharging}{\overset{Charging}{\rightleftarrows}} 3I_2 + 2e^- \tag{10.8}$$

$$I_2 + 6H_2O \underset{Discharging}{\overset{Charging}{\rightleftarrows}} 2IO_3^- + 12H^+ + 10e^- \tag{10.9}$$

The total capacitance of electrochemical supercapacitor is low in comparison to the positive electrode material. To increase the specific capacitance of ES, we can use two electrolyte materials for both electrodes. For positive electrode, 1.0 M KI and for negative electrode, $VOSO_4$ electrolytes are generally used in electrochemical supercapacitors [121]. To separate these electrolytes, glassy paper and Nafion 117 have been used. Specific capacitance, energy, and power of 500 F/g and 20 Wh/kg, and 2 kW/kg, respectively, have been reported for this system.

Heteropoly acids such as phosphotungstic acid (PWA) and silicotungstic acid (SiWA) are used as mediators for redox-active electrolyte. These mediators give high proton conductivity and multiple electron transfer in redox-active electrolytes. Mediators also help to increase the potential window of the supercapacitors. They contain metal ions so self-discharging can be increased due to the migration of the redox-active electrolyte between two electrodes. Redox-active electrolytes for pseudocapacitors have also been used; however, low cyclic stability remains a matter of concern. Commonly used redox-active electrolytes are KI and heteropoly acids such as hydroquinone (HQ) and p-Phenylenediamine (PPD). Mediators in redox-active electrolytes not only increase the specific capacitance but also speedup the Faradaic reaction.

10.6.1.1 Non-aqueous Redox-Active Electrolytes

To achieve good energy density, high capacitance and cell voltage, non-aqueous electrolytes are also studied based on organic and ionic liquid electrolytes. For example, $B_{12}F_xH_{12-x}$ in PC-DMC does effective contribution to increasing the pseudocapacitance performance of ESs. Again, [EMIM][I] in [EMIM][BF$_4$] increases the specific capacitance up to 50% in comparison to [EMIM][BF$_4$] [121].

10.6.1.2 Solid Redox-Active Electrolytes

Typically used redox additives in the solid electrolyte are iodide like KI, NaI, organic redox mediators like PPD, hydroquinone and mixture of both iodide and organic mediator. Normally, these redox-active electrolytes get dissolved into the gel polymer electrolyte, so this is similar to the organic and aqueous electrolyte. These gel polymer electrolytes in combination with redox-active mediator show very high specific capacitance in comparison to nonredox-active mediator GPEs. Polymers commonly used in gel polymer electrolytes are PVA, Nafion, and PEO. KI–VOSO$_4$ in combination with PVA/H$_2$SO$_4$ provides a high specific capacitance of 1232.8 F/g and an energy density of 25 Wh/kg [100].

10.7 Features of Electrolyte Materials

Any component of a device has one or more functions. For example, the component should support a load, hold pressure, transmit heat, and so forth. These specifications must be achieved. In designing such components, the designer has an objective: to make it as cheap as possible, perhaps, or as light, or as safe, or perhaps some combination of these. Table 10.7 lists the features of electrolytes used for supercapacitor.

Table 10.7 Features of an efficient electrolyte material for electrochemical supercapacitors

Primary features	Secondary features	Tertiary features
High potential window	Low viscosity	Environmental friendly
High ionic conductivity	Low cost	Easy to handle
Small size hydrated ions	High chemical stability	Easily available
High capacitance	Low volatility	Nontoxic
Large operating temperature range	Non-flammability	
	Optimum concentration	

10.8   Concluding Remarks

The basic requirements of electrolytes are good ionic conductivity, high potential window, and operating temperature range. H_2SO_4, KOH, and Na_2SO_4 are suitable as acid aqueous, alkali aqueous, and neutral electrolytes, respectively. In organic electrolytes, ionic conductivity and ion size play an important role. $TEABF_4$ salt with acrylonitrile solvent is used commonly. Ionic liquids are the organic liquid electrolytes at room temperature. Ionic liquid electrolytes are utilized in two ways, first is the solvent-free electrolytes and the second one is the mixture of organic solvent and ionic liquid electrolyte. The major concern remains as high viscosity and low ionic conductivity. To overcome these drawbacks, $[EMIM]^+[BF_4]^-$ has been utilized. Solid/quasisolid-state electrolytes are of great potential due to the good mechanical strength with no leakage. Gel polymer electrolytes have high conductivity and enough mechanical strength as well. Most commonly used gel polymer electrolytes are PVA/H_3PO_4 and PVA/H_2SO_4. Redox-active electrolytes are another important class of electrolytes used to improve pseudocapacitance of the supercapacitor by the redox reaction of electrolyte with the electrode material. The commonly used redox-active electrolytes are KI and heteropoly acids.

Acknowledgements The authors acknowledge the financial support provided by Department of Science and Technology, India (DST/TMD/MES/2K16/37(G)), for carrying out this research work. Authors are thankful to Ms. Tanvi Pal for drafting a figure.

References

1. C. Zhong, Y. Deng, W. Hu, D. Sun, X. Han, J. Qiao, J. Zhang, *Electrolytes for Electrochemical Supercapacitors*, 1st ed. (CRC Press, 2016)
2. J. Zhang, C. Zhong, Y. Deng, W. Hu, J. Qiao, L. Zhang, Chem. Soc. Rev. **44**, 7484 (2015)
3. M.L. Williams, Occup. Environ. Med. **53**, 504 (1996)
4. H. Tomiyasu, H. Shikata, K. Takao, N. Asanuma, S. Taruta, Y.Y. Park, Sci. Rep. **7**, 1 (2017)
5. R. Na, X. Wang, N. Lu, G. Huo, H. Lin, G. Wang, Electrochim. Acta **274**, 316 (2018)
6. S. Lehtimäki, A. Railanmaa, J. Keskinen, M. Kujala, S. Tuukkanen, D. Lupo, Sci. Rep. **7**, 46001 (2017)
7. M. Aslan, D. Weingarth, P. Herbeck-Engel, I. Grobelsek, V. Presser, J. Power Sources **279**, 323 (2015)
8. G. Nie, X. Lu, J. Lei, Z. Jiang, C. Wang, J. Mater. Chem. A **2**, 15495 (2014)
9. Y. Guo, X. Zhou, Q. Tang, H. Bao, G. Wang, P. Saha, J. Mater. Chem. A **4**, 8769 (2016)
10. Q. Tang, Y. Lin, M. Huang, F. Yu, L. Fan, J. Lin, J. Wu, H. Yu, Y. Li, J. Power Sources **206**, 463 (2012)
11. L. Demarconnay, E. Raymundo-Piñero, F. Béguin, Electrochem. Commun. **12**, 1275 (2010)
12. Q. Abbas, D. Pajak, E. Frąckowiak, F. Béguin, Electrochim. Acta **140**, 132 (2014)
13. K. Torchała, K. Kierzek, J. Machnikowski, Electrochim. Acta **86**, 260 (2012)
14. H. Liu, X. Chen, C. Xiong, T. Li, T. Zhao, H. Li, Y. Shang, A. Dang, Q. Zhuang, S. Zhang, Compos. Part B Eng. **141**, 250 (2017)
15. Y.J. Kang, H. Chung, C.-H. Han, W. Kim, Nanotechnology **23**, 065401 (2012)
16. Z. Jin, X. Yan, Y. Yu, G. Zhao, J. Mater. Chem. A **2**, 11706 (2014)

17. N.S.M. Nor, M. Deraman, R. Omar, E. Taer, Awitdrus, R. Farma, N.H. Basri, B.N.M. Dolah, AIP Conf. Proc. **1586**, 68 (2014)
18. F. Lufrano, P. Staiti, Electrochim. Acta **49**, 2683 (2004)
19. S. Yoon, H.I. Lee, K.-W. Jun, T. Hyeon, C. Jo, Y. Mun, J. Lee, J. Lee, S.-W. Hong, K.-S. Ha, S.-H. Lee, Carbon **64**, 391 (2013)
20. D. Jiménez-Cordero, F. Heras, M.A. Gilarranz, E. Raymundo-Piñero, Carbon **71**, 127 (2014)
21. Q. Chen, Y. Hu, C. Hu, H. Cheng, Z. Zhang, H. Shao, L. Qu, Phys. Chem. Chem. Phys. **16**, 19307 (2014)
22. L.-F. Chen, Z.-H. Huang, H.-W. Liang, H.-L. Gao, S.-H. Yu, Adv. Funct. Mater. **24**, 5104 (2014)
23. Y. Huang, S.L. Candelaria, Y. Li, Z. Li, J. Tian, L. Zhang, G. Cao, J. Power Sources **252**, 90 (2014)
24. X. Li, W. Xing, S. Zhuo, J. Zhou, F. Li, S.Z. Qiao, G.Q. Lu, Bioresour. Technol. **102**, 1118 (2011)
25. J. Qu, C. Geng, S. Lv, G. Shao, S. Ma, M. Wu, Electrochim. Acta **176**, 982 (2015)
26. W.-H. Qu, Y.-Y. Xu, A.-H. Lu, X.-Q. Zhang, W.-C. Li, Bioresour. Technol. **189**, 285 (2015)
27. J.H. Chen, W.Z. Li, D.Z. Wang, S.X. Yang, J.G. Wen, Z.F. Ren, Carbon **40**, 1193 (2002)
28. L.L. Zhang, X. Zhao, M.D. Stoller, Y. Zhu, H. Ji, S. Murali, Y. Wu, S. Perales, B. Clevenger, R.S. Ruoff, Nano Lett. **12**, 1806 (2012)
29. X. He, P. Ling, M. Yu, X. Wang, X. Zhang, M. Zheng, Electrochim. Acta **105**, 635 (2013)
30. B.E. Conway, *Electrochemical Supercapacitors* (Springer, US, 1999)
31. J. Cherusseri, K.K. Kar, Phys. Chem. Chem. Phys. **18**, 8587 (2016)
32. L. Deng, J. Wang, G. Zhu, L. Kang, Z. Hao, Z. Lei, Z. Yang, Z.H. Liu, J. Power Sources **248**, 407 (2014)
33. Q. Wang, J. Yan, Z. Fan, T. Wei, M. Zhang, X. Jing, J. Power Sources **247**, 197 (2014)
34. J. Zhang, J. Ding, C. Li, B. Li, D. Li, Z. Liu, Q. Cai, J. Zhang, Y. Liu, A.C.S. Sustain, Chem. Eng. **5**, 4982 (2017)
35. H. Wang, Y. Liang, T. Mirfakhrai, Z. Chen, H.S. Casalongue, H. Dai, Nano Res. **4**, 729 (2011)
36. L. Li, S. Peng, Y. Cheah, P. Teh, J. Wang, G. Wee, Y. Ko, C. Wong, M. Srinivasan, Chem. Eur. J. **19**, 5892 (2013)
37. C. Zhong, Y. He, G. He, Y. Deng, W. Hu, X. Han, J. Zhang, X. Zheng, Nanoscale **9**, 8623 (2017)
38. J.W. Lee, T. Ahn, J.H. Kim, J.M. Ko, J.D. Kim, Electrochim. Acta **56**, 4849 (2011)
39. Q.T. Qu, Y. Shi, L.L. Li, W.L. Guo, Y.P. Wu, H.P. Zhang, S.Y. Guan, R. Holze, Electrochem. Commun. **11**, 1325 (2009)
40. H. Xia, Y. Shirley Meng, G. Yuan, C. Cui, L. Lu, Electrochem. Solid-State Lett. **15**, A60 (2012)
41. S. Ishimoto, Y. Asakawa, M. Shinya, K. Naoi, J. Electrochem. Soc. **156**, A563 (2009)
42. J. Cherusseri, K.K. Kar, RSC Adv. **5**, 34335 (2015)
43. Y. Tian, J.W. Yan, R. Xue, B.L. Yi, Wuli Huaxue Xuebao/Acta Phys. Chim. Sin. **27**, 479 (2011)
44. U.M. Patil, R.R. Salunkhe, K.V. Gurav, C.D. Lokhande, Appl. Surf. Sci. **255**, 2603 (2008)
45. L. Su, L. Gong, H. Lü, Q. Xü, J. Power Sources **248**, 212 (2014)
46. X. Wang, A. Yuan, Y. Wang, J. Power Sources **172**, 1007 (2007)
47. X. Liu, P. Shang, Y. Zhang, X. Wang, Z. Fan, B. Wang, Y. Zheng, J. Mater. Chem. A **2**, 15273 (2014)
48. M.P. Bichat, E. Raymundo-Piñero, F. Béguin, Carbon **48**, 4351 (2010)
49. J. Cherusseri, R. Sharma, K.K. Kar, Carbon **105**, 113 (2016)
50. S.K. Singh, H. Prakash, M.J. Akhtar, K.K. Kar, A.C.S. Sustain, Chem. Eng. **6**, 5381 (2018)
51. Y. Tanaka, T. Kimura, K. Hikino, S. Goto, M. Nishimura, S. Mano, T. Nakagaw, in *Genetic Engineering—Basics, New Applications and Responsibilities* (Intech, 2012), p. 64
52. J.-Y. Choi, S. Kwon, A. Benayad, D. Lee, N. Jung, J.S. Park, D.-M. Yoon, Y.M. Park, Adv. Mater. **25**, 6854 (2013)

53. R. Francke, D. Cericola, R. Kötz, D. Weingarth, S.R. Waldvogel, Electrochim. Acta **62**, 372 (2012)
54. G.A. Ferrero, A.B. Fuertes, M. Sevilla, J. Mater. Chem. A **3**, 2914 (2015)
55. X. Zhang, H. Zhang, C. Li, K. Wang, X. Sun, Y. Ma, RSC Adv. **4**, 45862 (2014)
56. W. Hooch Antink, Y. Choi, K. Seong, J.M. Kim, Y. Piao, Adv. Mater. Interfaces **5**, 1701212 (2018)
57. A. Brandt, P. Isken, A. Lex-Balducci, A. Balducci, J. Power Sources **204**, 213 (2012)
58. E. Perricone, M. Chamas, L. Cointeaux, J.-C. Leprêtre, P. Judeinstein, P. Azais, F. Béguin, F. Alloin, Electrochim. Acta **93**, 1 (2013)
59. X. Yu, D. Ruan, C. Wu, J. Wang, Z. Shi, J. Power Sources **265**, 309 (2014)
60. G. Sun, K. Li, C. Sun, J. Power Sources **162**, 1444 (2006)
61. F. Markoulidis, C. Lei, C. Lekakou, Electrochim. Acta **249**, 122 (2017)
62. M. Mladenov, K. Alexandrova, N.V. Petrov, B. Tsyntsarski, D. Kovacheva, N. Saliyski, R. Raicheff, J. Solid State Electrochem. **17**, 2101 (2013)
63. Z. Boeva, R.-M. Latonen, T. Lindfors, Z. Mousavi, in *Electrochemical Nanofabrication* (Pan Stanford, 2016), pp. 417–471
64. C. Zheng, M. Yoshio, L. Qi, H. Wang, J. Power Sources **260**, 19 (2014)
65. W.H. Qu, F. Han, A.H. Lu, C. Xing, M. Qiao, W.C. Li, J. Mater. Chem. A **2**, 6549 (2014)
66. J. Cherusseri, K.K. Kar, RSC Adv. **6**, 60454 (2016)
67. H.E. Horne, K.N. West, N. Gouault, R.E. Barletta, M. Le Roch, G.E. Mullen, L. Chen, A.C. Stenson, J.M. Hendrich, C.G. Cassity, R.A. O'Brien, J.H. Davis, R.E. Sykora, K.R. Xiang, H.Y. Fadamiro, Angew. Chemie Int. Ed. **53**, 11762 (2014)
68. M. Galiński, A. Lewandowski, I. Stepniak, Electrochim. Acta **51**, 5567 (2006)
69. M. Shi, S. Kou, X. Yan, Chemsuschem **7**, 3053 (2014)
70. N. Handa, T. Sugimoto, M. Yamagata, M. Kikuta, M. Kono, M. Ishikawa, J. Power Sources **185**, 1585 (2008)
71. G.P. Pandey, S.A. Hashmi, Bull. Mater. Sci. **36**, 729 (2013)
72. K. Matsumoto, R. Hagiwara, J. Electrochem. Soc. **157**, A578 (2010)
73. A. Jänes, E. Lust, H. Kurig, K. Tõnurist, M. Vestli, J. Electrochem. Soc. **159**, A944 (2012)
74. Y. Zhu, Z. Tan, N. Quarles, L.L. Zhang, S. Murali, R.S. Ruoff, J.R. Potts, Y. Lu, Nano Energy **2**, 764 (2013)
75. T. Kim, G. Jung, S. Yoo, K.S. Suh, R.S. Ruoff, ACS Nano **7**, 6899 (2013)
76. B. kim, H. D. Jang, G. D. Moon, C. Lee, H. Chang, S. Cho, J.-H. Choi, Carbon **132**, 16 (2018)
77. H. Zhang, X. Zhang, X. Sun, Y. Ma, Sci. Rep. **3**, 1 (2013)
78. D. Rochefort, A.-L. Pont, Electrochem. Commun. **8**, 1539 (2006)
79. M.-T. Lee, C.-J. Su, Y.-S. Li, J.-K. Chang, I.-W. Sun, J. Mater. Chem. **22**, 6274 (2012)
80. A. Balducci, U. Bardi, S. Caporali, M. Mastragostino, F. Soavi, Electrochem. Commun. **6**, 566 (2004)
81. C. Arbizzani, M. Biso, D. Cericola, M. Lazzari, F. Soavi, M. Mastragostino, J. Power Sources **185**, 1575 (2008)
82. S. Sun, J. Lang, R. Wang, L. Kong, X. Li, X. Yan, J. Mater. Chem. A **2**, 14550 (2014)
83. A.B. McEwen, J. Electrochem. Soc. **144**, L84 (1997)
84. A. Orita, K. Kamijima, M. Yoshida, J. Power Sources **195**, 7471 (2010)
85. R. Lin, P. Huang, J. Ségalini, C. Largeot, P.L. Taberna, J. Chmiola, Y. Gogotsi, P. Simon, Electrochim. Acta **54**, 7025 (2009)
86. X.Y. Chen, Y.Y. He, H. Song, Z.J. Zhang, Carbon **72**, 410 (2014)
87. R. Kumar, H.J. Kim, S. Park, A. Srivastava, I.K. Oh, Carbon **79**, 192 (2014)
88. B.E. Francisco, C.M. Jones, S.-H. Lee, C.R. Stoldt, Appl. Phys. Lett. **100**, 103902 (2012)
89. K.K. Kar, *Composite Materials*, 1st edn. (Springer, Berlin, 2017)
90. A. Kumar, R. Sharma, M.K. Das, P. Gajbhiye, K.K. Kar, Electrochim. Acta **215**, 1 (2016)
91. B. De, A. Yadav, S. Khan, K.K. Kar, A.C.S. Appl, Mater. Interfaces **9**, 19870 (2017)
92. A. Yadav, B. De, S.K. Singh, P. Sinha, K.K. Kar, A.C.S. Appl, Mater. Interfaces **11**, 7974 (2019)
93. M. Kaempgen, C.K. Chan, J. Ma, Y. Cui, G. Gruner, Nano Lett. **9**, 1872 (2009)

94. M.L. Verma, M. Minakshi, N.K. Singh, Electrochim. Acta **137**, 497 (2014)
95. R.K. Kushwaha, A.B. Samui, P. Sivaraman, A. Thakur, D. Ratna, Electrochem. Solid-State Lett. **9**, A435 (2006)
96. S. Banerjee, K.K. Kar, J. Environ. Chem. Eng. **4**, 299 (2016)
97. S. Banerjee, K.K. Kar, High Perform. Polym. **28**, 1043 (2016)
98. S.Y. Kim, H.S. Shin, Y.M. Lee, C.N. Jeong, J. Appl. Polym. Sci. **73**, 1675 (1999)
99. F.H. Kuok, H.H. Chien, C.C. Lee, Y.C. Hao, I.S. Yu, C.C. Hsu, I.C. Cheng, J.Z. Chen, RSC Adv. **8**, 2851 (2018)
100. G. Lota, E. Frackowiak, Electrochem. Commun. **11**, 87 (2009)
101. Y.F. Huang, P.F. Wu, M.Q. Zhang, W.H. Ruan, E.P. Giannelis, Electrochim. Acta **132**, 103 (2014)
102. H. Gao, K. Lian, J. Electrochem. Soc. **160**, A505 (2013)
103. X. Li, K. Liu, Z. Liu, Z. Wang, B. Li, D. Zhang, Electrochim. Acta **240**, 43 (2017)
104. Z. Yang, J. Deng, X. Chen, J. Ren, H. Peng, Angew. Chemie Int. Ed. **52**, 13453 (2013)
105. C. Yuan, X. Zhang, Q. Wu, B. Gao, Solid State Ionics **177**, 1237 (2006)
106. H. Yu, J. Wu, L. Fan, K. Xu, X. Zhong, Y. Lin, J. Lin, Electrochim. Acta **56**, 6881 (2011)
107. T.M. Dinh, A. Achour, S. Vizireanu, G. Dinescu, L. Nistor, K. Armstrong, D. Guay, D. Pech, Nano Energy **10**, 288 (2014)
108. Y.J. Kang, H. Chung, W. Kim, Synth. Met. **166**, 40 (2013)
109. K. Zhang, H. Chen, X. Wang, D. Guo, C. Hu, S. Wang, J. Sun, Q. Leng, J. Power Sources **268**, 522 (2014)
110. K.K. Kar, J.K. Pandey, S. Rana, *Handbook of Polymer Nanocomposites. Processing, Performance and Application* (Springer, Berlin, 2015)
111. S. Banerjee, K. Kar, M. Das, Recent Pat. Mater. Sci. **7**, 173 (2014)
112. S. Banerjee, K. Kar, Recent Pat. Mater. Sci. **7**, 131 (2014)
113. K.F. Chiu, S.H. Su, Thin Solid Films **544**, 144 (2013)
114. X. Li, T. Zhao, Q. Chen, P. Li, K. Wang, M. Zhong, J. Wei, D. Wu, B. Wei, H. Zhu, Phys. Chem. Chem. Phys. **15**, 17752 (2013)
115. S. Wang, B. Hsia, C. Carraro, R. Maboudian, J. Mater. Chem. A **2**, 7997 (2014)
116. G. Ayalneh Tiruye, D. Muñoz-Torrero, J. Palma, M. Anderson, R. Marcilla, J. Power Sources **279**, 472 (2015)
117. Y.N. Sudhakar, M. Selvakumar, Electrochim. Acta **78**, 398 (2012)
118. S. Banerjee, K.K. Kar, Polymer J. **109**, 176 (2017)
119. K. Fic, E. Frackowiak, F. Béguin, J. Mater. Chem. **22**, 24213 (2012)
120. G. Lota, K. Fic, E. Frackowiak, Electrochem. Commun. **13**, 38 (2011)
121. C.M. Ionica-Bousquet, W.J. Casteel, R.M. Pearlstein, G. GirishKumar, G.P. Pez, P. Gómez-Romero, M.R. Palacín, D. Muñoz-Rojas, Electrochem. Commun. **12**, 636 (2010)

Chapter 11
Characteristics of Separator Materials for Supercapacitors

Kapil Dev Verma, Prerna Sinha, Soma Banerjee, Kamal K. Kar, and Manas K. Ghorai

Abstract Separator in a supercapacitor is used as a sandwich between two electrodes. The essential functions of separator materials remain the prevention of the device from short circuit, storage of electrolyte into its pores, and passage of ions during charging and discharging processes. Material selection for the separator also plays an important role in deciding the final performance of the supercapacitor devices. Ionic conductivity of the separator affects the power and energy density. For safety purpose, thermal stability of the separator is very important. For the flexible electronic devices, bendable separators of high mechanical strength are needed. Separator should have optimum porosity and high electrolyte uptake such that it can provide adequate electrolyte ions to the electrode. The example of the most commonly used separator is polyolefin materials. This chapter deals with the importance of the separator materials, the governing equations, and essential parameters.

K. D. Verma · P. Sinha · K. K. Kar (✉)
Advanced Nanoengineering Materials Laboratory, Materials Science Programme, Indian Institute of Technology Kanpur, Kanpur 208016, India
e-mail: kamalkk@iitk.ac.in

K. D. Verma
e-mail: kdev@iitk.ac.in

P. Sinha
e-mail: findingprerna09@gmail.com

S. Banerjee · K. K. Kar
Advanced Nanoengineering Materials Laboratory, Department of Mechanical Engineering, Indian Institute of Technology Kanpur, Kanpur 208016, India
e-mail: somabanerjee27@gmail.com

M. K. Ghorai
Department of Chemistry, Indian Institute of Technology Kanpur, Kanpur 208016, India
e-mail: mkghorai@iitk.ac.in

© Springer Nature Switzerland AG 2020
K. K. Kar (ed.), *Handbook of Nanocomposite Supercapacitor Materials I*,
Springer Series in Materials Science 300,
https://doi.org/10.1007/978-3-030-43009-2_11

315

11.1 Introduction

Supercapacitor is modern generation energy storage device, a high-capacity capacitor that bridges the gap between capacitor and rechargeable battery. An electrochemical capacitor is composed of two electrode materials disconnected by an ion permeable separator material also called the membrane and an electrolyte that is capable to connect these two electrodes of the device. Under application of an applied voltage, the electrodes get polarized and the ions present in the electrolyte lead to the formation of electrical double layers. The double layers so created have polarity opposite to the polarity of the electrode materials. Again, pseudocapacitor may emerge when the specially adsorbed ions out of the electrolyte permeates through the electrical double layer. Although separators do not participate in electrochemical reactions during energy storage in the supercapacitors, they play an important role in the whole phenomenon. Figure 11.1 shows the basic components of a supercapacitor device, a combination of four components, e.g., electrode, electrolyte, separator, and current collector. Out of these four components, separator is used in between two electrodes of opposite polarity.

Separator is used for the separation of the positive and negative electrode to avoid the short circuit, which is the primary cause of self-discharge. Although the separator is a excellent electrical insulator, it is a good ionic character means it allows electrolyte ions to pass through it. Generation of any mechanical damage or high temperature in the device has been sustained by the separator to ensure safety issues. Separator should have porous structure, lightweight, high permeability, good mechanical strength, hydrophilic nature (low contact angle), non-flammable, and high ionic conductivity for the development of high performance supercapacitors [1]. Ionic conductivity of the separator depends on the porosity and electrolyte uptake of the separator. For high ionic conductivity, separator should uptake large amount of electrolytes. This high ionic conductivity provides high energy and power densities in the device. On the other hand, mechanical strength of the separator depends on the porosity. As the porosity increases above a certain optimum value, tensile strength starts decreasing. Apart from these properties, separator should have high-dimensional stability, should not exchange any material with the electrode, and should be stable over the operating potential range. For the fabrication of the separator, different types of

Fig. 11.1 Basic components of a supercapacitor

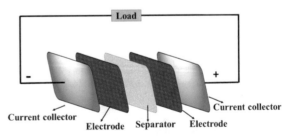

Electrolyte: 1M KOH

Table 11.1 Manufacturing companies and the respective separator materials [4]

Manufacturing company	Material	Separator design
Asahi Kasei chemicals	Polyolefin and ceramic-filled polyolefin	Biaxially orientated
Celgard LLC	PE, PP, and PP/PE/PP	Uniaxially orientated
Entek membranes	Ceramic-filled ultrahigh molecular weight polyethylene	Biaxially orientated
ExxonMobil/Tonen	PE and PE/PP mixtures	Biaxially orientated
SK energy	PE	Biaxially orientated
Ube industries	PP/PE/PP	Uniaxially orientated

materials are used such as non-woven fibers (cotton, nylon polyester, and cellulose), polymers [polypropylene (PP), polyethylene (PE), poly(tetrafluoroethylene) (PTFE), polyvinylidene difluoride (PVDF), polyacrylic acid (PAA), polyethylene terephathalate (PET), polyimide (PI)], and natural substrates (rubber, wood) [2]. During the short circuit, temperature inside the device reaches to 100 °C and above within a second, which can burst the device. Hence, for safety purpose, separator should sustain this much of high temperature. To overcome this problem, in recent days, separators are made with a combination of different polymers either during the fabrication of separator or afterward, during the formation of multilayer stacks [3]. This fabrication method is also useful to prevent thermal runways by improving the shutdown ability. At high temperature, polymers with low melting point get soften and close the pores of the separator and shut down the ion conduction process [3]. Solid electrolytes are also used as the separator in the supercapacitor applications. The use of solid electrolyte discards the need of an extra separator in the supercapacitor devices.

11.2 Commercially Available Separators

Commonly used separator material is polyolefin in which polypropylene and polyethylene are frequently used (Table 11.1).

11.3 Parameters of Interest for the Separator Used in Supercapacitors

The parameters, i.e., material properties of a separator material along with suitable combination of electrode and electrolytes decide the ultimate performance of a supercapacitor device. Separator should be cost-effective, thin, capable of high electrolyte uptake, and thermally stable [5]. Optimum values, which are required for the separator material, are listed in Table 11.2.

Table 11.2 Essential parameters of a separator material [3, 4, 6]

Parameter	Requirement
Sale price	<1.00 US$/m^2
Chemical stability	Stable more than 10 years
Thickness	<25 μm
MacMullin number	<11
Pore size	<1 μm
Wettability	Wet out in typical supercapacitor electrolyte
Puncture strength	>300 g/25.4 mm
Gurley (s/10 cm^3)	<35
Mechanical property	>1000 kg cm^{-1} (98.06 MPa)
Thermal stability	<5% shrinkage after 60 min at 90 °C
Purity	<50 ppm H$_2$O
Porosity	40–60%
Moisture Content	<50 ppm
Skew (mm/m)	<2
Dimensional stability	No curl up and lay flat
Shutdown temperature	100–110 °C
Pin removal	Easy removal from all major brands of winding machines
Melt integrity	≥200 °C

11.4 Physical and Electrochemical Characteristics of Separators

11.4.1 Porosity

Porosity of the separator should be optimum so that it can provide a sufficient amount of the electrolyte to the supercapacitor. Too high porosity leads to reduction in mechanical strength, and at high temperature, there will be a chance of shrinkage. On the other hand, if the porosity is too low, the separator cannot provide an adequate amount of electrolyte and internal resistance of the device will increase and thereby a reduction in power density is expected. Porosity of the separator should be uniform otherwise separator cannot provide the electrolyte to the entire electrode material, and hence, the final performance will again be decreased [6]. Porosity of the separator is defined by the ratio of void volume to geometric volume of the apparatus of the separator. Porosity can be calculated from (11.1)

$$\text{Porosity } (P) = \left(\frac{W - W_o}{\rho_e \times V} \right) \times 100\% \tag{11.1}$$

where P is the porosity, W is the weight of the separator after immerging in the electrolyte, W_o is the weight of the separator before immerging in the electrolyte, ρ_e is the density of the electrolyte, and V represents the volume of the separator.

11.4.2 Degree of Electrolyte Uptake

Degree of electrolyte uptake represents the percentage weight of the electrolyte absorbed by the separator. High electrolyte uptake is required such that separator can provide an adequate amount of electrolyte to the electrode material. For the calculation of the degree of electrolyte uptake, the first weight of the dry separator is measured and then the separator is dipped into the electrolyte for some time (say, for at least 20 min) and again the weight of the wet separator is measured [2, 7]. Degree of the electrolyte (DoE) uptake is calculated from (11.2)

$$\text{DoE uptake } (U) = \left(\frac{W - W_o}{W_o} \right) \times 100\% \tag{11.2}$$

where W is the weight of the separator after immerging in the electrolyte and W_o is the weight of the separator before immerging in the electrolyte.

11.4.3 Thermal Shrinkage

As the temperature of the device varies, the separator material starts changing its dimension. Thermal shrinkage of the separator should be low [8]. Thermal shrinkage is calculated from the following equation

$$\text{Thermal shrinkage } (\%) = \left(\frac{A_i - A_f}{A_i} \right) \times 100 \tag{11.3}$$

where A_i and A_f stand for the initial area of separator (before shrinkage) and final area of separator (after shrinkage), respectively .

11.4.4 Pore Size

Pore size of the separator should be small and uniform over the separator material. Small pore size prevents the short circuit caused by the migration of the electrode particles or dendritic growth [4]. Again, the contact angle of the electrolyte should be low. Low contact angle indicates a high electrolyte affinity of the separator. The pore size and contact angle of the separator can be calculated from (11.4)

$$d = -\frac{4T \cos \theta}{\Delta P} \tag{11.4}$$

where d is the pore diameter, T is the surface tension, θ is the contact angle, and ΔP is the pressure difference across the pore.

11.4.5 Permeability

Permeability is defined as the time for a given amount of the electrolyte to flow through the separator at a constant pressure. Permeability is characterized using the Gurley number. Permeability depends upon the porosity, fraction of the open pores, and tortuosity of the separator [4, 9, 10]. Permeability is calculated from (11.5)

$$B = \frac{l\mu v}{\Delta P} \tag{11.5}$$

where B is the permeability coefficient according to Darcy's law, l is the sample thickness, μ is the fluid viscosity, v is the fluid velocity, and ΔP is the pressure drop across the separator.

11.4.6 Mechanical Strength

Separator material should have high tensile strength and bendability. During the packaging of the supercapacitor device, separator material should sustain the load. For the flexible electronic devices, bendable separator is needed. Tensile strength is calculated from (11.6)

$$S = \frac{F}{t \times b} \tag{11.6}$$

where S, F, t, and b represent the tensile strength, force, thickness, and width of the separator, respectively.

11.4.7 Tortuosity

Tortuosity describes diffusion and fluid flow in the porous separator. Tortuosity is also represented as the ratio of the length of the streamline path between two points in the separator to straight line distance between the same points.

11.4.8 Ionic Conductivity

Separator of the supercapacitor device should have high ionic conductivity and electrical insulation. Ionic conductivity of the separator is calculated from (11.7)

$$\sigma = t/RA_0 \tag{11.7}$$

where σ is the ionic conductivity in mS/cm, t (cm) represents the thickness, R (ohm) is bulk resistance, and A_0 (cm^2) is the area of the electrode.

11.5 Separators for Supercapacitors

High internal resistance limits the performance of the supercapacitor device. Internal resistance of the supercapacitor depends on the contact resistance of electrolyte and electrode, electrolyte resistance, and molar conductivity of ions in the separator. Separator material is a semi-permeable material, which is responsible for ionic charge compensation between the electrodes. Therefore, the molar conductivity of the electrolyte ions in the electrolyte solution is important [11]. Separator material composition, structure, and thickness are the important factors to achieve high supercapacitor performance. Wettability and ionic conductivity of the separator should be high enough otherwise the power performance of the supercapacitor will be limited. Commercially used separators are generally prepared from cellulose and polypropylene [12].

11.5.1 Characteristics of Separators

Separator material selection has a major influence to decide the performance of the supercapacitor. Separator is used between two electrode materials to avoid the short circuit and compensate the ionic charge. It should have high molar ionic conductivity, high wettability, lightweight, high porosity, chemically stable, and high mechanical strength as mentioned in Table 11.3. Properties are mentioned in Table 11.4.

11.6 Concluding Remarks

Supercapacitor performance not only depends on the electrode and electrolyte material but also on the efficacy of the separator materials. Separator material should have high ionic conductivity, electrical insulation, high mechanical strength, high porosity, small pore size, high thermal stability, and low contact angle. Ionic conductivity

Table 11.3 Characteristics of separator materials for electrochemical supercapacitor

Primary characteristics	Secondary characteristics	Tertiary characteristics
High ionic conductivity	Minimal thickness	High corrosion resistance
High electrolyte wettability	High chemical inertness	Non-toxic
High mechanical strength	High thermal stability	Easily available
High porosity	Lightweight	Low cost

of the separator affects the power and energy density. Porosity of the separator should be optimum such that the mechanical strength remains unaffected. Again, the contact angle of the separator should be low such that separator gets wet easily and rapidly. According to the separator material selection analysis, best separator materials for the supercapacitor applications are cellulose, PET, polypropylene (PP), polyvinylidene difluoride (PVDF), polyethylene (PE), and polyimide (PI).

Table 11.4 Properties of different separator materials

Separator	Ionic conductivity (mS/cm)	Resistance (Ω)	Degree of electrolyte uptake (%)	Porosity (%)	Tensile strength (MPa)	Contact angle (°)	Thickness (μm)	References
Cladophora cellulose	0.82	–	–	44	137.6	–	10–40	[13]
Cellulose	1.74	–	340	75	–	–	25	[14, 15]
Polypropylene (PP) (Celgard)	0.503	2.15	134	41	139.16	56	20	[7, 16–18]
Polypropylene (PP)	1.05	–	120	55	–	82	25	[14]
Poly(vinylidene fluoride) (PVDF)	(0.3–3.7)	0.95	405.10	75	21.3	0	10	[6, 17, 19, 20]
PE (Tonen)	0.6	–	54	41.50	70.76	35	20	[8, 21–23]
PE	0.25	2.5	110.7	45	0.313	43.4	12	[24]
PE	0.59	1.65	100	41	–	49.2	–	[25]
Ceramic coating separator	1.10	–	71.20	41.20	–	–	24	[8]
PVA	0.52–1.27	–	225–260	–	45–70	–	100–120	[9, 26–28]
Egg shell membrane	–	19	81	–	6.59	–	14	[29]
Aquagel	72	–	–	–	–	–	125	[30]
Polyamide acid (PA)	0.80	1.58	2065	92.20	–	3.5	35	[31, 32]
Polyamide acid (PA)		1–2		38–87		17	200	[33]
polyethylene terephthalate (PET)	2.27	2.50	500	89	12	–	40	[34]
PI	2.5		2010	92.1		3.7	–	[31]
Polyimide (PI)	0.00173	–	250 ± 1.1	72.4	17.5 ± 1.4	–	39 ± 1.7	[35, 36]

(continued)

Table 11.4 (continued)

Separator	Ionic conductivity (mS/cm)	Resistance (Ω)	Degree of electrolyte uptake (%)	Porosity (%)	Tensile strength (MPa)	Contact angle (°)	Thickness (μm)	References
Polyacrylonitrile (PAN)	2–2.14	–	395–479	–	9	–	100	[6, 37]
PVA/5 wt%PVC	0.613	8–30	44.3	–	–	–	150–200	[20, 38]
10%PVA/PVP	0.79	1.94	226	63	–	8	–	[25]
Polytetrafluoroethylene (PTFE)/polyethylene (PE)	0.96	–	171.5	65–66	0.35	–	18	[24]
poly(butylene terephthalate) (PBT)	0.27	–	–	75	–	–	55	[39]
PMMA	2.8	–	300	57	–	–	85	[40, 41]

Acknowledgements   The authors acknowledge the financial support provided by Department of Science and Technology, India (DST/TMD/MES/2K16/37(G)) for carrying out this research work.

References

1. S. Banerjee, K.K. Kar, J. Environ. Chem. Eng. **4**, 299 (2016)
2. S. Banerjee, K.K. Kar, J. Appl. Polym. Sci. **133**, 42952 (2016)
3. T. Nestler, R. Schmid, W. Münchgesang, V. Bazhenov, J. Schilm, T. Leisegang, D.C. Meyer, *Separators—Technology Review: Ceramic Based Separators for Secondary Batteries* (2014)
4. X. Huang, J. Solid State Electrochem. **15**, 649 (2011)
5. N. Sykam, R.K. Gautam, K.K. Kar, Polym. Eng. Sci. **55**, 917 (2015)
6. H. Lee, M. Yanilmaz, O. Toprakci, K. Fu, X. Zhang, Energy Environ. Sci. **7**, 3857 (2014)
7. H. Cai, X. Tong, K. Chen, Y. Shen, J. Wu, Y. Xiang, Z. Wang, J. Li, Polymers **10** (2018)
8. C. Shi, J. Dai, C. Li, X. Shen, L. Peng, P. Zhang, D. Wu, D. Sun, J. Zhao, Polymers **9**, 10 (2017)
9. Y.S. Ye, M.Y. Cheng, X.L. Xie, J. Rick, Y.J. Huang, F.C. Chang, B.J. Hwang, J. Power Sources **239**, 424 (2013)
10. S. Banerjee, K.K. Kar, M.K. Ghorai, S. Das, High Perform. Polym. **27**, 402 (2015)
11. K. Liivand, T. Thomberg, A. Janes, E. Lust, ECS Trans. **64**, 41 (2015)
12. K.K. Kar, S.D. Sharma, S.K. Behera, P. Kumar, J. Elastomers Plast. **39**, 117 (2007)
13. B. Writer, *Lithium-Ion Batteries* (Springer International Publishing, Cham, 2019)
14. G. Ding, B. Qin, Z. Liu, J. Zhang, B. Zhang, P. Hu, C. Zhang, G. Xu, J. Yao, G. Cui, J. Electrochem. Soc. **162**, A834 (2015)
15. K.K. Kar, A. Hodzic, *Carbon Nanotube Based Nanocomposites: Recent Developments.*, 1st edn. (Research Publishing, 2011)
16. K. Tõnurist, T. Thomberg, A. Jänes, E. Lust, J. Electrochem. Soc. **160**, A449 (2013)
17. D. Wu, L. Deng, Y. Sun, K.S. Teh, C. Shi, Q. Tan, J. Zhao, D. Sun, L. Lin, RSC Adv. **7**, 24410 (2017)
18. K.K. Kar, S.D. Sharma, P. Kumar, Plast., Rubber Compos. **36**, 274 (2007)
19. S.K. Rath, S. Dubey, G.S. Kumar, S. Kumar, A.K. Patra, J. Bahadur, A.K. Singh, G. Harikrishnan, T.U. Patro, J. Mater. Sci. **49**, 103 (2014)
20. S. Banerjee, K. Kar, M. Das, Recent Patents. Mater. Sci. **7**, 173 (2014)
21. R.S. Baldwin, W.R. Bennett, E.K. Wong, M.R. Lewton, M.K. Harris, *Battery Separator Characterization and Evaluation Procedures for NASA's Advanced Lithium-Ion Batteries* (2010)
22. R. Kumar, S. Sahoo, E. Joanni, R.K. Singh, W.K. Tan, K.K. Kar, A. Matsuda, Prog. Energy Combust. Sci. **75**, 100786 (2019)
23. K.K. Kar, S.D. Sharma, P. Kumar, J. Ramkumar, R.K. Appaji, K.R.N. Reddy, J. Appl. Polym. Sci. **105**, 3333 (2007)
24. K. Zhang, W. Xiao, J. Liu, C. Yan, Polymers **10**, 1409 (2018)
25. W. Xiao, K. Zhang, J. Liu, C. Yan, J. Mater. Sci.: Mater. Electron. **28**, 17516 (2017)
26. A.A. Mohamad, N.S. Mohamed, M.Z.A. Yahya, R. Othman, S. Ramesh, Y. Alias, A.K. Arof, Solid State Ionics **156**, 171 (2003)
27. K.K. Kar, A. Hodzic, *Developments in Nanocomposites*, 1st edn. (Research Publishing, 2014)
28. A. Kumar, R. Sharma, M. Suresh, M.K. Das, K.K. Kar, J. Elastomers Plast. **49**, 513 (2017)
29. N.S.M. Nor, M. Deraman, R. Omar, E. Taer, Awitdrus, R. Farma, N.H. Basri, B.N.M. Dolah, AIP Conf. Proc. **1586**, 68 (2014)
30. S.T. Mayer, J.L. Kaschmitter, R.W. Pekala, 5402306 (1995)
31. X. Luo, X. Lu, G. Zhou, X. Zhao, Y. Ouyang, X. Zhu, Y.E. Miao, T. Liu, ACS Appl. Mater. Interfaces **10**, 42198 (2018)

32. K.K. Kar, J.K. Pandey, S. Rana, *Handbook of Polymer Nanocomposites, Processing, Performance and Application* (Springer, Berlin, 2015)
33. B. Szubzda, A. Szmaja, M. Ozimek, S. Mazurkiewicz, Appl. Phys. A Mater. Sci. Process. **117**, 1801 (2014)
34. J. Hao, G. Lei, Z. Li, L. Wu, Q. Xiao, L. Wang, J. Memb. Sci. **428**, 11 (2013)
35. C.E. Lin, H. Zhang, Y.Z. Song, Y. Zhang, J.J. Yuan, B.K. Zhu, J. Mater. Chem. A **6**, 991 (2018)
36. K.K. Kar, *Composite Materials* (Springer, Berlin, 2017)
37. N. Scharnagl, H. Buschatz, Desalination **139**, 191 (2001)
38. C.C. Yang, G.M. Wu, Mater. Chem. Phys. **114**, 948 (2009)
39. C.J. Orendorff, T.N. Lambert, C.A. Chavez, M. Bencomo, K.R. Fenton, Adv. Energy Mater. **3**, 314 (2013)
40. M. Yanilmaz, X. Zhang, Polymers **7**, 629 (2015)
41. K.K. Kar, S.D. Sharma, P. Kumar, J. Ramkumar, R.K. Appaji, K.R.N. Reddy, Polym. Compos. **28**, 637 (2007)

Chapter 12
Characteristics of Current Collector Materials for Supercapacitors

Kapil Dev Verma, Prerna Sinha, Soma Banerjee, and Kamal K. Kar

Abstract Current collector has a major role in electrochemical performance and cycle stability of supercapacitor. It collects electrons and supports the electrode material. Conductivity and contact resistance with the electrode material of a current collector have a direct influence on the power density and capacitance of a supercapacitor. Current collector should have high electrical conductivity, high mechanical strength/modulus, lightweight, high thermal stability, high electrochemical stability and low cost. Metal foam and metal foil type current collector are used in general, in which metal foam provides the highest performance. For flexible supercapacitor, carbon fiber is generally used as the current collector.

12.1 Introduction

Supercapacitor is one of the efficient energy storage systems, which has the capability of delivering high power and energy density. It is well known that supercapacitor delivers high power density as compared to batteries but fails to deliver high energy density. So, more efforts and research have been promoted towards the fabrication of high-performance supercapacitor devices. The performance of supercapacitor mainly depends upon its components, which are electrode, electrolyte, separator and current

K. D. Verma · P. Sinha · K. K. Kar (✉)
Advanced Nanoengineering Materials Laboratory, Materials Science Programme, Indian Institute of Technology Kanpur, Kanpur 208016, India
e-mail: kamalkk@iitk.ac.in

K. D. Verma
e-mail: kdev@iitk.ac.in

P. Sinha
e-mail: findingprerna09@gmail.com

S. Banerjee · K. K. Kar
Advanced Nanoengineering Materials Laboratory, Department of Mechanical Engineering, Indian Institute of Technology Kanpur, Kanpur 208016, India
e-mail: somabanerjee27@gmail.com

© Springer Nature Switzerland AG 2020
K. K. Kar (ed.), *Handbook of Nanocomposite Supercapacitor Materials I*,
Springer Series in Materials Science 300,
https://doi.org/10.1007/978-3-030-43009-2_12

collector [1, 2]. The charges are mainly stored in the electrode material, whereas electrolyte provides necessary ions to get adsorbed at the electrode electrolyte interfaces. Separator acts as semi-permeable membrane, which allows transfer of ions and prevents device from short-circuit. Current collector enhances the performance of the supercapacitor by the transportation of charge through external circuit. Along with the electrical conductivity of the electrode material, resistivity of the electrolyte, thickness of the separator material and the electrical conductivity of current collector play a significant role in determining internal resistance of the device [3, 4]. In accordance with this, each component of supercapacitor plays an important role to increase the performance of supercapacitor. So, selection of material for each component is required with respect to their associated properties for supercapacitor.

Present chapter deals with the material properties of current collector for supercapacitor application. Current collector collects electron from the electrodes and delivers to the external load. To achieve the high energy density, time constant (RC, resistor–capacitor, which is equal to the product of the circuit resistance (in ohms) and the circuit capacitance (in farads)) must be low such that resistance of the device should be as low as possible [5]. From supercapacitor perspective, time constant is the ability for how quickly a device can be charged and discharged. The term R in time constant (RC) is the equivalent series resistance (ESR) of the device. The performance of a current collector depends on its electrical conductivity and charge transfer resistance with electrode material. For the effective use of the current collector, it should have

(a) high electrical and thermal conductivity
(b) good bendability
(c) great compression strength
(d) high thermal stability
(e) thin and light in weight
(f) low contact resistance with the electrode
(g) high chemical and electrochemical stability
(h) high corrosion resistive
(i) environmental friendly,
(j) low cost, etc.

So, current collector is the important element of a supercapacitor device. The interaction between morphology of the electrode material and current collector decides the power delivery rate, efficiency and extraction of stored energy at the electrochemical interface of electrode and current collector [6, 7]. But the current rating depends on the current collector and electrode material loaded on it [8]. It is important to mention that electrode materials, synthesized from the same process, with different current collectors, show significant difference in the capacitance of the supercapacitor device [9, 10]. Current collector also affects the conductivity of the capacitor and chemical stability of electrolyte [8]. Another important point, i.e., functionality of the current collector is crucial for perfect designing of a flexible supercapacitor. Therefore, analysis of mechanical behavior in addition to electrical conductivity of a current collector is of immense importance in view of material selection. During charging and discharging cycle, current collector starts changing its color due to the electrochemical instability and low working temperature range. So, it is important to select appropriate material for current collector, which contributes for enhancing the cycle life of the supercapacitor. Also, modern generation demands lightweight and flexible electronic devices. It is possible by using lightweight and flexible current collectors as, during fabrication current collector weighs majority of the device weight. Along with various physical properties, the main function of current collector is to reduce the charge transfer resistance between the electrode material and current collector. This allows the easy passages of electron to flow from electrode surfaces toward current collector. To reduce contact resistance, it is essential to increase the contact surface area between current collector and electrode material [11]. Keeping these requirements of current collectors, various metals and carbon forms along with their properties are discussed in the next sections.

12.2   Metal Foil and Metal Foam-Based Current Collector

Electrode and current collector conjunction play an important role in the charge transfer resistance and power delivery rate of a supercapacitor. Electrode material is deposited over the current collector, and to minimize contact resistance, binders are used. Or in other words, it is a sandwich structure of electrode and current collector. Binders are of different types such as polytetrafluoroethylene (PTFE), poly-benzimidazole (PBI), polyvinylidene fluoride (PVDF) and bitumen paint, and their classification is much dependent on the underlined physical and chemical properties [8]. Selection of material for current collector should be done in such a way that it should not react with electrode and electrolyte during charging and discharging processes. Metal foil and metal plate are class of material, which are commonly used as current collectors for two electrode cells. To achieve high power performance, thickness of metal foil should be less with low contact resistance. Again, there is a need of lightweight supercapacitor cells for applications in portable devices. Also, in the foil-type current collector, whole material does not participate in the charge transportation, and hence, an increase in equivalent series resistance and decrease in efficiency are observed. According to recent studies, metal foam serves as better current collector than metal foil in order to achieve high performance. The advantages of metal foam over metal foil will be discussed in the succeeding sections of this chapter. Metal foam provides high contact area to electrode material, and charge transfer takes place under improved condition at the junction of current collector, metal frame and electrode material. Some electrode materials can be directly used with nickel foam current collector without using binder or conducting agent because of good mechanical adhesion and electronic connection with the current collector [12, 13]. The average diffusion distance of electron transfer decreases as the porosity of metal foam provides fast transport mechanism for electrode. On contrary, for metal foil, current collector exists only at the back side of the electrode material, which decreases contact area at electrode and current collector interface [14]. Impedance

Fig. 12.1 Comparison of
impedance curves for metal
foam (NiCrAl foam) and
metal (Al) foil. Reprinted
with permission from [14]

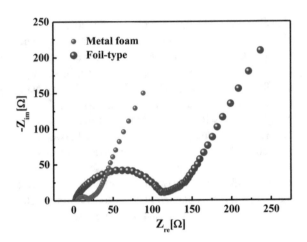

analyses of metal foil and metal foam as shown in Fig. 12.1 state that charge transfer
resistance of metal foam is much lower than that of metal foil. Although, there is no
significant difference in internal resistance, but charge transfer resistance of metal
foil is 7 times higher as compared to metal foams. From Fig. 12.1, it is clear that the
porous nature of metal foam provides even distribution of electrode material inside
current collector, which thereby increases the contact area and decreases charge
transfer resistance. On the other hand, contact area between metal foil and electrode
material is very less, which increases the charge transfer resistance. It is quite obvious that high charge transfer resistance of supercapacitor induces low power density
[14]. However, irrespective of all the above-stated advantages, major drawback of
metal foam is its low electrochemical stability, which deteriorates the performance of
supercapacitor at high potential window. Also, metal-based current collector limits
its usage in acidic electrolyte as these material gets corroded in acidic environment,
which affects the electrochemical stability of the device.

12.3   Dimensionally Stabilized Anode Current Collector

Dimensionally stabilized anodes are those materials, where a well-known form of
an electrode plays the role of both electrode and current collector. It consists of a
valve metal as electrode with film of platinum group metal deposited on the electrode
material. Metallic messes, plates, screens are also used as the current collector with
electrode materials. However, some drawbacks are also present with dimensionally
stabilized anode such as high fabrication cost and low corrosion resistance. As mentioned earlier, for fabrication of valves, metals like tantalum, niobium, titanium, etc.,
are used, which are quite costly. These materials form oxide layer over their surface,

which drastically reduces the electrical conductivity. To address this issue, it is necessary to coat the valve metal with non-oxide forming materials like platinum group metal, which further adds to the overall material cost [15, 16].

12.4 Current Collector for Flexible Supercapacitors

Rigidity of a storage device restricts its application in flexible and portable device. Flexibility plays an important role for fabrication of flexible and wearable electronics. For the flexible energy storage devices, development of flexible current collector is of uttermost importance. Critical challenges are to develop current collector with high electrical conductivity, ultrathin, lightweight, excellent mechanical properties like high bendability and compressibility [17]. Current collector material for flexible supercapacitor application is very limited. Commonly used current collector for flexible supercapacitor is carbon fiber-derived carbon fabrics. Carbon fabrics have high mechanical strength and good flexibility; however, they suffer from low electrical conductivity [18–20]. Other materials for flexible current collector include metal foil and polymer film with conducting coating. Nickel, titanium, aluminum foil, stainless steel fiber, etc., are used as a conducting substrate for flexible supercapacitor. These metal foils are of high conductivity; however, their large mass reduces mechanical flexibility and cyclic stability. Polymer substrate like polyimide shows weak adhesion with electrode material which further reduces cyclic stability [17, 18]. Metallic paper current collectors are also used for fabrication of flexible supercapacitor. Paper coated with metal substrate shows excellent flexibility, lightweight structure, cheap cost and hierarchical fiber distribution over the surface [17]. The fabrication of metallic paper-based current collector includes the coating of metal on paper via electrodeposition. However, the development of sufficient contact spots on the planer layer at the interface of current collector and electrode material seems difficult as contact spot decides the charge transfer efficiency. To overcome this problem, 3D metallic structure or nano-structured current collector is developed recently. Since, an improvement in the volumetric energy density with minimal contact resistance are possible [17, 21, 22].

12.5 Current Collector Materials

Metal foil, metal foam, polymer film and paper-coated metal substrate, carbon fiber-based current collectors are most commonly used for supercapacitor devices. Metals, for example, nickel, steel, aluminum, copper, titanium, platinum and gold are most frequently utilized as current collectors.

12.5.1 Nickel

Nickel-based current collector is the most commonly used for supercapacitor devices. It provides high mechanical strength, great electrical conductivity, low contact resistance with electrode material and of low cost [23]. Nickel is used both in foil and foam forms. Nickel foil of high electrical conductivity and mechanical strength is widely used depending on the type of the electrolyte [11]. But the nickel foam with concave surface and high contact area shows huge capacitive performance compared to nickel foil of flat surface [11]. Along with this, nickel foam containing activated carbon material composite sheet has also been used as electrode and current collector materials [11]. Cyclic voltammetry analysis demonstrates that nickel foam current collector shows faster redox reaction in comparison with the nickel foil [14]. 3D porous morphology of nickel foam as current collector accommodates large amount of electro-active material and provides low contact resistance between electrode and current collector. Also, the porous structure allows large amount of electrolyte ions to move over the electro-active material and electrode surfaces [24, 25]. Figure 12.2a shows the FESEM image of fresh nickel foam. Study reveals that during repeated electrochemical studies, nickel foam starts changing color, which indicates its instability due to the possible redox reactions in the nickel foam and contributes in the pseudocapacitance. Figure 12.2b shows the optical image of fresh nickel foam, which changes its color (as shown in Fig. 12.2c) after 2000 charge-discharge cycles due to reduction–oxidation reactions. Also, pretreated nickel foam shows some capacitive properties, which can bring some error in the capacitance value of the device [9]. This is due to the formation of oxide films at nickel foam surface, which act as active redox material and contributes in electrochemical processes. So, nickel foam should be properly treated with acidic solution prior to electrode deposition.

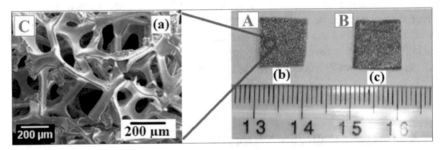

Fig. 12.2 FESEM image of fresh nickel foam (**a**), optical image of fresh nickel foam current collector (**b**) and nickel foam current collector after 2000 charge-discharge cycles (**c**). Redrawn and reprinted with permission from [9]

Fig. 12.3 **A** Stereo microscopy image of aluminum foam, **B** comparison of a bare Al foam and electrode material deposited on Al current collector. Redrawn and reprinted with permission from [30]

12.5.2 Aluminum

Aluminum is less costly, highly electrically conductive, high mechanical strength, light in weight and low contact resistive current collector. Although under poor working condition, electrode material gets peeled off from the aluminum foil, which results in high contact resistance. To overcome delamination of electrode from current collector and corrosion of aluminum surface, laser-treated aluminum and graphene-treated aluminum current collector are used [26–29]. Aluminum foam is also used as the current collector for supercapacitor applications. Aluminum foam provides 3D electronic conductive network with high mechanical strength in which large surface area for active material loading is available as shown in Fig. 12.3. Figure 12.3A shows stereo microscopy image of aluminum foam indicating 3D interlinked porous morphology. Figure 12.3B shows image of bare and electrode material-coated aluminum foam. The foam completely gets coated with electrode material, which indicated effective and even loading due to uniform porous structure. Supercapacitor fabricated using aluminum foam as current collector shows good cycle life and load capability [30].

12.5.3 Stainless Steel

Chemically stable and low cost stainless steel is widely used as a current collector in supercapacitor and battery applications [31]. Stainless steel also has high mechanical strength, good ductility, decent electrical conductivity and electrochemical stability [32, 33]. Apart from its excellent physical properties, it is easily available material. Electrochemical and chemical stability of stainless steel as current collector depend on the type of electrolyte used. In this vein, Wojciechowski et al. have studied the stability of stainless steel 316L in different electrolytes. Figure 12.4 shows the SEM micrograph of stainless steel 316L current collector after electrochemical test of

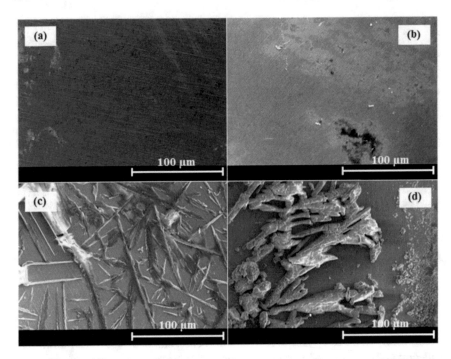

Fig. 12.4 SEM image of **a** bare stainless steel 316L current collector, stainless steel 316L current collector after electrochemical study with **b** 1 M H$_2$SO$_4$, **c** 1 M KOH and **d** 6 M KOH electrolyte. Reprinted with permission from [34]

supercapacitor device using 1 M H$_2$SO$_4$, 1 M KOH and 6 M KOH [34]. Figure 12.4a shows SEM images of bare stainless steel 316L and indicates plane surface. The presence of white spot might be due to organic pollutants. Figure 12.4b shows SEM micrograph of stainless steel 316L tested in 1 M H$_2$SO$_4$ electrolyte, where no appreciable change has been observed. For KOH-tested stainless steel 316L current collector shown in Fig. 12.4c, d, major change in morphology has been observed. This indicates the instability of stainless steel in KOH solution as electrolytic ion is getting dissolved with the elements of stainless steel 316L [34]. Also, during the repeated charge-discharge cycle, current collector may suffer from corrosion. Corrosion takes place at the electrode material. The polarization process may result in dissolution of steel current collector due to oxygen evolution corrosion reaction [31, 34–36]. To overcome corrosion problem, carbon-coated stainless steel is used, which shows better cyclic stability and rate capability compared to conventional stainless steel mesh and even titanium mesh-based current collectors [31].

12.5.4   Copper

Copper is reddish-orange color current collector with high electrical and thermal conductivity and ductile in nature. 3D porous structured copper foam is a new type of current collector, which provides high electrical conductivity. Pore size of the copper foam varies from micrometers to millimeters [37]. For high performance supercapacitor, along with high electrical conductivity, current collector should have high surface area and porosity. Ultrasonic technique is also used to increase the surface area of the copper current collector by producing mesoporous and microporous framework on the surface [37, 38]. Figure 12.5 demonstrates that, as the ultrasonication time increases, the development of nanopores on the surface of copper current collector also increases in 0.05% ammonia solution [37].

Some more current collectors like titanium, platinum, tantalum, carbon fiber, etc., are also used for energy storage applications. The selection of current collector material also depends upon the types of electrolyte used for supercapacitor applications. Most important point is that electrolyte material should not corrode the current collector to achieve best efficiency of supercapacitor.

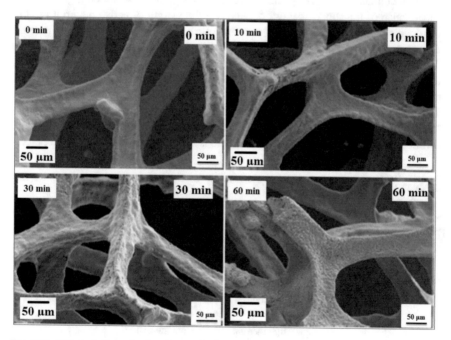

Fig. 12.5 Effect of ultrasonication time on the formation of nanopores on copper surface. Redrawn and reprinted with permission [37]

12.6   Current Collector Material Characteristics for Supercapacitors

Current collector directly affects the power density, flexibility and cycle life of the supercapacitor. The behavior of electrode and electrolyte material also depends upon the current collector material. Electrochemical properties of the electrode and electrolyte material are evaluated by placing or coating of electrolyte dipped electrode material between metallic foam and metallic foil current collectors. Current collector material is to be selected according to the property requirements for supercapacitor applications.

12.6.1   Characteristics of Current Collector Materials

Objective for current collector material selection is to improve the performance of supercapacitor. To meet this requirement, it should have high electrical conductivity and low contact resistance so that power and energy density can be high. Specific capacitance of the supercapacitor depends on the weight of active material used in the device, and hence, lightweight current collector is required to achieve high performance. It should have high thermal stability and great thermal conductivity so that heat produced during charging and discharging cycle cannot affect performance of the device. For flexible supercapacitor application, current collector should have high mechanical strength, high Young's modulus, high flexibility and less thickness. Table 12.1 represents primary, secondary and tertiary characteristics for current collector. Primary characteristic indicates the essential requirements, which should be fulfilled at any cost. Secondary characteristic is also required to achieve high perfor-

Table 12.1 Primary, secondary and tertiary characteristics for selection of a suitable current collector

Primary characteristics	Secondary characteristics	Tertiary characteristics
High electrical conductivity	High electrochemical stability	Low cost
High mechanical flexibility	High chemical stability	Environmentally friendly
Low contact resistance	Low density	Easy to fabricate
High thermal conductivity	High thermal stability	
High corrosion resistance	High compression strength	

mance. Different types of metal foam and foil current collectors are used for supercapacitor devices. Table 12.2 shows a list of current collectors with their electrical, mechanical and thermal properties achieved by several research groups.

12.7   Concluding Remarks

Along with electrode, electrolyte and separator material, current collector also plays a major role to achieve high performance of supercapacitor. Electrode material is deposited over the surface of the current collector. It collects electrons from the electrode and extends the same to the external load. Current collector should have high electrical and thermal conductivity, low contact resistance, lightweight, high electrochemical stability, extraordinary mechanical strength and bendability. Metal foam and metal foils are most commonly used as the current collector. Metal foam shows better results in comparison with metal foil because of low contact resistance and lightweight of current collector. The 3D interlinked porous morphology provides even distribution of electrode material. This not only reduces the contact resistance at electrode and current collector interfaces but also provides great exposure of electrode material for electrochemical processes. For flexible supercapacitor, carbon fiber is commonly used as current collector. Current collector is selected according to the properties required for the supercapacitor. Most promising current collector materials for supercapacitor application are carbon fiber, nickel, copper, stainless steel, titanium, platinum, tantalum and aluminum.

Acknowledgements The authors acknowledge the financial support provided by Department of Science and Technology, India (DST/TMD/MES/2K16/37(G)), for carrying out this research work.

Table 12.2 Current collector materials list with electrical, mechanical and thermal properties

Current collector material	Electrical conductivity (S/m) × 10^6	ESR (Ω)	Tensile strength (MPa)	Compression strength (MPa)	Young's modulus (GPa)	Thickness (mm)	Thermal stability (°C)	Density (g/cm^3)	Thermal conductivity (W/mK)	References
Nickel foam	15.80	0.07	–	0.25	200	0.5–40	1455	0.1–0.62	90.9	[11, 39]
Nickel foil	14.43	0.28	570	–	200	0.04	1455	8.9	90.9	[11, 40, 41]
Aluminum foam	39.2	–	3–15	3–17	70	4–50	660–800	0.25–0.75	205	[42–44]
Aluminum foil	35	3.3	105–120	–	70	0.016	660	2.70	205	[26, 45–48]
Stainless steel foam	–	–	2–3	0.25	190	1–25	1371–1537	7.65–7.94	15.1	[32, 49]
Stainless steel foil	1.4	–	1000–1200	–	190	0.15–0.36	–	7.98	15.1–19	[50, 51]
Copper foam	59.6	1.6	–	–	110–128	1–25	1085	0.7–1.1	401	[37]
Copper foil	59.6	0.39	318	–	110–128	0.25	1083	8.96	401	[52–54]
Titanium foil	2.38	0.94	140	–	116	0.25	1668	4.54	21.9	[55, 56]
Platinum foil	9.43	–	125–165	–	168	0.025–3	1772	21.45	71.6	[57]
Tantalum foil	7.63	–	–	–	186	–	3017	16.69	57.5	[58]
Carbon fiber	0.071	–	950–985	–	228	0.005	900–1500	2.267	119–165	[59–63]

References

1. P. Yang, W. Mai, Nano Energy **8**, 274 (2014)
2. A. González, E. Goikolea, J.A. Barrena, R. Mysyk, Renew. Sustain. Energy Rev. **58**, 1189 (2016)
3. R.S. Liu, L. Zhang, X. Sun, H. Liu, J. Zhang, Electrochem. Technol. Energy Storage Convers. **1–2** (2011)
4. B.K. Kim, S. Sy, A. Yu, J. Zhang, *Electrochemical Supercapacitors for Energy Storage and Conversion* (Wiley, Chichester, UK, 2015)
5. N. Tassin, G. Bronoel, J.-F. Fauvarque, I. Bispo-Fonseca, J. Power Sources **65**, 61 (1997)
6. K. Pandey, P. Yadav, I. Mukhopadhyay, RSC Adv. **4**, 53740 (2014)
7. K.K. Kar, A. Hodzic, *Carbon Nanotube Based Nanocomposites: Recent Developments*, 1st edn. (Research Publishing, 2011)
8. B. Panda, I. Dwivedi, K. Priya, P.B. Karandikar, P.S. Mandake, in *2016 Biennial International Conference on Power & Energy Systems: Towards Sustainable Energy* (IEEE, 2016), pp. 1–6
9. W. Xing, S. Qiao, X. Wu, X. Gao, J. Zhou, S. Zhuo, S.B. Hartono, D. Hulicova-Jurcakova, J. Power Sources **196**, 4123 (2011)
10. A. Yadav, B. De, S.K. Singh, P. Sinha, K.K. Kar, ACS Appl. Mater. Interfaces **11**, 7974 (2019)
11. J.-H. Kim, J.H. Park, Electrochemistry **69**, 853 (2001)
12. L. Li, J. Xu, J. Lei, J. Zhang, F. McLarnon, Z. Wei, N. Li, F. Pan, J. Mater. Chem. A **3**, 1953 (2015)
13. K.K. Kar, S.D. Sharma, P. Kumar, J. Ramkumar, R.K. Appaji, K.R.N. Reddy, J. Appl. Polym. Sci. **105**, 3333 (2007)
14. G.F. Yang, K.Y. Song, S.K. Joo, J. Mater. Chem. A **2**, 19648 (2014)
15. R.J. Lawrance, Hampstead, NH, Jan 1985
16. K.K. Kar, *Carbon Nanotubes: Synthesis, Characterization and Applications* (Research Publishing Services, Singapore, 2011)
17. Y. Li, Q. Wang, Y. Wang, M. Bai, J. Shao, H. Ji, H. Feng, J. Zhang, X. Ma, W. Zhao, Dalton Trans. **48**, 7659 (2019)
18. J. Yu, J. Wu, H. Wang, A. Zhou, C. Huang, H. Bai, L. Li, ACS Appl. Mater. Interfaces **8**, 4724 (2016)
19. H.A. Andreas, B.E. Conway, Electrochim. Acta **51**, 6510 (2006)
20. S. Banerjee, K.K. Kar, J. Appl. Polym. Sci. **133** (2016)
21. B. Yao, J. Zhang, T. Kou, Y. Song, T. Liu, Y. Li, Adv. Sci. **4** (2017)
22. K.K. Kar, A. Hodzic, *Developments in Nanocomposites* (Research Publishing Services, Singapore, 2014)
23. G. Li, X. Mo, W.-C. Law, K.C. Chan, J. Mater. Chem. A **7**, 4055 (2019)
24. B. Saravanakumar, S.S. Jayaseelan, M.K. Seo, H.Y. Kim, B.S. Kim, Nanoscale **9**, 18819 (2017)
25. S. Banerjee, K.K. Kar, J. Environ. Chem. Eng. **4**, 299 (2016)
26. Y. Huang, Y. Li, Q. Gong, G. Zhao, P. Zheng, J. Bai, J. Gan, M. Zhao, Y. Shao, D. Wang, L. Liu, G. Zou, D. Zhuang, J. Liang, H. Zhu, C. Nan, ACS Appl. Mater. Interfaces **10**, 16572 (2018)
27. M. Wang, M. Tang, S. Chen, H. Ci, K. Wang, L. Shi, L. Lin, H. Ren, J. Shan, P. Gao, Z. Liu, H. Peng, Adv. Mater. **29** (2017)
28. Y. Huang, Y. Zhao, Q. Gong, M. Weng, J. Bai, X. Liu, Y. Jiang, J. Wang, D. Wang, Y. Shao, M. Zhao, D. Zhuang, J. Liang, Electrochim. Acta **228**, 214 (2017)
29. K.K. Kar, J.K. Pandey, S. Rana, *Handbook of Polymer Nanocomposites. Processing, Performance and Application* (Springer Berlin Heidelberg, Berlin, Heidelberg, 2015)
30. M. Fritsch, G. Standke, C. Heubner, U. Langklotz, A. Michaelis, J. Energy Storage **16**, 125 (2018)
31. Y.H. Wen, L. Shao, P.C. Zhao, B.Y. Wang, G.P. Cao, Y.S. Yang, J. Mater. Chem. A **5**, 15752 (2017)
32. Y. Li, X. Zhang, L. Nie, Y. Zhang, X. Liu, J. Power Sources **245**, 520 (2014)

33. R. Sharma, S.S. Ahankari, K.K. Kar, A. Biswas, K. Srivastav, J. Elastomers Plast. **49**, 37 (2017)
34. J. Wojciechowski, Ł. Kolanowski, A. Bund, G. Lota, J. Power Sources **368**, 18 (2017)
35. M. Kakunuri, C.S. Sharma, ECS J. Solid State Sci. Technol. **6**, M3001 (2017)
36. K.K. Kar, *Composite Materials* (Springer Berlin Heidelberg, Berlin, Heidelberg, 2017)
37. H. Feng, Y. Chen, Y. Wang, Procedia Eng. **215**, 136 (2017)
38. C. Gayner, R. Sharma, I. Mallik, M.K. Das, K.K. Kar, J. Phys. D Appl. Phys. **49**, 285104 (2016)
39. Nickel Foam, https://www.americanelements.com/nickel-foam-7440-02-0. Accessed 2 Aug 2019
40. Electroformed Nickel Foil, http://www.specialmetals.com/assets/smc/documents/alloys/other/electroformed-nickel-foil.pdf. Accessed 5 Aug 2019
41. Nickel Foil | AMERICAN ELEMENTS, https://www.americanelements.com/nickel-foil-7440-02. Accessed 27 July 2019
42. Y. Feng, H. Zheng, Z. Zhu, F. Zu, Mater. Chem. Phys. **78**, 196 (2003)
43. D. Puspitasari, F.K.H. Rabie, T.L. Ginta, J.C. Kurnia, M. Mustapha, MATEC Web Conf. **225**, 01006 (2018)
44. Aluminum Foam, https://www.americanelements.com/aluminum-foam-7429-90-5. Accessed 30 July 2019
45. J. Butt, H. Mebrahtu, H. Shirvani, Prog. Addit. Manuf. **1**, 93 (2016)
46. S. Tóth, M. Füle, M. Veres, J.R. Selman, D. Arcon, I. Pócsik, M. Koós, Thin Solid Films **482**, 207 (2005)
47. R. Vicentini, L.H. Costa, W. Nunes, O. Vilas Boas, D.M. Soares, T.A. Alves, C. Real, C. Bueno, A.C. Peterlevitz, H. Zanin, J. Mater. Sci. Mater. Electron. **29**, 10573 (2018)
48. S. Banerjee, K.K. Kar, Polymer **109**, 176 (2017)
49. Stainless Steel Foam. https://www.americanelements.com/stainless-steel-foam-65997-19-5. Accessed 29 July 2019
50. C.-H. Chen, J.-T. Gau, R.-S. Lee, in *ASME 2008 International Manufacturing Science* and *Engineering Conference*, vol. 2 (ASME, 2008), pp. 457–463
51. Aluminum Vs. Steel Conductivity, https://sciencing.com/aluminum-vs-steel-conductivity-5997828.html. Accessed 29 July 2019
52. M. Valvo, M. Roberts, G. Oltean, B. Sun, D. Rehnlund, D. Brandell, L. Nyholm, T. Gustafsson, K. Edström, J. Mater. Chem. A **1**, 9281 (2013)
53. K.N. Kang, I.H. Kim, A. Ramadoss, S.I. Kim, J.C. Yoon, J.H. Jang, Phys. Chem. Chem. Phys. **20**, 719 (2018)
54. X.-Q. Yin, L.-J. Peng, S. Kayani, L. Cheng, J.-W. Wang, W. Xiao, L.-G. Wang, G.-J. Huang, Rare Met. **35**, 909 (2016)
55. R. Quintero, D.Y. Kim, K. Hasegawa, Y. Yamada, A. Yamada, S. Noda, RSC Adv. **4**, 8230 (2014)
56. Titanium Foil, https://www.americanelements.com/titanium-foil-7440-32-6. Accessed 31 July 2019
57. Platinum (Pt)—Properties, Applications, https://www.azom.com/article.aspx?ArticleID=9235. Accessed 31 July 2019
58. C. Shi, J. Dai, C. Li, X. Shen, L. Peng, P. Zhang, D. Wu, D. Sun, J. Zhao, Polymers (Basel). **9**, 10 (2017)
59. Carbon Fiber, https://www.americanelements.com/carbon-fiber-7440-44-0. Accessed 31 July 2019
60. D. Wentzel, I. Sevostianov, Int. J. Eng. Sci. **130**, 129 (2018)
61. G.E. Mostovoi, L.P. Kobets, V.I. Frolov, Mech. Compos. Mater. **15**, 20 (1979)
62. J.M.F. de Paiva, S. Mayer, M.C. Rezende, Mater. Res. **9**, 83 (2006)
63. S.K. Singh, M.J. Akhtar, K.K. Kar, ACS Appl. Mater. Interfaces **10**, 24816 (2018)

Chapter 13
Applications of Supercapacitors

Soma Banerjee, Bibekananda De, Prerna Sinha, Jayesh Cherusseri, and Kamal K. Kar

Abstract Supercapacitors are the most promising energy storage devices that bridge the gap between capacitors and batteries. They can reach energy density close to the batteries and power density to the conventional capacitors. Several researches have been carried out in the field of supercapacitors for the development of promising electrode and electrolyte materials as well as device fabrications to breakthrough in energy storage systems with diverse applications in electronics. They have a broad range of applications as they can deliver a huge power within a very short time. This chapter provides the detailed applications of supercapacitors in several sectors like consumer and portable electronics, transportation and vehicles, power backup, biomedical, military, aerospace, etc.

13.1 Introduction

Since the past decade, flexible/wearable/portable electronic devices are gaining significant importance in various sectors like mobility, consumer electronics, biomedical, sports, clean energy and environmental, and so on. Hence, due to the requirement

S. Banerjee · P. Sinha · K. K. Kar (✉)
Advanced Nanoengineering Materials Laboratory, Materials Science Programme, Indian Institute of Technology Kanpur, Kanpur 208016, India
e-mail: kamalkk@iitk.ac.in

S. Banerjee
e-mail: somabanerjee27@gmail.com

P. Sinha
e-mail: findingprerna09@gmail.com

B. De · J. Cherusseri · K. K. Kar
Advanced Nanoengineering Materials Laboratory, Department of Mechanical Engineering, Indian Institute of Technology Kanpur, Kanpur 208016, India
e-mail: debibek@iitk.ac.in

J. Cherusseri
e-mail: jayesh@iitk.ac.in

© Springer Nature Switzerland AG 2020
K. K. Kar (ed.), *Handbook of Nanocomposite Supercapacitor Materials I*,
Springer Series in Materials Science 300,
https://doi.org/10.1007/978-3-030-43009-2_13

341

of huge energy consumptions, these electronics need smart energy storage devices. Among the various energy storage systems, the supercapacitor is an important device that can provide high power density within a very short time by surface charge storage mechanisms [1–5]. Supercapacitor is a promising energy storage device that bridges the gap between capacitors and batteries. Supercapacitor is able to reach energy density close to the batteries and power density compared to the conventional capacitors. Several researches have been carried out in the field of supercapacitor for the development of promising electrode and electrolyte materials as well as device fabrications to a breakthrough in energy storage systems with diverse applications in electronics. Initially, supercapacitors are developed from high surface area carbon materials via the formation of electric double layer. They have excellent power density with very fast charge–discharge rates and long cycle life [2, 5–14]. However, later different types of pseudocapacitors are gaining interest due to the high energy density based on their Faradaic process as a result of the surface redox reactions [2, 5, 8]. Recently, to overcome the low energy density of carbon materials and poor conductivity, as well as surface area of pseudocapacitors, hybrid of both the materials are developed via utilization of the synergistic effect of both the components. Therefore, this is quite possible to achieve intermediate performance of conventional capacitors and batteries by improving overall energy and power densities.

At the initial stage, supercapacitors are traditionally used as additional energy backup systems in hybrid electric vehicles along with battery or fuel cell [2, 3]. Then, this attractive energy device has slowly progressed as a backup power supplier in complementary with batteries and fuel cells. Nowadays, the supercapacitors capture applications in several sectors like consumer and portable electronics, transportation and vehicles, power backup, biomedical, military and aerospace, and so on. This chapter mainly describes in detail several applications of supercapacitors, their advancement in modern technologies with concluding remarks and feature trends.

13.2 Applications of Supercapacitors

13.2.1 Portable and Flexible Electronics

Nowadays, we cannot imagine the world without portable electronics such as smartphones, smartwatch, laptops, cameras, and so many, to modernize our everyday lifestyles to fulfill a variety of new functions. However, these smart electronics need to power by smart energy storage systems. Supercapacitors play a significant role as an energy storage system along with batteries and fuel cells. Hybrid devices consisting of battery-supercapacitor hybrids are the best choices for the present and future mobile electronic devices to supply power. This combination enables lightweight battery devices to be fitted into small electronic portable gadgets such as watches, sensors, mobile, headphone, and so on, as shown in Fig. 13.1.

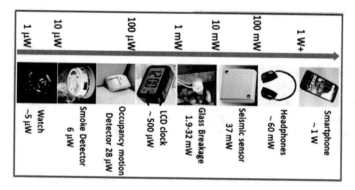

Fig. 13.1 Supercapacitor-enabled energy source in many smart and portable electronic devices. Reprinted with permission from [3]

Again, solid-state flexible supercapacitors have a number of applications in the present and next-generation flexible and wearable electronic devices. Flexible supercapacitors can easily be integrated with wearable textiles and function as power supplies for a variety of electronic devices such as mobile phones. The energy generated by the piezoelectric electric materials can be stored in a supercapacitor, and this energy can eventually be utilized for charging a mobile phone. One example is the supercapacitor-powered T-shirt "Sound Charge" developed by the Orange company, which is capable of generating electricity from pressure generated from sound waves. Tested at the Glastonbury festival, this T-shirt has generated enough electricity during the festive weekend to recharge two basic mobile phones. An image of activated carbon screen printed, knitted and woven with carbon fiber-based supercapacitor electrodes embedded in a long sleeve T-shirt has been displayed in Fig. 13.2.

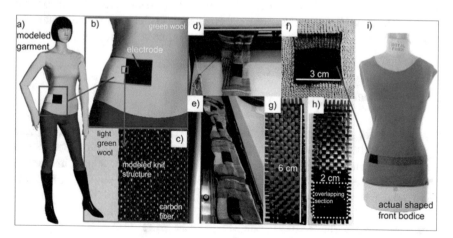

Fig. 13.2 Activated carbon screen printed, knitted and woven carbon fiber-based supercapacitor electrodes embedded in a long sleeve T-shirt. Reprinted with permission from [15]

13.2.2   Hybrid Electric Vehicles

Since 1990, supercapacitors have drawn attention after being utilized in hybrid electric vehicles along with batteries and fuel cells to deliver the required power for acceleration, and allow recuperating of brake energy [16, 17]. Supercapacitor and battery hybrids are suitable energy storage devices to supply power in different electric vehicles such as cars, boats, and buses. The importance of these alternative energy systems is that they are environment-friendly and devoid of any pollution, which is unavoidable in the case of conventional fuel-based systems such as diesel and petrol engines.

China has also produced the fastest charging electric bus equipped with CSR-CAP supercapacitors [18]. These supercapacitors are made by a Chinese company "Ningbo CSR New Energy Technology". The bus can be fully charged in just a few minutes and is also possible to recharge at stop stations, at the time when the passengers are getting on or off from the bus. Further, the bus consumes very little energy (30–50% less compared to other electric vehicles).

Another application of supercapacitors in hybrid electric vehicles is in the "brake energy regeneration" and "start–stop systems." Supercapacitors are found to be very effective in meeting this demand. The charge and discharge rate of hybrid supercapacitor are very rapid, almost without wear, and therefore, it can be considered as an ultra-fast and reliable battery. The first car is manufactured by PSA in 2010, using the concept of supercapacitors to optimize the start–stop system, were cut the engine if the car stops with gear lever in neutral for saving some fuel [19, 20]. Here, it is worthy to mention that around two times more mechanical and electrical energy is required to start a diesel compared to that of a gasoline car of the same size. During braking the vehicle, the recovered kinetic energy of such a system is converted into the electricity to power the alternator during the restart process. For example, in the case of new electrical architecture of its e-HDI cars, PSA thus integrates supercapacitors to rapid restart of the cars. Figure 13.3 shows a supercapacitor-battery-integrated system for "regenerative braking" as well as an extra energy source in hybrid electric vehicles.

13.2.3   Power Supply

Supercapacitors are also used in uninterrupted power supply (UPS) systems. These systems have advantages, serve as an emergency power device, surge protection, and also provide portable charging solutions. The UPS is designed to supply immediate backup to sensitive loads to bridge the start-up of a generator [21]. A supercapacitor-based uninterrupted power supply (UPS), 10A UPSU manufactured by inventlab, is shown in Fig. 13.4. This is one of the smallest industrial-/embedded-UPS solution in its power class within a voltage range of 12 or 24 V [22, 23].

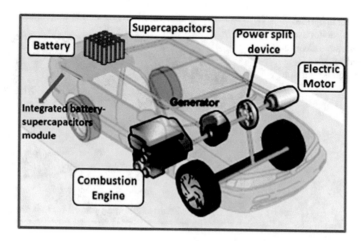

Fig. 13.3 Supercapacitor-battery-integrated system for "regenerative braking" as well as extra energy source in hybrid electric vehicles. Reprinted with permission from [3]

Fig. 13.4 A supercapacitor-based uninterrupted power supply (UPS), 10A UPSU manufactured by inventlab. Reprinted with permission from [22]

Fig. 13.5 An integrated
solar-supercapacitor system
to charge portable electronic
devices and vehicles.
Reprinted with permission
from [3]

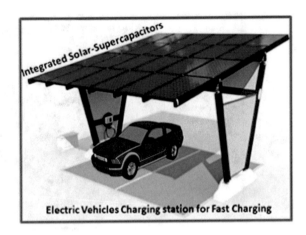

13.2.4   Wind/Solar Powering

Due to the intermittent nature of wind and solar technologies, supercapacitors are becoming unavoidable systems in the present and future energy storage devices. The energy converted by the solar cells during the daytime has to be stored by some means in order to utilize the same at night. Similarly, the energy converted by the wind turbines during the windy time has to be stored by some means in order to utilize the same when there is no wind present as source. These supercapacitor systems offer supplemental power during low-light or no-light conditions. These supercapacitors are available in cells and modules depending on the system voltage requirements and are ideal for energy storage and circuit-charging schemes in various solar power systems. Figure 13.5 represents an integrated solar-supercapacitor system that can be used as a charging stations at different places, like gardens, streets or parking areas to charge portable electronic devices and vehicles [3].

13.2.5   Implantable Healthcare

Piezoelectric supercapacitors are widely utilized in many implantable healthcare systems, where microwatts to milliwatts power is required. These supercapacitors are used in cardiac pacemaker, insulin pumps, healthcare applications, etc., as shown in Fig. 13.6.

Fig. 13.6 Application of supercapacitors in implantable medical devices. Reprinted with permission from [3]

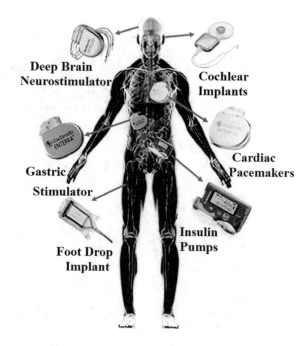

13.2.6 Defense and Aerospace

The ability of rapid power supply, long cycle life, and low temperature operation of supercapacitors marks them as a suitable device in various military and aerospace applications, like power backup in military vehicles and electronics, armored vehicles, fire control systems in tanks, black box on helicopters, backup power/memory hold-up for emergency handheld radios, GPS-guided missiles and projectiles, airbag deployment, and so on [3, 24]. They are also used in many aerospace applications, where high power density is required, like actuator systems for stage separation devices in launch vehicles, on-board systems of satellites and spacecraft. The military and aerospace vehicles where supercapacitors are utilized have been displayed in Fig. 13.7.

13.2.7 Other Electric Utilities

Supercapacitors also find use in various electric utilities such as power tools, electric transportation domain, pulsed laser, flashlights, cranes, elevators, and so on [3]. The supercapacitor enables rapid recharge and high power that lasts long. A supercapacitor-powered tool, a cordless screwdriver, is designed by Batavia as shown in Fig. 13.8. This lightweight flash cell screwdriver with new capacitor technology is capable to charge fully within a few minutes.

Fig. 13.7 Military and aerospace vehicles where supercapacitors are utilized as power backup systems. Reprinted with permission from [24] (SPEL)

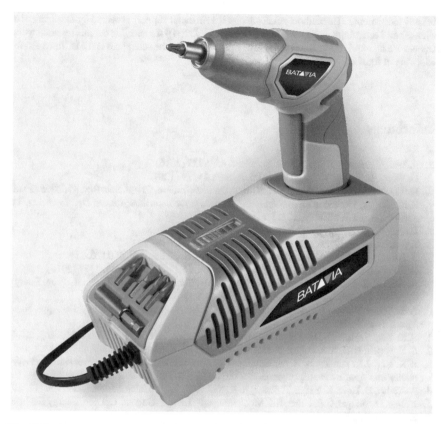

Fig. 13.8 A photograph of flash cell screwdriver with new capacitor technology designed by Batavia. Reprinted with permission from [25]

13.3   Concluding Remarks

Supercapacitors are intermediate energy storage systems between batteries and capacitors according to their electrochemical performance. They can provide power density like capacitors and deliver energy close to the batteries. The research is not only limited to the development of electrode or electrolyte materials and other components. Several new promising supercapacitor devices are also fabricated for advance and smart electronics. Excellent evolution has been made on wearable and flexible supercapacitors with sandwich, planar, wire, fiber, cable, etc., device designs. Research has also progressed on supercapacitors with novel functions like piezoelectric, shape-memory, self-healing, and so on to extend the practical applications in smart electronics. Therefore, supercapacitors have become an emergent technology to a breakthrough in energy storage systems with miscellaneous applications, like portable and wearable electronics, smart clothes, transportation and vehicles, power backup systems, implantable bioelectronics, military, aerospace, etc.

Acknowledgements The authors acknowledge the financial support provided by Department of Science and Technology, India (DST/TMD/MES/2K16/37(G)) for carrying out this research work. Authors are thankful to Mr. Kapil Dev Verma for rearranging the references and Ms. Tanvi Pal for redrawing of figures.

References

1. Y. Wang, Y. Song, Y. Xia, Chem. Soc. Rev. **45**, 5925 (2016)
2. G. Wang, L. Zhang, J. Zhang, Chem. Soc. Rev. **41**, 797 (2012)
3. D.P. Dubal, N.R. Chodankar, D.-H. Kim, P. Gomez-Romero, Chem. Soc. Rev. **47**, 2065 (2018)
4. K.K. Kar, A. Hodzic, *Carbon Nanotube Based Nanocomposites: Recent Developments*, 1st edn. (Research Publishing, 2011)
5. J. Cherusseri, K.K. Kar, J. Mater. Chem. A **4**, 9910 (2016)
6. J. Cherusseri, R. Sharma, K.K. Kar, Carbon **105**, 113 (2016)
7. S. Banerjee, P. Benjwal, M. Singh, K.K. Kar, Appl. Surf. Sci. **439**, 560 (2018)
8. D.P. Dubal, O. Ayyad, V. Ruiz, P. Gómez-Romero, Chem. Soc. Rev. **44**, 1777 (2015)
9. R. Kumar, S. Sahoo, E. Joanni, R.K. Singh, W.K. Tan, K.K. Kar, A. Matsuda, Prog. Energy Combust. Sci. **75**, 100786 (2019)
10. A.K. Yadav, S. Banerjee, R. Kumar, K.K. Kar, J. Ramkumar, K. Dasgupta, ACS Appl. Nano Mater. **1**, 4332 (2018)
11. K.K. Kar, A. Hodzic, *Developments in Nanocomposites*, 1st edn. (Research Publishing, 2014)
12. S. Banerjee, K.K. Kar, Polym. J. **109**, 176 (2017)
13. K.K. Kar, J.K. Pandey, S. Rana, *Handbook of Polymer Nanocomposites. Processing, Performance and Application* (Springer Berlin Heidelberg, Berlin, Heidelberg, 2015)
14. A. Yadav, B. De, S.K. Singh, P. Sinha, K.K. Kar, ACS Appl. Mater. Interfaces **11**, 7974 (2019)
15. K. Jost, D. Stenger, C.R. Perez, J.K. McDonough, K. Lian, Y. Gogotsi, G. Dion, Energy Environ. Sci. **6**, 2698 (2013)
16. R. Kötz, M. Carlen, Electrochim. Acta **45**, 2483 (2000)
17. K.K. Kar, *Composite Materials* (Springer Berlin Heidelberg, Berlin, Heidelberg, 2017)
18. World's fastest charging electric bus: 10 seconds thanks to supercapacitors, https://www.supercaptech.com/world-fastest-charging-electric-bus-10-seconds-thanks-to-supercapacitors. Accessed 12 Jan 2019
19. S.B. Barooah, FAME India Scheme set to be extended by up to 6 months (2017), http://www.autocarpro.in/news-national/fame-india-scheme-set-extended-months-24089. Accessed 30 July 2017
20. S. Banerjee, K.K. Kar, High Perform. Polym. **28**, 1043 (2016)
21. Supercapacitor, https://www.marathon-power.com/supercapacitor-ups.html. Accessed 12 Jan 2019
22. 10A UPSU is a 12V or 24V Embedded-/Industrial-UPS. It is available with Ultracapacitors or with Batteries, https://www.10a-upsu.com/. Accessed 02 Feb 2019
23. S. Banerjee, K.K. Kar, J. Appl. Polym. Sci. **133**, 42952 (2016)
24. Supercapacitors for defense & aerospace, http://www.capacitorsite.com/defence.html. Accessed 12 Jan 2019
25. Batavia Flash Cell 4.6V Duo Cordless Rechargeable Screwdriver 7061166—Brand New, https://www.brandedhousewares.co.uk/batavia-flash-cell-duo-cordless-rechargeable-screwdriver.html. Accessed 05 Feb 2019

Index

© Springer Nature Switzerland AG 2020
K. K. Kar (ed.), *Handbook of Nanocomposite Supercapacitor Materials I*,
Springer Series in Materials Science 300,
https://doi.org/10.1007/978-3-030-43009-2